T0329583

Government Policy and

FARMLAND

MARKETS

The Maintenance of Farmer Wealth

Government Policy and
FARMLAND
MARKETS

The Maintenance of Farmer Wealth

Charles B. Moss and Andrew Schmitz

Iowa State Press
A Blackwell Publishing Company

Charles B. Moss is Professor, Food and Resource Economics, University of Florida, Gainesville, and co-editor of the *Journal of Agricultural & Applied Economics* 1999-2001.

Andrew Schmitz is the Ben Hill Griffin Eminent Scholar and Professor, Food and Resource Economics, University of Florida, Gainesville.

Iowa State Press
2121 State Avenue, Ames, Iowa 50014

Orders:	1-800-862-6657
Office:	1-515-292-0140
Fax:	1-515-292-3348
Web site:	www.iowastatepress.com

First edition, 2003

Library of Congress Cataloging-in-Publication Data

Government policy and farmland markets: The maintenance of farmer wealth/edited by Charles B. Moss, Andrew Schmitz.
    p. cm.
    Papers presented at a conference entitled: Government policy and farmland markets implications for the new economy held in May 2002 in Washington, D.C.
    Includes bibliographical references and index.
    ISBN 0-8138-2329-3
    Farms—Valuation—United States—Congresses. 2. Farms—Prices—United States—Congresses. 3. Farms—Government policy—United States—Congresses. 4. Land use, Rural— Government policy—United States—Congresses. 5. Agriculture and state—United States—Congresses. 6. Farmers—United States—Economic conditions—Congresses. 7. Urbanization–United States—Congresses. I. Moss, Charles B. (Charles Brit) II. Schmitz, Andrew.

HD256.G29 2003
338.1'0973—dc21

                                                                  2003056586

The last digit is the print number: 9 8 7 6 5 4 3 2 1

Contents

Section III: Capital Markets and Farmland Values

Section IV: Transaction Costs and Farmland Values

Section V: Urbanization, Environmental Quality, and Farmland Values

Section VI: Regional and International Dimensions

Contributors

Charles Barnard
United States Department of Agriculture
Economic Research Service
Washington, DC, USA

Peter J. Barry
Department of Agricultural and Consumer Economics
University of Illinois at Urbana-Champaign
Urbana, Illinois, USA

Vince Breneman
United States Department of Agriculture
Economic Research Service
Washington, DC, USA

Jared Carlberg
Department of Agribusiness and Agricultural Economics
University of Manitoba
Winnipeg, Manitoba, Canada

Jean-Paul Chavas
Department of Agricultural and Resource Economics
University of Maryland
College Park, Maryland, USA

Willard W. Cochrane
Professor Emeritus, Department of Applied Economics
University of Minnesota
St. Paul, Minnesota, USA

Xinshen Diao
International Food Policy Research Institute
Washington, DC, USA

Kenneth Erickson
United States Department of Agriculture
Economic Research Service
Washington, DC, USA

Allen M. Featherstone
Department of Agricultural Economics
Kansas State University
Manhattan, Kansas, USA

Hartley Furtan
Department of Agricultural Economics
University of Saskatchewan
Saskatoon, Saskatchewan, Canada

Bruce Gardner
Department of Agricultural and Resource Economics
University of Maryland
College Park, Maryland, USA

Barry K. Goodwin
Department of Agricultural, Environmental, and Development Economics
The Ohio State University
Columbus, Ohio, USA

Jeffrey Hopkins
United States Department of Agriculture
Economic Research Service
Washington, DC, USA

Elena G. Irwin
Department of Agricultural, Environmental, and Development Economics
The Ohio State University
Columbus, Ohio, USA

Warren Johnston
Professor Emeritus, Department of Agricultural Economics
University of California at Davis
Davis, California, USA

Richard E. Just
Department of Agricultural and Resource Economics
University of Maryland
College Park, Maryland, USA

Sergio H. Lence
Department of Economics
Iowa State University
Ames, Iowa, USA

Lawrence W. Libby
Department of Agricultural, Environmental, and Development Economics
The Ohio State University
Columbus, Ohio, USA

Sabrina J. Lovell
Environmental Protection Agency
Washington, DC, USA

Lori Lynch
Department of Agricultural and Resource Economics
University of Maryland
College Park, Maryland, USA

Douglas J. Miller
Department of Agricultural Economics
Purdue University
West Lafayette, Indiana, USA

Ashok K. Mishra
United States Department of Agriculture
Economic Research Service
Washington, DC, USA

Charles B. Moss
Department of Food and Resource Economics
University of Florida
Gainesville, Florida, USA

Susan Offutt
United States Department of Agriculture
Economic Research Service
Washington, DC, USA

François N. Ortalo-Magné
London School of Economics
London, United Kingdom

Philip M. Raup
Professor Emeritus, Department of Applied Economics
University of Minnesota
St. Paul, Minnesota, USA

John E. Reynolds
Department of Food and Resource Economics
University of Florida
Gainesville, Florida, USA

Terry Roe
Department of Applied Economics
University of Minnesota
St. Paul, Minnesota, USA

Andrew Schmitz
Department of Food and Resource Economics
University of Florida
Gainesville, Florida, USA

Bruce J. Sherrick
Department of Agricultural and Consumer Economics
University of Illinois at Urbana-Champaign
Urbana, Illinois, USA

J.S. Shonkwiler
Department of Applied Economics and Statistics
University of Nevada-Reno
Reno, Nevada, USA

Agapi Somwaru
United States Department of Agriculture
Economic Research Service
Washington, DC, USA

Calum Turvey
Department of Agricultural, Food, and Resource Economics
Rutgers University
New Brunswick, New Jersey, USA

Luther Tweeten
Professor Emeritus, Department of Agricultural, Environmental, and
Development Economics
Ohio State University
Columbus, Ohio, USA

Keith Wiebe
United States Department of Agriculture
Economic Research Service
Washington, DC, USA

Acknowledgements

The editors would like to thank all those who participated in and attended the conference entitled "Government Policy and Farmland Markets: Implications for the New Economy", which was held in May 2002 in Washington, DC. We thank all the participants who presented papers at the conference, and especially those who contributed chapters for the present book. We also thank the authors who contributed additional chapters to this volume after the conference was completed.

We particularly thank the agencies that contributed financially to both the conference and to the publication of this volume. These agencies include the United States Department of Agriculture through the National Research Initiative and the Economic Research Service of the United States Department of Agriculture. In addition, we would like to thank the Farm Foundation for their generous support of both the conference and the publication costs of this volume. The Ben Hill Griffin, Jr., Endowed Chair at the University of Florida; the Swank Chair at Ohio State University, and the Department of Food and Resource Economics at the University of Florida also provided additional funding.

The editors would like to especially thank H. Carole Schmitz for her tireless efforts organizing the conference in Washington DC, as well as for her invaluable administrative work editing this volume and managing its production for publication. Special thanks also to Jason Snart formerly of the University of Florida for his expertise as editor, research specialist, and technical advisor in the production of this volume. In addition, we would like to thank Kjerstin Terry for her assistance in the final stages of the preparation of the edited volume.

Foreword

Susan Offutt

United States Department of Agriculture, Economic Research Service

THE SIGNIFICANCE OF THE VALUE
OF FARMLAND

It is not necessary to be a farmer, or even an economist, to figure out that land is the foundation of the farm enterprise. For most of this nation's history, the value of any particular piece of farmland was largely determined by how productive it was. Proximity to markets mattered, too, but less so as transportation, processing, and storage improved. In the 20th century, the initiation of the New Deal farm programs and acceleration in the growth of urban and suburban areas have presented two additional, important determinants of the price of farmland.

Understanding how government policies and urbanization affect farmland value matters because of the centrality of farmland to the requirements of the physical production of food and also to farm household wealth. Improvement in the well being of American farm households, an important policy objective, is too often seen solely in terms of commodity price levels. But income and wealth, together two important components of financial health, are better indicators of well being. Just as it is worth studying the impact of farm policy on farm prices, it is worth understanding the effects on farmland value. Pressure on land values created by competition for nonfarm land uses in urban and suburban areas can be strong in places, and more and more often the effects of urbanization are mediated by state, local, and Federal policies aimed at farmland preservation.

Land is the most important asset in the farm business and in the farm household-investment portfolio. Nationally, in 2001, real estate accounted for almost three-fourths of farm business assets. For the average investment portfolio of the farm household, worth about $660,000 in 2001, real estate comprised 60 percent of its value. For landowners who operate farms, the value of land as a business asset lies in its worth as collateral for operating and expansion loans. For landowners that do not operate farms, the revenue stream from renting farmland often represents an important component of income. Retired farmers appear to represent a sizable part of this class of nonoperator-owners. National USDA surveys show that about one-half of nonoperator-owners are age 65 or older (compared to about one-third of operator-owners) and that 85 percent live within 50 miles of the land they rent. For retired self-employed farmers without the benefits of standard pensions, the appreciation in farmland value that is realized on its sale or the stream of income from its rental may be significant to their financial security.

Productivity in agricultural uses, the level of farm-program benefits, and proximity to urban and suburban development represent major determinants of farmland value. In the West, access to water is also obviously important, and other regional constraints may matter, too. So, local farmland markets are conditioned by a variety of factors whose importance varies regionally and over time. Of special interest, though, is the role of Federal policy in determining land values because of its explicit aim to improve the lot of farmers. How well is this objective achieved, as measured by or reflected in the value of farmland?

Federal farm policy influences figure most in those areas where supported commodities are grown. In 2001, government payments went to about 40 percent of the nation's farms, mainly those growing the eight program crops (wheat, corn, soybeans, sorghum, cotton, rice, barley, and oats) which are geographically concentrated in the middle of the country. Economists have understood for some time that the value of Federal program benefits are capitalized in land values as these payments become a component of expected future returns. USDA's Economic Research Services estimates that, nationally, about one-quarter of the value of farmland is attributable to the expected continued benefits of farm programs tied to the farming operation. With the trend toward use of direct payments, as opposed to supply controls and price interventions, the translation of payments into benefits for landowners has become more transparent. For farmers who own the farm, the appreciated value of the land is realized when it is sold. For landowners who do not operate farms but rent out acreage, any payments made to tenants can be captured through adjustment in lease terms. Because 40 percent of farmers rent a part or all of the land they work, there may be a disconnect between the political desire to assist active farmers and the distribution of benefits to nonfarmer landlords. However, to the extent that landowners who do not farm are retired farmers, the outcome may not be inconsistent with aims to improve the well being of farm households. In any case, Federal farm policy is a direct contributor to farmland value appreciation, and the distribution of these benefits is an important matter for study.

The chapters in this volume address the range of economic issues concerning the value of farmland. Beginning with a historical perspective, the focus progresses to the interaction between government policies and farmland values, then to the mechanics and behavior of the markets for land, the major farm business asset. These markets have special characteristics that make transaction costs a significant component of a land purchase, and these aspects are also considered. The more contemporary influences of urbanization and also environmental subsidies and regulations are examined next, with concluding chapters about special circumstances in regional farmland markets. Taken together, these chapters represent a most comprehensive and timely look at farmland values. The future prospects for trade liberalization and domestic policy reform, for ongoing farmland conversion, and for management of environmental quality mean that farmland values, with their undeniable importance to farm businesses and households, will continue to be of key interest.

Section I:
Historical Perspectives
on Farmland Values

Chapter 1

Farmland Markets in the Development of U.S. Agriculture

Willard W. Cochrane
Professor Emeritus
University of Minnesota

Abundant land—cheap or free, distributed with or without corruption—served as an important stimulus to the overall development of this nation. Land was the magnet that drew the first settlers to English colonies, once the bubble of instant riches had been pricked. It was the magnet that continued to draw them to these shores for almost three centuries. And it was the magnet that drew settlers into the wilderness, over the Appalachian, and across the continent in one century following the Revolutionary War. To the landless and land-hungry people of Western Europe the pull of cheap or free land in North America was overwhelming. They came in droves and they suffered untold misery to make that land their own.[i](Cochrane 1993: 173)

How that land was distributed is the subject of this essay: the role of government, the role of private markets, and the more common role of some combination of government and markets. And we will observe these different and changing roles in a historical context—first in colonial times, then in period after period as the nation developed.

THE COLONIAL PERIOD

The Virginia Colony founded in 1607 barely survived its first twenty years of existence. The mortality rate from disease and starvation among the colonists was extremely high. The colony was rejuvenated annually with an infusion of settlers and supplies. The settlers kept coming, not in droves, but in increasing numbers as they learned from the Native Americans how to produce corn and tobacco. The government of colonial Virginia attracted settlers by giving them land free of charge. Throughout the 1600s, land in Virginia was distributed through a headright system (that right given by a colony to each head of a family for the ownership of a tract of public land). Thus, a colonist who transported himself to Virginia received a 50-acre grant of land as his headright. He also received a grant of fifty acres for each additional individual that he transported to the colony. This system was adopted in some of the other southern colonies.

The Massachusetts Bay Colony, founded in 1630, had a relatively successful colonizing experience. Its members came from a dissenting religious group in England, the Puritans, who sought with great zeal to establish a theocracy of their particular religious persuasion in the wilderness, and they succeeded. As a result of this tight cohesiveness, the colony followed a land-distribution path very different from that of Virginia. In the usual case, an organized group— usually a church congregation—would petition the colonial government for a grant of land. If approved, that group would receive a grant of land for the use of the group and for distribution among investing members of the group. The tract of land on which settlement was to be made was known as the town grant, and its size varied from four to ten square miles. Near the center of the land grant a village was laid out that contained a site for a village green, a meetinghouse, and a house lot for each settler. Next, the arable lands and meadows were subdivided. Large fields of several hundred acres were surveyed and divided into strips, which were then distributed among investing settlers by lot. In the distribution of the arable lands and meadows, two criteria were used: the settler's investment in the original enterprise, and his ability to use the land. The allotments of arable lands and meadows were not fenced separately, but were surrounded by a common fence; hence they came to be known as proprietors' commons. Each settler cultivated his own lot or lots, subject to restrictions imposed by the proprietors' commons dealing with kinds of crops and dates of planting and harvesting. In addition, there were town commons that could be used by all inhabitants of the town; they included all land of the original grant not yet divided among the original proprietors or allotted to latecomers. On such commons any inhabitant might pasture his animals and take wood, stone, or earth, subject again to the regulation of the town authorities.

The New England land-distribution system had important social and political implications. It developed habits of group action and tended to foster a compact cohesive social life. It provided an effective and equitable method for distributing large areas of land to cultivators in rough proportions to their ability to use land. In New England, during the years when the system of town grants was in operation, land speculation was unknown—that is, up to 1725.

William Penn, who founded the Colony of Pennsylvania in the 1680s, pursued still a different land-distribution and colonizing route. He offered religious toleration to prospective colonists (who often were from continental Europe) and offered land for sale at low prices and on easy terms. This proved to be a successful colonizing arrangement.

These unique systems of land distribution gave way to land speculation in the 1700s. Untold amounts of unoccupied land lay at the edge of settlements in all of the colonies (to white Europeans, the fact that native Americans roamed those lands was unimportant). Friends of kings, associates of colonial governors, or successful military men could and often did obtain large land grants on the frontier, which they hoped to sell in smaller parcels to the next wave of settlers. The way in which blocks of land on the frontier were obtained varied over time, but land speculation did not change. It was the number one game on the frontier in which all participated—the squatter, the pioneer settler, the railroad companies, Eastern moneymen, military heroes, and politicians of every stripe.

THE 1783 TO 1860 PERIOD

With the victory of the colonists over Great Britain in the American Revolution, the nation that was to become the United States was a solid land mass extending from the Atlantic Coast in the East to the Mississippi River in the West, and from the Great Lakes in the North—to but not including Florida and West Florida in the South. When the Congress of Confederation accepted the cession by Virginia, 1784, to its claims to most of the land lying north of the Ohio River, the new nation acquired a large public domain. When five other states joined Virginia in ceding western lands claimed under their colonial charters to the United States between 1785 and 1802, the new nation had acquired a huge public domain consisting of what are now the states of Ohio, Indiana, Illinois, Michigan, Wisconsin, Tennessee, Alabama, and Mississippi. That mass of land under the control of the United States expanded to the Pacific by 1860 through the Louisiana Purchase in 1803, the annexation of the Republic of Texas as a state in 1845, the acquisition of the Oregon Territory by a treaty with Great Britain in 1846, and the acquisition of California and much of the Southwest by military conquest of Mexico in 1848.[ii]

The new nation's expansive land mass prompted the question of how this land was to be distributed. One knotty question had been resolved by the individual colonies during the Revolution when the colonial governments eliminated the essential elements of a feudalistic tenure system, including quit rents and the right of primogeniture, and stimulated private enterprise by placing land in the hands of private individuals as rapidly as possible. Thus, a freehold land-tenure system was well established in the new nation by the end of the Revolutionary War period.

The Ordinance of 1785 established the procedure to use when distributing lands within the public domain and the procedure to use when conducting public-domain land sales. Under the provisions of the Ordinance of 1785, government surveyors were to establish horizontal lines called base lines and vertical lines called meridians on all unoccupied lands of the public domain. The first meridian was laid off in what is now the state of Ohio, and the first survey covered land north of the Ohio River, excluding the reserves—the areas of land reserved for the different states for specific purposes. From the established meridians and base lines, tiers of townships were laid out; within each township thirty-six sections were also laid out. Eventually all the land in the United States was included in the survey except the thirteen original states, Vermont, Kentucky, Tennessee, parts of Ohio, and all of Texas.

The Ordinance of 1785 also set the terms of the sale of federal lands in public auctions to private persons: Land was to be sold in lots of 640 acres at a per-acre price of not less than $1.00—cash payments only. Within each township, five sections of land were to be reserved; four sections for the United States and one section for the support of public schools.

With respect to sales policy, two opposing views were addressed. Advocates of the "Conservative view" favored selling public lands for cash in large tracts at high prices and for cash. Supporters of this view were seeking revenue for the new government and large tracts of land for speculation. Advocates of the "Liberal view" favored putting land in the reach of everyone by selling it in small parcels at low prices and on easy credit terms. This was the Jeffersonian position. The policy result was a compromise that leaned strongly toward the Conservative view. The sale of land in the public domain at public auction in minimum lots of 640 acres at a price of not less than $1 per acre was in fact a harsh, self-defeating sales policy; almost no pioneering family in the frontier could come up with $640 in cash.

The sale of public lands in the Northwest Territory proceeded very slowly after the passage of the ordinance of 1785 for a number of reasons. First, there were serious Indian troubles in the territory that were not resolved until after the victory over the Indians at the battle of Fallen Timbers in 1794. Second, states like New York, Pennsylvania, and Kentucky were engaged in enterprising schemes of land disposal within their borders in direct competition with the federal program. Third, and by far the most important, the minimum purchase that a settler could make ($640 for 640 acres) exceeded by far what the typical impoverished pioneer could afford to expend for land. The general state of poverty among pioneers of this period reduced the disposal of land under the Ordinance of 1785 to a mere trickle.

Further, it was a fact of life that on the frontier, the impoverished pioneer could always move beyond the land survey, squat on some unsurveyed land, and hope that at some future time legislation would enable him to hold on to it— which was very often the case. Land squatters bedeviled the land-sales program of the federal government for the next 50 years.

Dissatisfaction with the sales results under the Ordinance of 1785 led to passage of a new land act by the Congress of the United States in 1796. But this act represented another victory for those with a conservative view of land disposal. The minimum per-acre price was increased to $2.00 and the minimum purchase of land remained 640 acres. Not surprisingly, the pace of land sales from the public domain remained slow under the Act of 1796. Agitation started immediately by representatives in Congress from the agrarian South and West to liberalize the provisions of the land-sales program. Thus, a new act was passed in 1800 that kept the $2.00 per-acre minimum sales price but reduced the minimum acreage requirement to 320 acres and initiated a credit provision that gave the purchaser up to four years to pay for the land. These more liberal provisions stimulated land sales, but not sufficiently to satisfy the Southern and Western agrarians. Consequently, new land acts were passed in 1804, 1820, 1841 and 1854, and each one further liberalized the sales provisions of the land-disposal programs.

These continued efforts to liberalize the land-sales program of the federal government had a positive effect. The federal government sold almost 20 million acres of land from the public domain between 1800 and 1820. (In 1818 alone, some 3.5 million acres were sold.) But these land-disposal programs did not achieve the success that many had hoped. Some 167 million acres of public lands had been surveyed by 1835, but only 44.5 million acres had actually been sold, leaving some 122 million acres unsold. This continued to be the case through the early 1850s. Numbers like these led to the passage of the Graduation Act in 1854, which lowered the price of unsold public lands. Under the provisions of this Act, the per-acre price of land that had been on the market for 10 years was reduced to $1.00. If the land remained unsold for another 5 years, its per-acre price would decline to 75 cents, and so on down to 12.5 cents after 30 years. As might be guessed, the leading beneficiaries of this act were knowledgeable speculators and mining and timber companies. These companies knew the location of the parcels of land that fitted into their operations, and when the prices of those parcels dropped low enough, they gobbled them up.

This land law was not well received by the public, and almost immediately agitation began to further liberalize the distribution of public lands. This time it took the form of the passage of homestead legislation to distribute lands in the public domain to settlers free of charge. Interestingly, the pressure for free land in the 1850s did not come from agrarian Westerners and Southerners, but rather it came from Eastern reformers, who sought ways of improving conditions of recent immigrants to the United States. Before we turn to the many ramifications of the Homestead Act of 1862, we need to consider two other forces bearing on land markets between 1800 and 1860.

On the frontier there were vast amounts of unoccupied land, and an increasing number of pioneer families with little or no cash. Consequently, pioneer families squatted on parcels of land and carved farms out of the wilderness. When the land surveyors caught up with the squatters and land in these areas was put up for sale at public auction, chaos broke out. Squatters formed protective associations and turned up at the public land auctions with their rifles. Often they were successful in discouraging buyers from bidding on land upon which the squatters had settled. As the West gained political power in Congress, that body passed 33 pre-emption laws between 1799 and 1830 giving squatters the right to purchase their land at a minimum price without having to bid for it at a public auction. These pre-emptive laws, however, were mainly local in application. But the right of pre-emption became a burning political issue in the 1830s, and Congress passed five general pre-emptive laws that forgave squatters already on public lands for their illegal trespass, and also gave them the right to buy their land at the minimum price. But Westerners wanted more, and they got it. They wanted the right of prospective pre-emption that would sanction squatting on public lands before the lands were thrown open for sale, which they won in the Pre-emption Act of 1841. The pressure to obtain land in the West, free or cheap, legal or illegal, was tremendous in the period leading up to the Civil War.

The railroads became the leading players in the land market of the nineteenth century. The federal government made direct grants of some 131 million acres of land to different railroad companies to help them finance the construction of their routes. If land grants by states were accounted for, the gift of land to the railroads by all levels of government totaled at least 183 million acres. With the help of conniving politicians, the railroads also obtained much of the best land from the federal government—the future state of Iowa. Thus, when my great grandparents emigrated from Northern Ireland to Southwestern Iowa in the 1850s, they bought their 100 acres of land from a railroad company. They were lucky, because rising farm-product prices during the Civil War period enabled them to survive and even to prosper in a modest way.

With the railroads unloading their millions upon millions of acres as fast as they could to get the funds to finance the construction operations of their ever-expanding routes, and with the squatters who either legally or illegally outwitted the efforts of the federal government to dispose of the public domain, it is not surprising that land-sale efforts of the federal government lagged. Further, it is not surprising that farmland prices rarely rose above the minimums established by the federal government between 1800 and 1860. There was a great and growing demand for farmland beyond the Appalachian Mountains, but there was even a greater supply of land, albeit part of it illegal, to meet that demand. Thus, farmland prices rode on the federal government's established minimums over that 60-year period.

THE 1860 TO 1900 PERIOD

The Westerners with the help of Eastern reformers finally got what they wanted in 1862—free land. The Homestead Law enacted on 20 May 1862 made it possible for any person to file for 160 acres of unappropriated public land if he or she:

1. Was an American citizen or had filed his or her papers of intention to become a citizen.
2. Was 21 years of age, or had served 14 days in the armed forces of the United States.
3. Had never fought against the United States (a condition intended to exclude persons who had served in the Confederate Army—a condition that was dropped in 1866).

The man or woman filing on 160 acres of land would obtain a fee-simple title to that land when he or she:

1. Had resided on or farmed the claim for five consecutive years;
2. Had become a citizen of the United States; and
3. Paid the requisite fees (which in the early years amounted to $10).

The enacting of the Homestead Act of 1862 was historic and important to the development of American agriculture, but it did not play as important a role in that development as many people had come to believe for two reasons. First, most of the productive agricultural lands east of the Western boundaries of Minnesota, Iowa, and Missouri were already gone by 1862, because the federal government had already sold it or had given it away to the states, railroads, and veterans of the various wars. Thus, the bulk of the filings over the next 30 years occurred in Kansas, Nebraska, and the two Dakotas. Further, much of this land was located in arid areas. For these areas, the 160-acre limitation was a huge mistake, because a homestead of 160 acres was too small a unit to be economically viable even for crop farming, and was far too small for grazing operations in the arid West.

Second, between 1862 and 1900 the federal government disposed of about 500 million acres of the public domain. Of that total only about 80 million acres were distributed under the Homestead Act. During the same period, some 108 million acres were sold at auction or through other disposal programs (four to be noted later). That left some 300 million acres that were given as grants to states and railroads, which in turn sold these lands to speculators who then sold these lands to farmers, ranchers, and others. The distribution of public lands under the Homestead Act was important in absolute terms, but not in relative terms.

The Congress passed four major land-disposal acts in the 1870s, largely at the behest of special-interest groups. They were the Timber Culture Act of 1873, the Desert Land Act of 1877, the Timber and Stone Act of 1878, and the Timber Cutting Act of 1878.

The first of these acts was passed to help small homesteaders on the Plains enlarge their units. But Congress could not achieve through legislation what nature could not do—namely, grow trees on the Plains. Thus, the Timber Culture Act achieved little in the way of helping homesteaders enlarge their units. The remaining three land acts were clearly land-grabbing acts much used by the special-interest groups and little used by farmers.

In the 1880s more and more people began to realize that the land resources of the nation were being squandered in a reckless fashion, so political pressure developed to tighten the many and varied legal provisions for the disposal of the public domain. The General Revision Act of 1891 closed several of the land-disposal loopholes that were part of the previous pieces of legislation. The Act of 1891 repealed the Timber Cutting Act; it amended the Desert Land Act by specifying that a definite plan of irrigation had to be submitted with an application to take up land under that act; and finally, it repealed the Pre-emption Act of 1841. The new law further authorized the U.S. President to set aside forestland as public reservations. The Act of 1891 may be viewed as the first concrete step of the Conservation movement that gained strength under the leadership of President Theodore Roosevelt, and one that is still battling to protect the natural resources of the United States more than 100 years later.

The principal land-disposal programs of the federal government in the nineteenth century and their numerous shortcomings have been described and discussed, but little has been said to this point about the positive aspects of those programs. In the judgment of this writer, despite all the corruption and speculation associated with those disposal programs and the policy mistakes involved in those programs, the results in terms of the national economy and the agricultural economy were good, not perfect, but good. The land donated to the railroads to help finance the rapid construction of a national railway network certainly contributed to the growth of the national economy, and the food surpluses developed in the second half of the nineteenth century contributed to the industrialization of the nation by providing an abundant low-priced food supply.

The agricultural sector at the close of the nineteenth century was primarily composed of small freehold units in which families had a vital stake in the productivity of those units and an important stake in the well being of their communities. To an important degree, the Jeffersonian dream had become a reality by 1900.

But there were losers, too. Native Americans were the big losers. They lost their lands, they lost their cultures, and a high proportion of them lost their lives—some were killed in battle, but most died of the White man's diseases—smallpox and cholera.

THE 1900 TO 1933 PERIOD

In the period 1900 to 1933, we came as close to a free-market economy in American agriculture, both with respect to product prices and land prices, as we have ever come over the almost 400-year period, 1607 to 2003. The federal government in that 33-year period was almost out of the land-disposal business, and it had not yet gotten into the farm-price and income-support business for commercial agriculture.[iii] The free market generated the "Golden Age of American Agriculture," 1910 to 1914, and the economic hell running from 1921 to 1933.

The economic dimensions of this period were as follows. The farm product-price level rose persistently, if not steadily, between 1900 and 1915, then shot skyward between 1915 and 1919. Farmland prices followed the product-price level: farmland prices in Iowa, for example, doubled between 1900 and 1910, and more than doubled between 1910 and 1920. For the period 1920 to 1933, the farm product-price level fell sharply in 1920/21, recovered modestly in the mid-1920s, and then fell sharply once again between 1929 and 1933. Over the 13-year period, the farm product-price level fell from an index value of over 200 in 1919 to less than 70 in 1933 (1910 to 1914 = 100). Once again the value of farmland followed the farm product-price level; the average per-acre price for the nation as a whole fell more than 50 percent. For Iowa, the per-acre average of farmland fell from U.S. $227 in 1920 to U.S. $60 in 1933.

If one is to judge the economic performance of this free-market period from the perspective of the farmer participants, it might look as follows. If you were a young farmer getting started in farming in 1900, rode the rising product-price level to its crescendo in 1919, sold out in 1920, retired to Long Beach, California, and lived happily ever after, it was the best that could be imagined. But if you were a young man back from the Great War who bought the farm mentioned above, rode the farm product-price level down, found that it was impossible to meet the mortgage payments on the high-priced land of that farm, was foreclosed out-of-business sometime between 1925 and 1933, and ended up in town driving a school bus (if you were lucky), it was the worst that could be imagined. Farming in a free-market situation can be good to you or it can kill you; that is the case where a high proportion of your financial resources are tied up in the asset, land.

THE 1933 TO 2003 PERIOD

The farmland market has been a free market in form from 1953 to 2003. Subject to local zoning laws, a landowner has been free to sell any part of his or her land for whatever price he or she could get for it. And a buyer has been free to purchase land for whatever legal purpose he or she had in mind for it, for any price that he or she was willing to pay. The mechanics of buying and selling farmland have been as free of government control, or regulation, as any free-market ideologue could wish for. *But the prices at which those buying and selling transactions occurred were not true free-market prices, or values.* Since 1953, federal price and income-program subsidies to commercial farmers have held farm incomes above free market equilibrium levels in most years. These above-free-market equilibrium levels of income have enabled farmers to bid up the price of land to levels above the long-run equilibrium; how much above the free-market equilibrium-price level is open to debate. Obviously this price-level discrepancy varied with the magnitude of price and income subsidies flowing to commercial farmers at any particular time. Perhaps the econometricians in one of the later chapters of this book will provide us with some useful estimates.

Until then, it is my firm judgment that if the level of price and income subsidies flowing to commercial farmers since 1996 (and gives promise of continuing into the future) was suddenly terminated, the price level of farmland would fall sharply and importantly. Every farmland owner, rural banker, farm-machinery dealer, fertilizer salesman and Congressman from a farming district from Western Ohio to the Rocky Mountains and from the Canadian border to the Rio Grande *knows* that the major hunk of wealth, in the form of land values, in that broad sweep of land would simply evaporate if major price and income subsidies to commercial farmers were shut down. That is why all of the players noted above fight so hard, and in the past, so successfully, to keep price and income subsidies flowing.

But this is not the end of the story. In the 1933 to 2003 period we had the following situation: rapid farm-technological advance, a free land market, and a fixed supply of land. In this situation, an enterprising few were quick to adopt the new, improved technologies, get their production costs down and reap increased profits. Each enterprising farmer then sought to increase his or her profits further by expanding his or her farming operation by acquiring more land. But where was each enterprising farmer to get more land, when the total supply of land was fixed? They got land from their neighbors, of course, by buying it from them. As more and more enterprising farmers pursued this course of action, bidding for the scarce resource, land, forced the price of that land upward. At the same time, the output of those firms was increasing. As the price of land was rising, production costs to all farmers were rising, and as the total supply of products was increasing, the prices of those products were falling. The treadmill was running.

The profit positions of the early enterprising farmers were now moving back toward zero, and Mr. Average Farmer and all the laggards were feeling the price-cost squeeze. As the financial positions of all these U.S. farmers worsened, they cried out to their government for help, and, as we know, their government has regularly provided them with price-income assistance. We also know that most of this help went to the farmers who produced the most product, thereby putting them in a position to buy more land and produce even more. The consequences of this additional bidding for land has been to increase the price of land still more. Here I leave it to the ingenuity of the econometricians to determine how much of the price of land, say in 2003, is attributable to farm-technological advances and to increased productivity, and how much is attributable to the price and income subsidies to farmers from the federal government.

We do not, however, need an econometric analysis to describe for us the extent to which productive resources in American agriculture have been concentrated in a relatively few hands over the past 70 years. Census data do that for us. In the late 1990s, some 160,000 large to very large production enterprises (I hesitate to call them farms) produced between 60 percent and 70 percent of the total agricultural product of the nation; the remainder was produced on a declining number of full-time family farms, a growing number of part-time farms, and a smattering of Asian truck farms, organic farms, and retirement farms. Over the last 70 years in the context of a free land market, rapid technological advances and generous price and income support, the American farm economy has moved a long way down the road to the Latin-American model, namely, one in which the best land is occupied by large estates, the work on those estates is done by a floating labor force (i.e., peons), and the little people are pushed back into the hills and mountains to live as best they can on subsistence plots of land. The U.S. agricultural economy is not there yet, but it is coming closer every year that passes.

REFERENCES

Carstensen, V. (ed). (1968) *The Public Lands: Studies in the History of the Public Domain.* Madison: University of Wisconsin.
_____. (2003) *American Agricultural Abundance: Solutions for the Future.*
Cochrane, W.W. (1993) *The Curse of American Agricultural Abundance: A Sustainable Solution.* Lincoln: University of Nebraska Press.
Michener, J.A. (1974) *Centennial.* New York: Random House.
Raban, J. (1996) *Bad Land: An American Romance.* New York: Pantheon Books.
Roberts, E. (1930) *The Great Meadow.* New York: Viking.
Shannon, F.A. (1945) *The Farmers Last Frontier: Agriculture, 1860–1897.* New York: Farrar and Rinehart (Chapter III).
USDA/ERS (U.S. Department of Agriculture, Economic Research Service). (1993) *Farm Real Estate, Historical Series Data: 1850–1970.* Washington, DC: USDA/ERS (June).

ENDNOTES

[i] Cochrane, W.W. (1993) *The Development of American Agriculture: A Historical Analysis,* Second Edition. Minneapolis: University of Minnesota Press (173).

[ii] All of the land acquisition of the United States by treaties and purchases are not listed— only those that took the U.S. land mass to the Pacific Ocean.

[iii] Some homesteading took place in Montana from 1900 to 1910. Although that was important in the development of Montana in terms of the national agricultural economy, it was rather insignificant.

Chapter 2

Disaggregating Farmland Markets

Philip M. Raup
Professor Emeritus
University of Minnesota

INTRODUCTION

When we speak of the market for farmland, we invoke a terminology that invites comparison with other markets that are national in scope and include the stock market, the bond market, the grain markets, the markets for used cars, and the oil and gas markets. The key characteristics of markets other than farmland are the specific identification of what is being traded, and their portability.

Shifting this terminology to the market for land introduces confusion. Land is site-specific. Although many tracts are similar and their values can be compared, they are still unique because they are immovable.

This creates limitations on our ability to aggregate the values placed on specific parcels of land as revealed in market transactions. Much effort has been expended and much progress has been made specifying the characteristics of land that give it value. Soil surveys are tedious, labor intensive, expensive, and widely though not yet universally available, within the United States. These soil surveys have been supplemented in recent decades by data on erodiblity, subsurface characteristics, water-retention capacity, and many related physical and chemical properties. It almost seems that, in the most valuable soil and climate regions, everything relevant to the value of a land parcel that can be measured has been measured, or soon will be.

The problem of specifying what is being traded in developing an effective and efficient market for farmland is much closer to solution than ever before. And yet, we are far from developing a basis that enables us to speak of a farmland market in the sense of a national forum that attracts buyers and sellers whose only linkage involves an agreement on price. The farmland market is different and these differences have not diminished over time; they have expanded in ways that will be explored in this chapter.

THE ROLE OF TRANSPORTATION COSTS

The expanding concentric waves generated by a pebble when it is dropped in a pond provide the dominant image that guides our understanding of forces affecting land values. This has remained true throughout the evolution of modern economic analysis. The most explicit early development of this imagery was by J.H. von Thuenen, in *Der isolierte Staat* (The Isolated State), written in the 1820s and 1830s, and first published in complete form in 1842. He assumed an isolated and featureless plain of uniform fertility in which he located a city. He then explored the variety of choices available to landowners when deciding what uses to make of the land, and by implication, what values the various segments of the land would have.

His result was a series of concentric rings of land-use intensity, centered on the city (the market). Perishable garden and vegetable crops occupied the first rings with crops and products of progressively lower ratios of value to weight confined to successively more distant rings. The exception was forestry. The demand for fuel wood and its relatively high cost of transport dictated a forested and woodlot belt near the city. Grazing and pastureland uses were in the outer rings.

von Thuenen's treatise is a remarkable early use of the modeling of a hypothetical situation to promote economic analysis. The driving force of the model, of course, is the cost of transport. It is a precursor of what today constitutes an entire family of gravity models of land use and industrial location, with market access as the key precondition for transportation costs to exercise their determining effects.

The von Thuenen model of determining transportation-cost effects on land use and land value included a river linking his isolated state to the wider market. He used the river to show how proximity to this transportation route distorts and realigns his concentric rings of land-use intensity by reducing transportation costs for the products of nearby land.

The pre-railroad era of north Germany in the 1830s provides a usable model for subsequent analysis of the effects of transportation-cost reduction on land use and land value. It is a model that remains durable to the present day. His river was quickly augmented by the railroad and, in our day, by the autobahn and the super highway. The focus of land use and valuation remains throughout history on transportation costs in time and money, but with major changes.

The concentric-ring model has been altered beyond recognition. First the railroad and now the highway have converted rings of market influence into a visual pattern for which a better image is a starburst. Long streamers of energy explode into the countryside, reordering land use and land value in a hub-and-spoke pattern. The limited goal of this chapter is to highlight some of the ways in which transportation costs and other forces shape the rural land market.

FARMLAND MARKETS: ARE THEY LOCAL?

Tip O'Neal, the sometime speaker of the U.S. House of Representatives, left an enduring legacy in his remark that "all politics is local." Recognizing the hyperbole in this judgment, we can adapt it to argue that the farmland market is overwhelmingly local.

Supporting data are not easily assembled. Land prices are widely reported, but data on who is selling and who is buying are elusive and expensive to collect. In the United States, the property tax is the driving force supporting most systematic efforts to collect and report farmland-market statistics. Throughout history, the desire for public revenue from land use has energized the expenditure of effort to compile data on land ownership, land value, and tax-paying capacity. Historical data—from Egyptian papyri, Sumerian clay tablets, Greek stone property markers, the Doomsday book that followed the Norman Conquest, and modern-day efforts to complete soil surveys and land-classification schemes—confirm this. How much can the ruler, the conqueror, the Crown, or the state expect to collect in taxes? When do marginal tax yields turn negative? The search for answers to these questions has supported the assembly of land-market statistics.

Motivation is weaker for a more detailed analysis of land-market data. In several U.S. states, episodic data exist on who buys and sells land. Time-series data are rare. One exception is available from Minnesota that covers forty years from 1953 through 1992 (Brekke et al. 1993). In this annual survey, a question was asked about the distance in miles between the residences of farmland buyers and their tracts of purchased land. From the results of this survey, stable farmland-buying patterns emerge. Roughly two-thirds of the purchases of farmland have been made by buyers who lived less than 10 miles from the tracts of land they purchased. This percentage rose to 75 percent or 80 percent in areas that had the highest price of land. Purchases by non-local buyers were clustered in areas of higher risk, which were more remote from markets or had lighter soils. Given the proximity to the sales of available farmland and the knowledge of its agricultural value, it seems apparent that the locals exercised a bidding advantage for the better-quality land.

The frequency of sales of farmland was also inversely related to prices paid. The turnover rate in areas of the highest-priced farmland was much lower than it was in the areas that had medium-priced land.

It is unclear whether or not one-state data are representative of land markets outside the midwestern grain and dairy regions. It is also doubtful that the time span involved can be projected into the future. But the data do suggest that the trend toward farm-size enlargement depends primarily on demand for land from the neighbors.

Note some of the implications of these data. It is axiomatic that the best measure of the value of a tract of land is the price a willing buyer will pay a willing seller in an arms-length transaction, which is a sale that is not forced by foreclosure of a mortgage or by some other unusual circumstance. Aggregating sales prices to derive a countywide, regional, or statewide average price of farmland, however, can be very misleading.

The highest priced farmland, which usually has the best quality, sells infrequently. When consolidating sales prices to derive an average price, however, the lower priced lands are over-represented because they sell more frequently. This effect can be moderated by judiciously weighting the data, but this can become highly subjective.

Sales prices are useful in a local setting. This is reflected in their widespread use by appraisers in their search for comparable sales in a specific area to bolster their appraisals of other farmlands in that area. Sales prices lose relevance when they are aggregated. Yet sales prices undoubtedly remain uppermost in the judgment of individuals when they are asked or required to report the value of land for reasons including tax assessments or as respondents to land-market surveys.

Research workers in land economics know this and attempt to compensate by shifting their focus from land prices to land values. The implication is that the data sought should apply to all land, especially to land that was not sold. These are the data most useful when determining a tax base, or when judging the significance of trends when determining debt burdens that are secured by farmland.

THE MELDING OF FARMLAND AND HOUSING MARKETS

Historically, the local nature of farmland markets has been dictated by natural conditions of soil, climate, topography, and location, with the importance of location providing the greatest source of variability. Good soils in remote locations may be lowly valued, whereas poorer soils near active markets attract the investments needed to convert them into virtual capital goods. Thus, location is not the only factor that influences farmland markets, but it dominates.

This dominance of location has been expanded greatly within the United States in the past four decades. The interstate-highway program initiated in 1956, and largely finished by the 1980s, completed the process of opening up the interior of the country that had begun a century earlier with the railroad. But the interstate-highway program did much more because it changed the nature of cities.

By stimulating the construction of feeder roads and commuting corridors nationwide, the interstate-highway revolution extended the influence of cities on land use far into the urban hinterland. It is now quite customary for some workers in central cities to commute by automobile from residences 50 miles to 70 miles distant from their workplaces.

As the periphery of these commuting zones or commuter sheds expands, they do so in an irregular fashion. They are not the concentric rings postulated by von Thuenen. Instead, they follow the high-speed highways.

Transportation and travel-time-to-work studies have shown for over a century a clustering around 20 minutes to 30 minutes to be a tolerable commute. This held true when walking to work was necessary and also when riding to work during the streetcar era was the latest form of transportation. This 20-minute to 30-minute commuting time reappeared in the automobile era with the extension that, when driving on the interstate, tolerable travel time elongated to an hour or more.

This puts an enormous area of farmland inside the potential commuting sheds of central cities. Other chapters in this volume explore the measurement of the areas of several regions (e.g., Barnard et al. 2003). In this chapter, my focus is on the changes these expanding commuting sheds bring to the farmland market.

A most noticeable change of expanding commuting sheds to the farmland market, and especially to the once predominantly farming country, is the reversal of the relative values of buildings on farms. An aphorism in rural America has been the observation that, in a good farming community, the barns usually were better built, were more distinctive, and were more valuable than the residences. In a rapidly increasing area of rural America, these relationships are now inverted, and the residence is becoming the core of farmland value.

In the most explicit agricultural areas that lie outside zones of urban influence, a residence on the land adds nothing to the value of a tract of land. The residence can even detract from the price of land; especially if the residence is the typical 5-acre to 8-acre farmstead with aging outbuildings that reduces the area of tillable land.

In sharp contrast, if the farmland in question is located in a commuting shed, a usable residence with a water source and a septic tank may be the major component of the land's value. This evolution is converting the market for farmland into a dual market that incorporates pluses and minuses from the urban housing market. Farmland and housing markets are increasingly difficult to separate.

Several dimensions of this melding of the two markets deserve emphasis. One dimension is the element of housing-market risk that is being added to the conventional sources of risk when valuing farmland. There are current and widespread concerns that the housing market is exhibiting "bubble" symptoms. That is, prices reflect an abnormal combination of secular high demand; simultaneously they reflect low costs of housing finance. As an example, the aggressive action of the U.S. Federal Reserve Bank taken when lowering short-term interest rates in 2001/02 supported a similar, but not a sharp, fall in real-estate mortgage rates. The effect of an effort to use interest-rate policy to prevent or to reverse an economic recession has stimulated credit markets, particularly the real-estate markets.

The easing of mortgage-interest rates affects farmland markets directly in the same way it affects housing markets, but with the added prospect that a deflation in the housing market could feed deflation into the farmland markets in important rural regions, including the states of the Northeast, the Mid-Atlantic, the Midwest, the Pacific, and the Southwest. A softening in farmland markets triggered by a housing-market weakness is now a more realistic possibility than ever before. The two markets moved more or less independently during the real-estate market collapse in the 1980s; today they are more tightly intertwined.

The insertion of housing-market risk into farmland markets is obvious and measurable. What is less easily perceived is the greater risk resulting from the unusual structure of the 2002/03 U.S. financial market. We now have low interest rates for real-estate credit relative to our own history over the past half century, and relative to other developed countries. At the same time, we have a record of large negative balances of international trade, both in merchandise terms and in the current accounts.

The 2002/03 U.S. financial market is inherently unstable. It is maintained only because other countries are willing to cover our trade deficit by investing their positive balances from trade with the United States into the United States. The nominal rate of return to foreign U.S. investments tells us only a part of the story: The decisive variable is their risk-adjusted rate of return. For most of the past decade this has moved in favor of the United States, but this could change. Whenever interest rates begin to climb, or faith in the stability of the U.S. economy weakens, the volatility of our trade-based role as a debtor could explode painfully. Real-estate-based credit would be among the first sectors to feel the reversal because of its sensitivity to interest-rate trends.

We have little direct foreign investment in U.S. farmland. The effect of a flight from the dollar by foreign creditors will not flood the farmland market with land for sale. But the indirect effects on financial markets could be severe. The direct exposure of U.S. farmland markets to an increased risk in housing markets has been stressed above. A potentially larger risk is from our uncovered international trade deficit.

Another dimension of the urban influence on farmland markets is the reordering of amenity values associated with farmland. A tract of fertile land has always had an attraction that is not fully measured by the capitalized and discounted value of the products it can produce. It promises a food supply, it may define identity, it offers security, and much else. These and many other hard-to-quantify elements enter into the calculation of the value of farmland. This is not new.

What is new in our time is the enlargement of these amenity values to include recreational and environmental values that have grown in prominence with the triumph of urban culture (Libby and Irwin 2003). This, of course, is a part of the explanation for the intrusion of the housing market into the valuation process once considered confined to agricultural productivity. It is now not only its attractiveness as a residential site that can give value to former farmland but also it is its capacity to yield environmental and aesthetic rewards.

It is no accident that the most rapid expansion of forestland acreage in recent decades has been in the heavily urbanized Northeastern states and in the Atlantic- and Gulf-coastal corridors (Smith et al. 2001). Former farmland has passed into the hands of owners who have little incentive to use the land for agriculture purposes but who often hesitate to sell the land, so they leave it idle. Nature takes charge, and it reverts the land to brush and eventually to trees.

This process can also be seen in the revaluation of water and trees. Wetlands that once detracted from farmland value when viewed through farming eyes can become eco-friendly opportunities for wildlife support, species diversity, and water-regime management when viewed by urban buyers. The spread of urban-oriented demand into land markets once thought to be agricultural can introduce complete reversals in the ordering of the components of farmland value.

DIVERGING TRENDS IN FARM SIZE AND LAND-OWNERSHIP UNITS

One of the best-publicized dimensions of American agriculture is the expanding size of farms, especially in the grain and other-cash-crop regions. Between the Appalachian and Rocky Mountains, from Canada to the Gulf of Mexico, the average size of farm-operating units has more than doubled in a single generation. This dramatic change has overshadowed a compositional shift that is less noticed but may be more profound in its long-run effects: The average size of farmland ownership units is decreasing.

A land market comprises ownership units, not operating units. Not long ago these were often identical, but no more. It once was quite common to find statistics on farmland sales that included intact operating units, complete with farm residence and buildings. Into the 1970s, the modal size of farmland sales in the Midwest and Great Plains states fluctuated around 160 acres (the quarter section), which was the acreage of land tracts available under the Homestead Act of 1862 and became the transaction unit used when disposing of much of the public domain.

The decline in the average acreage of sales tracts of what had been classified as farmland is a major indicator of the internal shifts within land tenure in U.S. agricultural regions. The large majority of the children of farm families are leaving agriculture as an occupation, but fewer of them are surrendering ownership of their inherited land. Here again, as with the identity of buyers and sellers, reliable statistics are scarce and tend to be the product of case studies, rather than the product of systematic collection efforts. Available data are revealing.

In the majority of counties that have agricultural land, plat maps or plat books exist that contain maps of ownership tracts compiled from ownership records in county offices. The U.S. Census of Agriculture reported a peak in the number of farms in 1935 for most states. If we use that year as a base and compare plat books for the mid-1930s with contemporary records for the 1990s, one fact stands out: There are many more tract owners today than there were in the 1930s. Making this comparison outside of urban commuting sheds in selected counties of the Middle West and Great Plains states suggests a 50 percent increase in the number of owners, and a decrease in the size of ownership tracts by 33 percent or more. This estimate, however, may be too conservative.

Simultaneously, with an unprecedented increase in the size of farm-operating units, the United States is experiencing a parceling of ownership units. Migration out of agriculture has obviously facilitated farm-size enlargement, because those owners that remain add to their existing farm size the tracts of land of those who depart. The key point in this process of succession in the control of cultivated cropland is that farm-size enlargement is occurring to date primarily through the land-rental market, not through the outright transfer of ownership. The United States has fewer farmers but more farmland owners than ever before in our history (USDA/NASS 2001).

Farm-size enlargement has contributed to an enormous increase in transaction costs in farmland management. It is not uncommon to find farm operators in the corn, soybean, and wheat states who rent their land from a dozen or more landowners. We are approaching levels of fragmentation in farmland ownership that resemble the strip-farm patterns of Napoleonic Europe, where equal division of land among heirs became a symbol of rebellion against the monarchy and against control by a landed aristocracy. With a few isolated exceptions, we have no cultural tradition of primogeniture in the United States. But we are building up a land-ownership pattern that is slanted toward excessive fragmentation of ownership.

Other chapters in this volume (Gardner 2003; Goodwin et al. 2003; Roe et al. 2003; Schmitz and Just 2003; Tweeten and Hopkins 2003) deal with the tendency of government-support programs in agriculture to yield benefits that are capitalized directly into land values and thus flow immediately to landowners. When evaluating the validity of this argument, it is important to keep in mind that generational succession processes are constantly increasing the number of landowners, and land-sales processes are not keeping pace. It is arguable that one of the unintended consequences of government-instigated farm programs has been to interrupt the generational-transition process in farmland ownership.

Although statistics that confirm this tendency would be impossibly expensive to collect, it is likely that today we have our largest-ever population of involuntary farmland owners. Apart from the Eastern and Southern seaboards, most American farmland is in the hands of third-generation to fifth-generation descendants of the original settlers. Again, without benefit of statistical support, it is also probable that we have our largest-ever population of farmland owners whose status reflects a decision not to sell their inherited farmland, rather than a decision to buy farmland. This is relevant to the analysis of choice processes that affect farmland markets. It is also very relevant to the support that can be generated for the family farm. With no possibility of statistical verification, it is also likely that the majority of those who count themselves as supporters of the concept of the family farm do not live on farms. The majority of farmland owners almost surely live in towns and cities. This gives another dimension to problems of interpreting aggregated statistics on farmland markets.

These observations highlight the fact that studies repeatedly show that sales of farmland for farming purposes are triggered primarily by death or retirement. In this regard, they are similar to decisions made by doctors, dentists, or accountants when they sell a professional practice: It is not only income-earning property rights that are being sold but also jobs that define the individuals involved. This is one of the most important ways in which markets for farmland differ from other asset markets. And it forms the basis for emphasizing the cultural components of farmland markets.

The United States is studded with rural communities that shape their attitudes toward the sale or transfer of land through their religious beliefs or practices. These typically are related—but only weakly—to any conventional calculation of profit or loss. Prominent examples are the rich variety of farming based religious settlements, from Pennsylvania westward and from Mexico into Canada, including Amish, Mennonite, and Hutterite. These do not dominate farmland markets in large regions, but they are clearly influential in many local markets.

Sharply high rates of natural reproduction, a commitment to a rural lifestyle, and an ability to mobilize family capital make the religious settlements formidable players in markets for agricultural land. When taking an inventory of the ways in which farmland markets should be differentiated, religious-based communities provide outstanding examples of economic exceptionalism. Conventional approaches to appraisal or to valuation of land in these cases may be irrelevant.

REGIONAL CHANGES IN FARMLAND MARKETS ARISING WITHIN AGRICULTURE

One of the great changes affecting farmland markets has been the virtual disappearance of mixed-enterprise farming. Going back in time by only one generation, we find that the majority of farms had livestock, produced forages and crops for cash sale, and followed land-use patterns that included a variety of different crops. This can be illustrated with data from Minnesota, which historically has been one of the most diversified agricultural states in the United States (Raup 2002, 1994).

In 1950, corn or maize, wheat, and soybeans were grown on 40 percent of the total land cultivated in Minnesota. This increased to 75 percent from 1990 to 1994, and these crops today account for over 85 percent of all acres of harvested cropland. Oats in 1950 occupied almost five times the acreage planted to soybeans, and the acreage in flax exceeded the acreage in soybeans. Today, oats and flax have become minor crops on Minnesota farms (Minnesota DOA 2002). The state's agricultural economy is heavily dependent on corn, soybeans, and wheat, each of which is highly sensitive to export markets. This shift has been repeated throughout all regions in which corn, soybeans, or wheat predominate.

One consequence of the increased farm-operating-unit size—those of which have specialized in the production of fewer products or have been coupled with the concentration of animal production in large batteries, feedlots, or dairies—has been the relative revaluation of grasslands. This is prominent in a swath of northern lands from the Great Plains to the Atlantic, which includes major parts of Minnesota, Wisconsin, Michigan, Pennsylvania, New York, and the New England states, but is not confined to this area. Revaluation is also apparent in some southern and intermountain states, and in the Pacific coastal states.

The major driving force for this revaluation of forage-producing lands has been the substitution of feed concentrates (primarily grain and soybeans) for grasses that were used previously to feed ruminant animals (especially dairy cows). When grasses were a major feed source, the family dairy provided a superior opportunity for the investment of family labor in the creation of value-added products in farming. Production of feedstuffs could be combined with initial processing into animal products to provide more nearly year-round employment.

The separation of animal production from the production of their feeds has had two major consequences. First, it has restricted the usefulness and reduced the relative value of lands unsuited to the production of non-forage grain and oilseed crops, but capable of producing a forage crop that can be converted into meat, milk, or wool in the ruminant stomach.

Second, it has increased the economic exposure of farms on lands producing grains and feed concentrates to the vagaries of climate. In a drought year they have little shock-absorbing capacity in the form of alternative enterprises. We can see the result in the current incessant pressure for drought and disaster relief for farms in the grain belt. When measured in terms of annual employment, many of them have become part-time farms, but even this opportunity has been sharply reduced.

The government farm-program payments from the beginning in the 1930s focused on price supports for cash crops, for which export markets were expanding. They had the unintended consequence of transforming animal agriculture. The value profile of the national stock of farmland was reconfigured. Where grassland utilization confronted the option of a land-use shift to residential or to related urban-oriented uses—including horse farms, sod farms, and hobby farms—the transformation has not diminished the market for farmland. On the contrary, a country area of near-defunct dairy farms can be the focus of a land developer's dream and can become a major component of the farmland market.

But where there has been little incentive to shift grasslands to urban-driven uses, grasslands and forage-producing lands have lost relative value. There are many reasons for this shift. The opportunity cost of family labor has risen sharply. Research into cash crops has pushed the limits of sown-crop production ever further into the boundaries once set by rainfall and frost. Advances in irrigation have supported dairying in sometime semi-arid regions and have promoted a dominant concentration of beef feedlots in the southwestern Great Plains states. Rice and cotton now occupy lands once utilized as pasture for cattle ranches. The investment of research effort and human capital to accomplish these shifts can be attributed in good part to government farm programs, as distorting as they may seem. Whether or not the consequential reordering of values in land is desirable or durable remains an unanswered question.

The decline of mixed-enterprise farming has increased the regional characteristics of farmland markets in the contiguous 48 states in other important ways. One way is that it has sharpened the definition of regions that are explicitly exposed to trends in international trade. Historically, U.S. agriculture has always been involved in international markets. Apart from early colonial periods, this sensitivity was largely confined to a few products, primarily wheat, cotton, tobacco, and pork. Up to the mid-twentieth century, the regional dependence on foreign markets centered on areas that produced these products.

Change in this pattern, however, accelerated during and after World War II. Today, we export roughly one-half of our annual wheat production; just under one-half of our rice, sorghum, and soybean output; and one-fifth of our corn. The big pattern change has been in the added export dependence of feed-grain producing and soybean-producing regions, primarily in the Middle West (USDA/FAS 2002).

With the exception of wheat, land values in this region responded primarily to the domestic market until well after 1950. Recall that this region was home to the America First movement of the 1930s and 1940s, with its isolationist overtones. Now agriculture in this region is heavily dependent on export trade. Land-use decisions over a much-expanded area of the nation have been internationalized, with a consequent increase in risk exposure.

This added risk introduced by concentration on one or two crops, and on animal production in factory-like units, has three dimensions for both plants and animals: price risk, climate risk, and disease risk.

These risks have always plagued agriculture, but their potential for agricultural disaster has been multiplied. As a result, the stability of markets for agricultural land has lost some of its aura of permanence and security that has been historically an attribute of investment in food-producing land. Farmland markets in the future are unlikely to vary with the volatility of stock exchanges, but the chance of a wipeout of values has clearly increased, with regional impacts more explicitly defined.

REFERENCES

Barnard, C., K. Wiebe, and V. Breneman. (2003) See Chapter 18, this volume.

Brekke, J., H.L. Tao, and P.M. Raup. (1993) *The Minnesota Rural Real Estate Market in 1992.* Economic Report ER 93-5 and annual preceding reports, 1953 to 1992. St. Paul: University of Minnesota, Department of Applied Economics (July).

Gardner, B. (2003) See Chapter 5, this volume.

Goodwin, B.K., A. Mishra, and F.N. Ortalo-Magné. (2003) See Chapter 6, this volume.

Libby, L. and E. Irwin. (2003) See Chapter 19, this volume.

Minn DOA (Minnesota Department of Agriculture and USDA). (1850 to 2002) *Minnesota Agricultural Statistics 2002.*

Raup, P.M. (2002) "Reinterpreting Structural Change in U.S. Agriculture." In M. Canavari, P. Caggiati and K.W. Easter, eds., (185–200), *Economic Studies on Food, Agriculture, and the Environment.* Norwell/Dordrecht: Kluwer Academic/Plenum Publishers.

____. (1994) "Four Decades of the Minnesota Rural Real Estate Market." *Minnesota Agricultural Economist* (675).

Roe, T., A. Somwaru, and X. Diao. (2003) See Chapter 7, this volume.

Schmitz, A. and R. Just. (2003) See Chapter 4, this volume.

Smith, W.B., J.L. Vissage, D.R. Darr and R. Sheffield. (2001) *Forest Resources of the United States, 1997.* U.S. Department of Agriculture, Forest Service, GTR-NC-219. St. Paul: North Central Research Station.

Tweeten, L. and J. Hopkins. (2003) See Chapter 8, this volume.

USDA/FAS (U.S. Department of Agriculture, Food and Agricultural Service). (2002) *Grain: World Markets and Trade.* USDA/FAS Report FG 12-02. Washington, DC: USDA/FAS (December and preceding issues).

USDA/NASS (U.S. Department of Agriculture, National Agricultural Statistics Service). (2001) *Agricultural Economics and Land Ownership Survey (1999),* Vol. 3, Special Studies, Part IV, 1997 Census of Agriculture, AC 97-SP-4. Washington, DC: USDA/NASS (December).

von Thuenen, J.H. *Der isolierte Staat,* Part I, first published in Hamburg in 1826, revised and published with Part II in Rostock in 1842. Perhaps the most readily accessible edition is that published in Jena by Gustav Fisher in 1910 as Vol. XIII of the *Sammlung Sozialwissenschaftlicher Meister.* A comprehensive review of von Thuenen's significance for contemporary thought is in Sonderheft 213, *Berichte Ueber Landwirtschaft,* Muenster (2000), and in Sonderheft 215, *Berichte Ueber Landwirtschaft,* Muenster (2002).

Chapter 3

Farmland Markets:
Historical Perspectives and Contemporary Issues

Bruce J. Sherrick and Peter J. Barry
University of Illinois

INTRODUCTION

Debate about future farm policy is underpinned with historical evidence about the performance of the sector and about major asset classes within the sector. This chapter provides a general overview of farmland markets in the United States including a summary of farm real-estate values, farm structure, and control arrangements, turnover rates, and market frictions. The performance of agricultural real estate as an asset class is examined vis-à-vis other investment alternatives, and relative to its specific risks. The chapter then offers a list of issues that deserve attention in future research and in debates about the effects of government policy on farmland markets.

Farmland occupies a uniquely important role in the performance of the agricultural sector because of its dominance on the balance sheet of agriculture (Tables 3.1a and 3.1b). In aggregate, farm real estate accounts for roughly 75 percent of the total value of assets in the U.S. agricultural sector. The share of agricultural-sector debt supporting real estate is not proportional, however, it accounts for just over 50 percent of total-sector debt. By comparison to other sectors, the aggregate debt-to-asset (D/A) ratio of the sector is low, at roughly 16 percent. The data in Tables 3.1a and 3.1b show that while the nominal values of assets, liabilities, and equity have changed greatly for the past five decades, the relative composition of the balance sheet has remained fairly constant through time. Debt shares are included, as these have shifted among lenders, although there have been only small movements between debt and equity.

Table 3.1a Balance Sheet of U.S. Agricultural Sector (1960, 1970, and 1980)

Item	1960	1970	1980
Assets:	Thousand U.S. dollars		
Real estate	123,280,105	202,417,225	782,819,263
Livestock and poultry	15,607,597	23,705,761	60,633,203
Machinery and equipment	19,068,447	30,363,932	80,347,162
Crops	6,359,991	8,700,574	32,831,043
Purchased inputs	0	0	0
Financial assets	10,033,998	13,676,837	26,674,177
Total farm assets	174,350,138	278,864,329	983,304,848
Liabilities:			
Real estate debt held by:			
Farm Credit System	2,222,301	6,420,357	33,224,684
Farm Service Agency	623,895	2,179,873	7,435,059
Commercial banks	1,355,733	3,328,876	7,765,058
Life insurance companies	2,651,587	5,122,291	11,997,922
Individuals and others	4,408,545	10,308,264	27,813,346
CCC storage and drying loans	47,523	146,276	1,456,359
Total real estate debt	11,309,593	27,505,932	89,692,429
Non-real-estate held by:			
Farm Credit System	1,509,061	5,303,115	19,749,504
Farm Service Agency	369,168	699,478	10,029,347
Commercial banks	4,716,945	10,491,172	29,985,795
Individuals and others	4,540,901	4,752,996	17,366,580
Total non-real-estate debt	11,136,073	21,246,762	77,131,229
Total farm liabilities	22,445,666	48,752,694	166,823,658
Farm equity	151,904,472	230,111,635	816,481,190
Selected ratios:	Percent		
Debt/equity	14.8	21.2	20.4
Debt/assets	12.9	17.5	17.0
Real estate value/equity	81.2	88.0	95.9
Real estate value/assets	70.7	72.6	79.6
Real estate debt to total debt	50.4	56.4	53.8

Source: Authors' compilation from USDA/ERS (2002) for 1960 to 1980.

Table 3.1b Balance Sheet of the U.S. Agricultural Sector (1990 and 2000)

Item	1990	2000
Assets:	Thousand U.S. dollars	
Real estate	619,149,242	929,453,515
Livestock and poultry	70,856,057	76,825,534
Machinery and equipment	86,297,864	92,038,510
Crops	23,178,199	27,932,428
Purchased inputs	2,807,525	4,895,713
Financial assets	38,319,683	57,114,380
Total farm assets	840,608,570	1,188,260,080
Liabilities:		
Real estate debt held by:		
Farm Credit System	25,924,490	31,766,396
Farm Service Agency	7,639,490	3,657,025
Commercial banks	16,288,128	31,835,855
Life insurance companies	9,703,958	11,825,503
Individuals and others	15,169,299	18,388,795
CCC storage drying loans	6,506	0
Total real estate debt	74,731,876	97,473,581
Non-real-estate held by:		
Farm Credit System	9,848,060	16,757,205
Farm Service Agency	9,374,181	3,873,724
Commercial banks	31,267,499	44,572,769
Individuals and others	12,739,994	21,300,301
Total non-real-estate debt	63,229,734	86,504,003
Total farm liabilities	137,961,610	183,977,584
Farm equity	702,646,960	1,004,282,496
Selected ratios:	Percent	
Debt/equity	19.6	18.3
Debt/assets	16.4	15.5
Real estate value/equity	88.1	92.5
Real estate value/assets	73.7	78.2
Real estate debt to total debt	54.2	53.0

Source: Authors' compilation from USDA/ERS (2002) for 1990 to 2000.

In addition to its dominance in the agricultural sector, farm real estate has other features that deserve special consideration. Among its distinguishing characteristics are its (1) nondepreciability, (2) historically large capital-gains component relative to current income in total return (Barry and Robison 2002), (3) its low correlation with returns to other (financial) asset classes, (4) irreversible development potential and nonagricultural-use influence in value, (5) the capitalization of government payments into land values (USDA/ERS 2001a, 2001b), and (6) fixed supply.

As the residual claimant of farm income (see the discussion of residual rents in Erickson et al. 2003) the impact of government payments on farm income is of particular importance. Understanding linkages between income flows, nonsector macro influences, and market frictions becomes increasingly important when assessing the impact of government policy on farmland values. While an understanding of the historic features of the farmland market provides neither a direct vision of the future nor a recipe for government policy related to farmland, it does provide an important foundation and context for future research and debate.

Average land values through time provide important summary evidence of the performance of farmland as an asset and also suggest other relationships that likely influence farmland values. Figure 3.1 shows average aggregate farmland value per acre in the United States from 1960 to 2001, with farmland's share of total farm assets on the right-hand axis. The familiar run-up in values in the late-1970s and early-1980s followed by the precipitous drop in the mid- and late-1980s breaks an otherwise long progression of capital gains. While there are pronounced regional differences in levels of farmland values, the patterns of change are similar throughout most of the major agricultural regions of the United States.

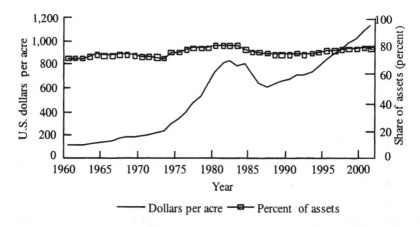

Figure 3.1 Average U.S. Farmland Value per Acre and Average Percent Share of Assets (1960 to 2001)

Since 1960, the continuously compounded nominal capital-gains rate has averaged 5.6 percent per year (geometric average). Taken since 1970, the capital-gains rate has averaged nearly the same at 5.7 percent. But from 1980 to 2001 alone, the capital-gains rate has been just over 2 percent. Thus, the data period used to implicitly represent the structure of agriculture is critically important to conclusions about its financial performance.

Estimates of income to farmland investments vary depending on the approach used, but taking cash rent less property taxes as a share of value results in estimates of rates of return to land in the 3-percent to 5-percent range (Lins et al. 1992). Combining returns with capital gains under this approach yields estimates of total returns in the 5-percent to 10-percent range. Alternatively, estimating returns by the allocation of shares of farm income to asset classes that are based on their relative asset shares and depreciability minus their property taxes (Miller and Sherrick 1994), and combining this with their capital gains results in estimates of total rates of return to farmland from 6.5 percent (from 1980 to 2001) to 9.97 percent (from 1970 to 2001). These estimates depend heavily on the time period under consideration. Under both approaches, however, relatively low current returns illustrate the inherent non-liquidity of leveraged investments in farmland in which interest rates on borrowed funds may exceed the rate of earnings from the land and can result in debt-service requirements that exceed the cash flow generated.

THE ROLE OF LEASING IN FARMLAND MARKETS

Complicating the assessment of factors that have influenced the value of farmland is the continued evolution in the structure of ownership of farmland and the control of farm units. As shown in Figure 3.2, the number of farms has continued to decline through time from nearly 5.6 million farms in 1950 to approximately 2 million in 2001. The definition of a farm unit has changed as well.

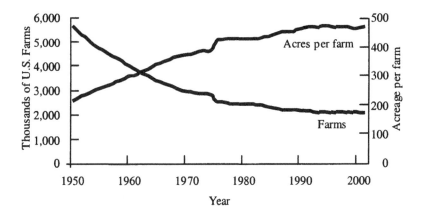

Figure 3.2 U.S. Farms and Average Acres per Farm (1950 to 2001)

Acres farmed in the United States have declined at a rate of just under .5 percent per year over the past five decades, and have fallen from 1.2 billion acres in 1950 to 948 million acres in 2001 (Figure 3.3). As a result, the average farm size has increased correspondingly from just over 200 acres per farm in 1950 to 470 acres per farm in 2001. Again, these aggregate numbers mask considerable regional variation, but nonetheless they demonstrate important trends that show no sign of stopping or reversing. For U.S. agriculture as well as for many other countries, farmland markets have two major dimensions—ownership and control of the farmland asset. The ownership dimension is defined as farmers who own the land they operate. The control dimension is defined by the leasing of farmland under cash rent, share rent, a combination of cash and share rent, or by the hiring of farming services.The leasing market is substantial. Nationwide, U.S. farmers lease about 45 percent of the land they operate (Table 3.2) while in Illinois, farmers lease about 65 percent of the land they operate.

Among larger, commercial-scale family farms, the incidence of leasing is still higher. In Illinois, for example, the approximately 6,500 farmers who participated in the Farm Business Farm Management Association in 2001 reported tenure ratios (acres leased to total acres operated) that averaged 78 percent for Northern Illinois, 86 percent for Central Illinois, and 75 percent for Southern Illinois.

It is important to consider both ownership and leasing dimensions when analyzing farmland markets. In ownership transactions, the focus is on the value of the land, which often is determined by farm real-estate appraisers using the comparable-sales method of land valuation, or the income-capitalization approach. In leasing transactions, however, the focus is on the determination of the rent (both its level and its form) for the acreage involved. At the aggregate-market level, value and rent come together through the capitalization process

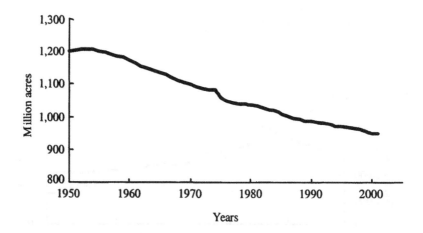

Figure 3.3 Acres of Farmland in the United States (1950 to 2000)

$$V = \frac{P(1+g)}{i-g}, \qquad (1)$$

where P is the current rent, i is the real cost of capital, g is a real growth rate, and V is the value per acre.

An important point to remember about the valuation of farmland is that farmland-leasing markets determine the returns (rent) to the farmland asset. Moreover, the growing reliance on cash leasing (e.g., in the Midwest region of the United States where share leasing is widespread) results in a more explicit determination of the level of those returns because the value of rent is more easily determined under cash leasing than it is under share leasing. Understanding the factors affecting rents is therefore fundamental to understanding land values. Improving the availability of farmland-rental information will in turn enhance the efficiency of farmland markets. Among the factors influencing rent are government payments, land quality, financing costs, farm-expansion pressures, and local-market practices. Nonfarm effects on farmland (e.g., urban and commercial development) also influence the magnitude and variability of future land values, which are more difficult to determine than are present-day rents.

A review of the history of the leasing market can be insightful, although changes in leasing arrangements make contemporary and forward-looking analysis essential to farmland-market evaluations. Historically, leasing of

Table 3.2 1999 Percentage of U.S Farmland Owned, Leased, and Type of Lease by Subregion

Subregion	Owned[a]	Leased[b]	Type of lease[c]			
			Cash	Share	Cash/Share	Other
			Percent			
United States	54.7	45.3	59.4	24.0	10.8	5.9
Northeast	69.2	30.8	80.0	2.6	3.4	14.0
Lake	60.5	39.5	79.4	8.0	7.4	5.2
Corn Belt	49.5	50.5	45.4	34.0	18.7	1.9
Northern Plains	47.4	52.6	54.7	28.5	13.9	2.9
Appalachian	63.7	36.3	56.4	16.5	12.0	15.1
Southeast	72.8	27.2	82.2	3.7	3.5	10.6
Delta	47.7	52.3	51.0	34.6	9.8	4.5
Southern Plains	51.2	48.8	65.8	18.6	6.6	9.0
Mountain	61.7	38.3	61.3	25.9	8.1	4.7
Pacific	52.2	47.8	65.6	19.9	6.2	8.2

[a]Source: Authors' compilation from USDA/NASS (2001: Table 3)
[b]Source: Authors' compilation from USDA/NASS (2001: Table 3)
[c]Source: Authors' compilation from USDA/NASS (2001: Table 99)

farmland has occurred in largely informal markets under relatively simple contracts. Market boundaries are unclear, turnover rates of leased tracts of land in the past have been low, and information about rental prices and rates of return have been limited. Lease contracts often are oral and short-term, reflecting practices in local markets and the reputations of the leasing parties involved, although the short-term contract often becomes a long-term business relationship due to close ties between landowner and farmer (Allen and Leuck 1992). Culture and tradition also have significant influences on leasing markets and rental arrangements (Young and Burke 2001).

Two long-standing trends in land-leasing markets have been (1) the expansion of absentee ownership of farmland whereby owners may reside at greater distances from the farm or be institutional investors concerned about rates of return from their land investment, and (2) the greater reliance on cash-leasing arrangements relative to share-leasing arrangements (Barry et al. 2000). For many farmers, the cash-leasing arrangement is believed to provide simpler, more flexible bidding opportunities for expanding leased acreage (Barry et al. 1999). Greater reliance on the cash-leasing arrangement enables the farmer to have greater management autonomy, especially when multiple landlords and the production of specialty crops are involved. Under cash leases, landlords may also achieve more stable returns over time, adjust more easily to ownership costs of changing land values, eliminate sharing of operating expenses, avoid management participation and crop-marketing decisions, remain eligible for social security payments at retirement, and bid more readily for government payments (under share-leasing arrangements, government payments tend to be allocated in the same proportions that the crop is shared).

Increased cash-leasing arrangements could also reflect inequities in share levels due to rigidities in the traditional sharing arrangements (Barry et al. 2000). The 50/50 crop-share lease that is common in much of Midwestern agriculture is an equal, yet not necessarily equitable, approach to the allocation of returns between farmers and landowners. Equal shares may overcompensate landowners that have low-quality soil and may under-compensate those that have high-quality soil. The spread between the values of share rent and cash rent may not be large enough to compensate the landowner for the greater risks and agency costs he has under share leasing (Barry et al. 2000). Econometric analysis of the cash lease as opposed to the share lease in Illinois suggests that a greater likelihood of the cash lease is related significantly to higher income variability, smaller tracts of leased acreage, shorter contract relationships between landlords and farmers, and farmers with larger net-worth positions and higher debt-to-asset ratios (Sotomayor et al. 2000). The level of cash rents in the lease contracts under study in this chapter are associated primarily with differences in soil productivity, tract size, and net worth. Generally, the growth factor of farms is more influential than is the risk consideration in the choice of leasing contracts and the level of cash rent.

According to the Agricultural Economics and Land Ownership Survey (AELOS) (USDA/NASS 2001) much of the recent surge in cash leasing has come about because of increases in a combination of cash/share leases under which some fixed payment is combined with returns dependent on production (Table 3.2). Nationwide, the acreage under a cash/share lease was 10.8 percent in 1999, up from 3.2 percent in 1988 (the latter figure is not shown in the table). The incidence of cash/share leasing was highest in the Corn Belt region (18.7 percent) followed by the Northern Plains region (13.9 percent).

For individual states, the cash/share lease arrangement comprised 26.5 percent of lease numbers and 26.0 percent of acreages in Indiana for 1999, which was up from that of 1.8 percent and 2.1 percent, respectively, in 1988. Similar magnitudes of increase occurred in Illinois (from 2.3 percent and 3.6 percent in 1988 to 24.4 percent and 26.9 percent in 1999). Increases up to the 13-percent to 16-percent range occurred in Ohio, Kansas, and Nebraska. Iowa, Missouri, and the Dakotas experienced increases in the 9-percent to 12-percent range (USDA/NASS 2001). Much of the 1999 increase in cash/share-leasing combinations reflected the decline in share leasing, while the number of cash leases alone remained relatively steady, even though average cash-lease size declined.

Leasing arrangements in U.S. agriculture are becoming more sophisticated. However, determining equitable lease pricing remains a problem. Greater precision in lease pricing is possible albeit demanding in terms of data needs, determining rates of return to contributed assets, and negotiating strength between farmers and landowners (Barry et al. 2000). Nonetheless, improved understanding of lease-pricing alternatives can enhance the equitability of leasing contracts, expand the range of contract choices, promote mutually compatible incentives for the parties involved, and heighten the efficiency of leasing markets through greater standardization of leases. Improved information can also lead to relationships between rental rates and land values that are more clear. Natural roles for government include facilitating and enforcing contract laws, and improving the provision of, and access to, public information that is related to leasing land.

OWNERSHIP TURNOVER RATES AND MARKET EFFICIENCY

Detailed aggregate data on farmland acreage sold for farm and ranch use, or for other agricultural uses, are available periodically from the Census of Agriculture and the AELOS follow-up surveys. While historic rates of rural land sales have been reported at 3.5 percent for all rural land (USDA/ERS 1989), specific turnover data indicate that farmlands sold for farm and ranch purposes have occurred at a much lower rate. According to aggregate AELOS 1999 data, just over .55 percent of the total farmland in the United States was sold annually and nearly 18 percent of those sales have been for nonfarm or non-ranch purposes. These 1999 values contrast with rates of 1.4 percent total and 7.8 percent for nonfarm purposes from the comparable 1988 AELOS data. Average turnover rates through time have been more problematic to construct, and regional differences in nonfarm development seriously impact evidence of turnover rates of farmland sold for continued farming or of farmland sold that has relatively little development potential. Thus we turn to more specific data from Illinois for evidence about the turnover rates of farmland sold that could be considered open market, or available for acquisition by either a competing farmer or an investor.

The University of Illinois has collected and developed a farm-level transfer database that utilizes Illinois Department of Revenue transfer declarations. The database contains records for all parcels in the state of Illinois greater than 5 acres that were sold during the 1979 to 1999 period. Complete records exist for more than 68,000 parcels, and represent just over 4.2 million acres of the state total acreage of approximately 27.7 million acres.

From these University of Illinois title-transfer data, we calculate a series of variables to determine if the sales were considered arm's length or if other considerations (e.g., equipment) were involved in the title transfer that determined a net price for each real-estate sale. We exclude estate transfers to related parties from those sales considered to be arm's length and available to a public investor, but we include contract-sales units in our data that contain information describing the terms of financing. In addition, the title-transfer data we use for this study include separate indicators of building-tax assessments; information about whether the real estate was improved, unimproved, or had other-use categories; and information about the terms of financing if seller financing was involved. Further, we segment the data into regions to control for issues such as differing population pressures on land values.

To calculate annual Illinois turnover rates, we sum the acreage from all sales in a year classified as arm's length, all farm-use classes, and all locations by date and divide by farm acres in the state as calculated by the U.S. Department of Agriculture (USDA) National Agricultural Statistics Service (NASS). The data in Figure 3.4 reveal the resulting turnover rates. The statewide Illinois farmland turnover rate ranged from a low of .38 percent in 1982 to a high of 1 percent in 1994. The average over all periods was .73 percent, or just under three-quarters of one percent. Our results reflect a relatively unrestricted definition of farm acreage. These results would decline further if sales were screened to control for small acreage, high-valued parcels sold either for future development or sold in regions whose long-run farming prospects were not high (e.g., Chicago-collar counties).

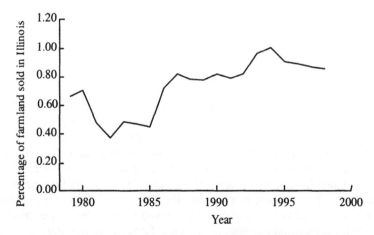

Figure 3.4 Farmland Turnover Rates in Illinois (1979 to 1999)

The thinness of the farmland market is indicated by low turnover rates and is important to consider when evaluating the performance of farmland as an investment. While not shown, we find a parcel-size-to-price effect in the data, such that parcels less than 20 acres sold for significantly more per acre than did larger tracts of land. In empirical studies, parcel size likely proxies for development potential (residential or otherwise).

The increasing reliance on leasing by farm operators discussed earlier is also related to the low turnover rates observed in farmland markets. If little farmland is sold, then per-farm acreage increases must take place through additional leased acreage. On one hand, increased demand for leasing to overcome liquidity and debt-servicing difficulties associated with farmland ownership could explain low turnover rates. Alternatively, other factors leading to low turnover (including high capital gains, incentives to transfer real estate at death, limited divisibility, and high transaction costs) limit the ability of a producer to acquire farmland through purchase. Under these circumstances, the only way for a producer to increase his amount of acreage is to enter into lease agreements.

SPECIALIZED INSTITUTIONAL FEATURES OF FARMLAND MARKETS

Making direct comparisons between farmland markets and other asset markets is complicated by farmland's unique characteristics and by the numerous specialized institutions related to farmland markets. Despite the value represented in farmland markets, there is no well-functioning equity market for trading shares of farmland or claims on farm income. There have been a few limited-partnership funds (e.g., Westchester Farmland Limited Partnerships, and farmland mutual funds proposed in the 1970s), and there have been limited (and as yet unsuccessful) attempts to develop farm real-estate investment trusts (REITs) containing farmland (e.g., *harVestco*). Thus, land ownership is still very much parcel-specific.

On the debt side, specialized credit institutions exist that have dedicated missions related to the provision of debt capital to agriculture [e.g., Farm Credit System (FCS), Farmer Mac, and Farm Services Administration (FSA)]. In contrast to the absence of equity markets for farmland, both FCS and Farmer Mac pool farm mortgages and issue mortgage-backed securities and/or debt securities to financial market investors. (FCS and Farmer Mac are government-sponsored enterprises, and FSA is a government agency.) Other than housing, few other sectors have similarly dedicated lending enterprises.

Seller financing represents another unique arrangement within agricultural real-estate markets. While seller-financing arrangements do not account for a large share of lenders, they do serve a unique financial role in the sale of farmland. Historically, seller financing has represented as much as one-third of farm real-estate lending, but its share has been declining over time as farm real-estate lending by financial institutions increased and as changes in income tax laws made contract sales less attractive for sellers.

Another feature of farm real estate is its frequent property and estate-transfer taxation at artificially low values rather than at market values. Many states have adopted favorable property-tax treatment (i.e., use-valuation assessment) for farmland in response to the argument that landowners bear disproportionately high shares of the tax base in rural communities. Other states have adopted preferential-tax treatment to encourage continued farm usage and to discourage land-development growth through recapture of use-valuation tax benefits when the land is sold. The federal government also offers a use-value option to value farmland in estate transfers if various eligibility conditions are met. A plausible response to preferential-tax treatment is that market values increase to equilibrate the effect of favorable tax treatment, with a resulting (government-policy induced) transfer to landowners.

Environmental amenities (and externalities) associated with farmland and its usage in customary farming practices serve as a source of increasing potential conflict with nonfarm populations (Libby and Irwin 2003). Restrictions through use-zoning and environmental compliance (either incentive versions or punitive versions) can affect materially the value of specific farmland parcels. Likewise, preservation groups have become more visible in recent years through their efforts to sequester farmland in permanent farm-use forms of ownership (Libby 2003, Lynch and Lovell 2003). Other headlines have been generated related to takings—the appropriation of property rights by governments—when use by others or in other forms is viewed as being of greater public value than the displaced private-property rights of ownership. Put more directly, the potential influences on farmland values of issues related to environmental issues, natural amenities, good-neighbor practices, water rights, zoning additions, takings, and green practices, to list a few of the more prominent phrasings, are likely to continue to increase in the future. As each of the above represents a conflicting interpretation of property rights, the role of government in the resolution of each will be central.

Government farm programs influence both the level and variability of the incomes of those owning and/or using farmland. Most visible are the traditional commodity programs [including loan deficiency payments (LDPs) and marketing loan programs] and transition payments [including Agricultural Market Transition Act payments (AMTAP), production flexibility payments (PFCP), and market loss assistance payments (MLAP)] associated with the 1996 Federal Agricultural Improvement and Reform Act (FAIR) (Goodwin et al. 2003). Other forms of government participation in agriculture include disaster payments, subsidization of crop-insurance programs, and conservation reserve program payments.

While the ostensible intent of FAIR was to limit the involvement of government in agriculture and to be a transition to lower levels of government support, direct payments to agriculture remained a large share of net farm income and will continue under the terms of the 2002 Food, Security, and Rural-Investment Act (FSRI). According to U.S. Department of Agriculture data, direct-government payments between 1970 and 2001 represented 23.9 percent of the aggregate net income from farming operations, and represented 32 percent of farm income exclusive of direct government payments [i.e., direct government payments divided by net income prior to government payments (USDA/ERS

2002)]. Since 1990, total direct government payments have represented 27 percent (including government payments) and 37 percent (prior to government payments) of net farm income. Since 1996 these figures have increased to 32 percent and 47 percent, respectively.

The simple correlation between government payments and aggregate net farm income from 1970 to 2001 (including livestock components) is positive at .16, since much of the payment stream is tied to agricultural production activities rather than to direct countercyclical payments (although the effect at the individual producer level is more likely countercyclical). As a result, the standard deviation of net farm income through time is actually higher when it includes government payments (U.S. $12.1 billion) than it is when it excludes government payments (U.S. $9.5 billion). Of course, the average level of net farm income including government payments is also much higher, so it would not be fair to classify government payments as risk-increasing in this regard. Combining these measures, the coefficient of variation of net farm income over this sample period is .37 when it includes direct farm payments, and is .38 when it excludes direct government payments.

While the long-term prospects for government participation in agricultural income support and stabilization programs are uncertain, historic levels of the agricultural sector's income have ranged from 20 percent to 30 percent. If capitalization of income holds as a basis for asset valuation, farmland values reflect expected levels of future incomes. If future income expectations reflect the historically high and relatively stable share of income provided to the sector by direct government payments, the implications are equally direct in that changes in policies that affect farm income will have commensurate effects on farmland values. Further, to the extent that government programs are income stabilizing or risk reducing, the same directional effect results in that government programs reduce the required capitalization rate through lower risk, and thereby result in higher land values. Removal of these effects through the elimination of government programs would result in the decline of farmland values.

PERFORMANCE OF FARMLAND INVESTMENTS

Numerous studies have considered different attributes of farmland values, rates of return to farmland investments, and farmland market structure and performance. A seminal contribution by Melichar (1979) establishes the relationship between growth in the earnings of farmland and growth in land values, with the attendant feature of unrealized capital gains representing a significant portion of the total economic return to farmland. The impact of the level of farm income on land values has been clearly established (Castle and Hoch 1982; Alston 1986). Impacts of financial costs, credit availability, and financing properties have been identified and validated (Shalit and Schmitz 1982). The positive impacts on land values by government support of farm incomes and commodity prices have been conceptualized and measured (Harris 1977; Boehlje and Griffin 1979; Barrickman and Herriges 1992; USDA/ERS 2001a and 2001b). Relationships between economic and accounting rates of return to farmland have been clarified (Barry and Robison 2002). The applicability of present-value models to the estimation of land values has been delineated and qualified (Falk 1991; Clark et al. 1993; Just and Miranowski 1993). The dynamics of land prices have been measured in terms of leads and lags in market determinants, and in terms of informational components (Phipps 1984; Burt 1986; Featherstone and Baker 1987; Falk 1991; Moss 1997; Falk and Lee 1998). The impact of transaction characteristics, such as tract size and location, has been empirically analyzed (Palmquist and Danielson 1989; Stewart and Libby 1998; Hardie et al. 2001). The value of rights to future land development has been demonstrated (Plantinga and Miller 2001). Spatial econometrics has accounted for the use of unobservable variables that are correlated with farmland values and location (Hardie et al. 2001).

From a market-portfolio perspective, previous studies have assessed the relationship between farmland returns and various market proxies, and have found little evidence of systematic (market) risk in farmland investments across a wide array of approaches. Pioneered by the Barry (1980) article, capital asset pricing model (CAPM) applications consistently indicate that farmland adds little systematic risk to the market portfolio, and that risk premiums would be close to zero on investments by well-diversified investors. These studies reflect a wide variety of refinements, including explicit consideration of inflation (Irwin et al. 1988), arbitrage-pricing theory and the range of agricultural assets (Bjornsen and Innes 1992; Arthur et al. 1988; Collins 1988), market segmentation (Lins et al. 1992; Shiha and Chavas 1995) and commodity pricing (Bjornsen and Carter 1997). Each study addresses farmland risks and risk premia relative to nonagricultural investments. Only Chavas and Jones (1993) find a significant relationship between land values and the variability of farm income, while Chavas and Thomas (1999) find significant relationships between land values, risk aversion, and transaction costs. Krause and Brorsen (1999) also find a significant inverse relationship between revenue variability and cash rents on farmland.

To summarize, under CAPM-like formulations, farmland βs are generally low or insignificantly different from zero over many periods, over different market proxies, and with various other control variables included. Little evidence exists of strong systematic risk components in returns to farmland. While findings on the degree of excess or superior returns are more varied, most authors find neutral positive measures of excess return. When adjusted for market frictions and other linkages, the excesses over risk-adjusted returns sometimes disappear, but the bulk of the literature applying traditional models of financial performance to farmland puts farmland investments in a good light.

We present evidence about the relative performance of farmland as an investment by comparing risk and returns across a wide variety of competing assets. We collected data from the National Association of Real Estate Investment Trusts (NAREIT) on all publicly traded REITs to provide an alternative real-estate benchmark. For market indexes, we collected returns data on the Dow Jones Index, S&P 500 Index, NYSE cap-weighted, the Moody's North American Index, and Europe, Asia, and Far East Total Returns Index (EAFE) (Morgan Stanley). Also, we collected returns on corporate bonds rated Aaa to Baa, as well as on commercial paper to provide representative corporate debt investments (Morgan Stanley, Federal Reserve). We compiled various Treasury series including yields on 3-month, 1-year and 10-year constant maturity series published by the Federal Reserve. We developed inflation indicators including the Consumer Price Index (CPI) and the Producer Price Index (PPI) (Bureau of Labor Statistics). We developed farmland returns by calculating farmland's annual portion of net farm income based on asset values and depreciation rates of other farm assets (Miller and Sherrick 1994), less property taxes, plus capital gains, divided by USDA value per acre. We developed a comparable series using cash rent less property taxes, but it could not be extended to the full period due to unavailable data after 1996.

Table 3.3 contains summary correlations between U.S. farmland and the other listed asset classes and indices from 1970 to 2000. The correlations in returns between farmland and market indexes (EAFE, Dow, S&P, and N. American) are all negative, while the correlations with the inflation measures are highest at .63 when we used CPI and .72 when we used PPI. These results vary greatly by sample period, and are only presented here to demonstrate the historic evidence on systematic risk and inflation risk.

Table 3.3 Correlation between Returns to U.S. Farmland and other Selected Investment Indices (1970 to 2000)

Asset/Index	Correlation with farmland
REITs (real-estate investment trusts)	-0.133
EAFE[a] (Far East total returns index)	-0.472[a]
Dow[b]	-0.429[b]
S&P 500	-0.342[c]
North American Index	-0.330[c]
10-year treasuries	-0.226
3-month treasuries	0.036
Commercial paper	0.087
Baa bonds	-0.294
Aaa bonds	-0.300
CPI (consumer price index)[a]	0.630[a]
PPI (producer price index)[a]	0.722[a]
T-Bills	0.008

[a]Significant at 0.01 level,
[b]Significant at 0.05 level, and
[c]Significant at 0.10 level.
Source: Authors' compilation.

When run in a traditional CAPM formulation using the 1-year treasury rate as the risk-free rate, the resulting estimate of excess return is positive and significant relative to either the Dow or the S&P as the market proxy (with excess returns approximately 2.2 percent and significant). In both cases, β is insignificantly different from zero. These results illustrate the type of analysis and findings common in the literature, and reinforce the low correlations with returns to broad-based market indexes. Continuing from this perspective, we interpreted low systematic risk by an investor to exert downward pressure on the required risk-adjusted discount rate and thus to lead to an increase in land values.

Figure 3.5 extends the U.S. farmland investment analysis within the context of other investments. This graph contains the results from solving for the minimum-risk portfolio across a parametrically varied set of mean returns. We constructed Figure 3.5 by considering risk-efficient combinations of farmland, REITs, S&P 500, 10-year Treasury securities, 3-month Treasury securities, and commercial paper as the allowable investment universe. The risk-efficient frontier (commonly termed the E-V frontier or the maximum expected value-variance frontier return E for each given level of risk V) was solved subject to restrictions that farmland would receive no more than 50-percent weight, and each of the debt securities would receive no more than 25-percent weight. These restrictions and classes of competing investments put farmland in a good light in that the set of assets chosen for comparison were those with the most favorable correlations and returns relative to farmland.

Expected value-variance frontier

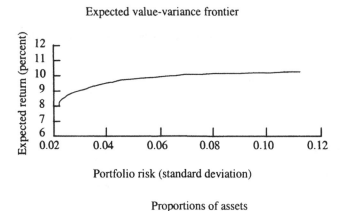

Portfolio risk (standard deviation)

Proportions of assets

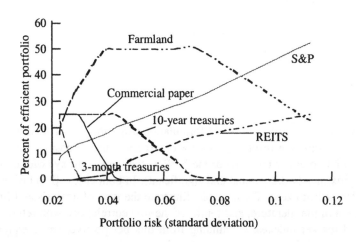

Portfolio risk (standard deviation)

Figure 3.5 Risk-Efficient Asset Combinations and Shares versus Risk

The top panel of Figure 3.5 displays the E-V frontier, while the bottom panel identifies shares of assets present at each point along the E-V frontier. At low levels of risk, the portfolio is dominated by Treasury securities, and farmland begins to substitute primarily for these risk-free assets as returns are required to increase. Moving to higher levels of risk, farmland continues to substitute for debt securities, but eventually it is replaced by equity-index investments. As we expected, the shares committed to equities and REITs increased as the required rate of return increased. Of significance is the high relative share represented by farmland along the E-V frontier (despite the fact that few investors hold such high relative farmland positions). The results do not reflect the relatively high transaction costs of acquiring farmland. Instead, the results emphasize that low correlations with other asset classes have led to historically favorable performance of farmland as an asset class in aggregate.

Despite these findings, it may be difficult for either individuals or institutional investors to capture these effects through direct investment. Individual farmer-investors generally need to keep their tracts of land in a small geographic area (Raup 2003). Thin markets limit the speed with which size adjustments can be made, and institutional investors require a management infrastructure to access farmland as an investment. Transaction costs, which were ignored in all the returns series, are much higher for farm real estate than they are for any of the other classes of assets. (Including transaction costs would reduce the relative attractiveness of farmland as an asset.) Further, from an individual farmer's perspective, concentrated investment in a specific set of agricultural assets can involve substantial idiosyncratic risks. Thus, excess returns relative to market risks may still be needed to compensate for specific risks if there are sufficient market impediments to holding the assets in a well-diversified portfolio.

Within the class of agricultural investments, we also determine adjustments in returns due to differences in risk across different farmland investments. To assess this issue, we develop two additional data series to reflect the cost of control and revenue level and to reflect variability among different farmland locations. The first series utilizes Illinois Farm Business-Farm Management (FBFM) data to develop average cash rents from 1996 to 2000 by county for 92 of the 102 counties in Illinois (excluding the Chicago area and a few counties with only a few farms). The second series represents gross crop revenue reflecting local cash basis, cropping patterns, and harvest dates. To construct this series, we compile county-level data from USDA/NASS (2001) on crop yields for corn, soybeans, and wheat, along with their acreage by county. These data are combined with Illinois Ag-Marketing Services basis data and historic harvest-progress data to generate harvest data gross revenue weighted by crop and adjusted for local cash-price basis information (i.e., the average gross revenue per acre in a county that reflects local crop weights, harvest-date prices, and local cash-market-basis details). (Details of basis calculations and yield data are available at http://www.farmdoc.uiuc.edu.) We then construct the fraction of average gross revenue paid in cash rent for each county.

By direct analogy to certainty-equivalent arguments, we test the hypothesis of whether or not cash rent as a fraction of expected revenue will be discounted more (have a higher risk premium) for higher-risk circumstances. Algebraically, the certainty equivalent is

$$y_{ce} = \mu_y - \lambda\sigma_y^2, \tag{2}$$

where y_{ce} is the certainty equivalent, or the amount one would pay for a risky return with mean μ_y and variance σ_y^2, and λ is a measure of the decision-maker's risk-aversion coefficient. We view cash rent cr as the fixed payment made in exchange for the risky returns from the crop, and if production costs are relatively fixed proportion k of average returns, then dividing by the mean and rearranging relates the fraction of mean gross return paid in cash rent to the measure of the risk per unit mean return, or

$$\frac{cr}{\overline{\mu}_y} = k - \lambda\left(\frac{\sigma_y^2}{\overline{\mu}_y}\right). \tag{3}$$

To test this relationship, the share of average revenue paid in cash rent by county was regressed against the per-acre premiums charged for Actual Production History (APH) crop insurance by county (Table 3.4). The APH premiums were for 85-percent election levels on a farm in each county with the same average yield as the county. Relative premiums by county represent an independent assessment of risk, but on a specific crop and not reflecting local crop-mix or harvest-date differences. The results are negative and significant for both corn and soybean premiums (soybean results are not shown). This analysis demonstrates that rental rates do adjust to reflect differences in risk within farmland markets.

Table 3.4 Regression Statistics for Cash-Rent on Average Production History (APH) Crop-Insurance Premiums

Variable	Estimated coefficients
Constant	0.60[*]
	(0.03)[a]
Average Production History Crop-Insurance (APH) Premium	–0.010[*]
	(0.002)
Goodness of fit	
R^2	0.23
Adjusted R^2	0.22

[*]Denotes statistical significance at the 0.01 level of confidence.
[a]Numbers in parenthesis denote standard errors.
Source: Authors' computations from the Illinois Farm Business-Farm Management data.

OTHER ISSUES IN FARMLAND MARKETS

"Past performance is no guarantee of future returns." This sentiment is particularly relevant to farmland markets in light of several changing conditions in agriculture, and in the face of increased pressures for fundamental reform in agricultural policy. This section identifies some of the more prominent contemporary forces affecting farmland markets and suggests issues appropriate for consideration in future policy debates.

Government Payments and Farmland Values

The future role of government in agriculture is a major source of uncertainty affecting agricultural production and farmland markets. While the agricultural lobby remains powerful by most observers' assessments, the farm agenda is increasingly evolving toward a food agenda, and is under ever-increasing pressures to recognize linkages to environmental facets of agricultural production. Potential future linkages of production and green payments could fundamentally alter the degree of stability of direct agricultural payments to producers.

Additionally (although it is difficult to detect in recent data), the policy rhetoric suggests that the government intends to develop risk-management programs that have strong private-market linkages and incentive structures to replace the government's traditional role in income stabilization through production-linked payments. The cornerstone of these efforts is reflected in crop-insurance programs and in the intent by the government to transfer payments in the form of subsidized premiums to attract producer participation. Although it is unlikely that the government will ever fully eliminate ad hoc disaster payments, development of private or quasi-private risk-management markets in agriculture will be important first steps.

Changing Lease Arrangements

Changing lease arrangements and increasing separation of ownership and use represent other forces that will change the nature of farmland markets in the future. Agricultural policy may be usefully targeted to develop other forms of lease/risk-sharing contracts that substitute for other risk-bearing market mechanisms, and backfill for the absent (liquid and transparent) equity markets in agriculture. In addition, the farmland market of the future could involve a more complete separation of property rights, and could develop new markets for previously unseparated rights (e.g., specific natural amenities, carbon sequestration, and hunting rights), which would be in addition to the existing production-use value.

Nonfarm Demand for Land

Continuing nonagricultural development and the slow removal of farmland from agricultural uses will increase in importance as the supply of farmland shrinks and population pressures continue to grow. Long-run usage plans are appropriate roles for governments to consider (not necessarily with the intent to force a particular outcome, but to insure that the market signals are appropriately and completely transmitted so that prices contain appropriate information about alternative uses). In any case, nonfarm pressures will only increase in the future. Complete and transparent information is an important ingredient to use when trying to avoid market failures and frictions.

Market Efficiency

Many of the characteristics of farmland markets suggest that existing markets are not highly efficient. In particular, high transaction costs, low turnover, low debt-service capacity, and other indicators of inefficiency in land markets including wide bid-ask spreads could be reduced with better information. Perhaps less attention should be paid to land values per se, and more attention should be paid to the development of more extensive and higher-quality publicly available information about prices, leasing rates and practices, location differentials, and contract terms. Government policy could play a role in enhancing market efficiency in farmland markets through information channels just as it has for other types of assets (e.g., disclosure, reporting requirements, grades, and standards). Doing so would require resolving competing interests between those holding private information, who may benefit from specialized information (e.g., farm managers), and the public, whose interest is distributed among all existing and potential investors in farmland.

Risk and Farmland Values

Whether or not rates of return are appropriate measures for the risk represented in farmland ownership is central to the formation of sound government policy that subsequently impacts farmland markets. Historic governmental support of agriculture, whether direct or otherwise, has been an important component in understanding the structure of agricultural-asset markets and particularly the asset of farmland itself. Further evidence of the relationship between farmland value and returns to risk bearing is needed to better understand how farmers in particular and landowners in general are compensated for risks in agriculture.

CONCLUSION

History suggests that the major attributes of farmland markets (rent and values) have been important indicators of the wealth position and financial performance of the agricultural sector. In part, these roles reflect the dominance of farmland among the assets of the farm sector. Farmland investments have also been shown to provide favorable risk characteristics to well-diversified investors. Moreover, the effects of various government programs have been directly reflected in these farmland-market attributes through their effects on the level and stability of the returns to farm assets.

When looking to the future, there is little reason to expect that these fundamental relationships will change significantly. However, important policy issues and research needs will continue, especially regarding the distributional implications of farmland markets, risk and return relationships, the efficiency of leasing and ownership markets, equity-capital markets, effects of environmental-use and land-use regulation, and the delineation of ownership, financing, and managerial control over farmland. The future research agenda is clearly a rich one.

REFERENCES

Allen D. and D. Leuck. (1995) "Risk Preferences and the Economics of Contracts." *Journal of the American Society of Farm Managers and Rural Appraisers* 85: 447–51.

_____. (1992) "The 'Back-Forty' on a Handshake: Specific Asset Reputation, and the Structure of Farmland and Contracts." *Journal of Law, Economics, and Organization* 8: 366–76.

Alston, J.M. (1986) "An Analysis of Growth of U.S. Farmland Prices, 1963–82." *American Journal of Agricultural Economics* 68(1): 1–9.

Arthur, L., C. Carter, and F. Abizadeh. (1988) "Arbitrage Pricing, Capital-Asset Pricing and Agricultural Assets." *American Journal of Agricultural Economics* 83: 359–65.

Barnard, C.H., G. Whittaker, D. Westenbarger, and M. Ahearn. (1997) "Evidence of Capitalization of Direct Government Payments into U.S. Cropland Values." *American Journal of Agricultural Economics* 79(5): 1642–50.

Barrickman, N. and J. Herriges. (1992) "The Implicit Value of Corn-Base Acreage." *American Journal of Agricultural Economics* 74: 50–8.

Barry, P.J. (forthcoming) "The Cash/Share Combination in Farmland Leasing." *Journal of the American Society of Farm Management and Rural Appraisers.*

_____. (1980) "Capital Asset Pricing and Farm Real Estate." *American Journal of Agricultural Economics* 62(3): 549–53.

Barry, P.J. and L.J. Robison. (2002) "Agricultural Finance: Credit, Credit Constraints, and Consequences." *Handbook of Agricultural Economics.* New York: Elsevier Publishing.

Barry, P.J., N.L. Sotomayor, and L.M. Moss. (1999). "Professional Farm Managers' Views on Leasing Contracts and Land Control: An Illinois Perspective." *Journal of the American Society of Farm Managers and Rural Appraisers* 62: 15–9.

Barry, P.J., L.M. Moss, N.L. Sotomayor, and C. Escalante. (2000) "Lease Pricing for Farm Real Estate." *Review of Agricultural Economics* 22(Spring/Summer): 2–16.

Bjornsen, B. and C. Carter. (1997) "New Evidence of Agricultural Commodity Return Under Time-Varying Risk." *American Journal of Agricultural Economics* 79(August): 918–30.

Bjornson, B. and R. Innes. (1992) "Another Look at Returns to Agricultural and Nonagricultural Assets." *American Journal of Agricultural Economics* 74(1): 109–20.

Boehlje, M. and S. Griffin (1979) "Financial Impacts of Government Support-Price Programs." *American Journal of Agricultural Economics* 61(2): 285–96.

Burt, O.R. (1986) "Econometric Modeling of the Capitalization Formula for Farmland Prices." *American Journal of Agricultural Economics* 68(1): 10–26.

Castle, E.N. and I. Hoch. (1982) "Farm Real-Estate Price Components." *American Journal of Agricultural Economics* 64(1): 8–18.

Chavas, J.P. and A. Thomas. (1999) "A Dynamic Analysis of Land Prices." *American Journal of Agricultural Economics* 81(4): 772–884.

Chavas, J.P. and B. Jones. (1993) "An Analysis of Land Prices Under Risk." *Review of Agricultural Economics* 15: 351–66.

Clark, J.S., M. Fulton, and J.T. Scott, Jr. (1993) "The Inconsistency of Land Values, Land Rents, and Capitalization Formulas." *American Journal of Agricultural Economics* 75(1): 147–55.

Collins, R. (1988) "The Required Rate of Return for Publicly Held Agricultural Equity." *Western Journal of Agricultural Economics* 13(2): 163–69.

Erickon, K., A. Mishra, and C.B. Moss. (2003) See Chapter 12, this volume.

Falk, B. (1991) "Formally Testing the Present-Value Model of Farmland Prices." *American Journal of Agricultural Economics* 73(1): 1–10.

Falk, B. and B. Lee. (1998) "Fads Versus Fundamentals in Farmland Prices." *American Journal of Agricultural Economics* 80(4): 696–707.

Featherstone, A.M. and T.G. Baker. (1987) "An Examination of Farm Sector Real Asset Dynamics: 1910–85." *American Journal of Agricultural Economics* 69 (3): 532–46.

Goodwin, B.K., A. Mishra, and F.N. Ortalo-Magné. (2003) See Chapter 6, this volume.

Hanson, S.D. and R.J. Myers. (1995). "Testing for a Time-Varying Risk Premium in the Returns to U.S. Farmland." *Journal of Empirical Finance* 2(3): 265–76.

Hardie, I.W., T.A. Narayan, and B.L. Gardner. (2001) "The Joint Influence of Agricultural and Nonfarm Factors on Real-Estate Values: An Application to the Mid-Atlantic Region." *American Journal of Agricultural Economics* 83(1): 120–32.

Harris, D. (1977) "Inflation-Indexed Price Supports and Land Values." *American Journal of Agricultural Economics* 59: 489–95.

Irwin, S., L. Forster, and B. Sherrick. (1988) "Return to Farm Real Estate Revisited." *American Journal of Agricultural Economics* 70: 580–87.

Just, R.E., and J.A. Miranowski. (1993) "Understanding Farmland Price Changes." *American Journal of Agricultural Economics* 75(1): 156–68.

Koenig, S., C. Dodson, and J. Ryan. (1996) "Leased Capital *v* Debt Capital in U.S. Agriculture." *Journal of Agricultural Lending* 9(Summer): 16–21.

Krause, J.H. and B.W. Brorsen. (1999) "The Effect of Risk on the Rental Value of Farmland." *Review of Agricultural Economics* 17: 71–6.

Libby, L. (2003) See Chapter 17, this volume.

Libby, L. and E. Irwin. (2003) See Chapter 19, this volume.

Lins, D.A, B.J. Sherrick, and A. Venigalla. (1992) "Institutional Portfolios: Diversification through Farmland Investments" *American Real Estate and Urban Economics Journal* 20: 549–71.

Lynch, L. and S.J. Lovell (2003) See Chapter 16, this volume.

Melichar, E. (1979) "Capital Gains Versus Current Income in the Farming Sector." *American Journal of Agricultural Economics* 61 (5): 1085–92.

Miller, L.H. and B.J. Sherrick. (1994) "An Examination of Farm-Asset Returns." *Journal of Agribusiness* 12(1): 15–35.

Moss, C.B. (1997) "Returns, Interest Rates, and Inflation: How They Explain Changes in Farmland Values." *American Journal of Agricultural Economics* 79 (4): 1311–18.

Palmquist, R. and L.E. Danielson. (1989) "A Hedonic Study of the Effects of Erosion Control and Drainage on Farmland Values." *American Journal of Agricultural Economics* 71: 55–62.

Phipps, T.T. (1984) "Land Values and Farm-Based Returns." *American Journal of Agricultural Economics* 66(4): 422–29.

Plantinga, A.J. and D.J. Miller. (2001) "Agriculture Land Values and the Value of Rights to Future Land Development." *Land Economics* 77(1): 56–67.

Raup, P.M. (2003) See Chapter 2, this volume.

Shalit, H. and A. Schmitz. (1982) "Farmland Accumulation and Prices." *American Journal of Agricultural Economics* 64: 710–19.

Shiha, A.N. and J.P. Chavas. (1995) "Capital Segmentation and U.S. Farm Real Estate Pricing." *American Journal of Agricultural Economics* 77(2): 397–407.

Sotomayor, N.L., P.N. Ellinger, and P.J. Barry. (2000) "Choice Among Leasing Contracts in Farm Real Estate." *Agricultural Finance Review* 60: 71–84.

Stewart, P.A. and L.W. Libby. (1998) "Determinants of Farmland Values: The Case of DeKalb County, Illinois." *Review of Agricultural Economics* 20: 80–95.

Young, H.P. and M.A. Burke. (2001) "Competition and Custom in Economic Contracts: A Case Study of Illinois Agriculture." *American Economic Review* 91 (June): 559–73.

USDA/ERS (U.S. Department of Agriculture, Economic Research Service). (1989) *Rural Land Transfers in the United States*, Agricultural Information Bulletin 574, U.S. Department of Agriculture, Washington, DC.

_____. (2001a) *Government Payments to Farmers Contribute to Rising Land Values* Agricultural Outlook, U.S. Department of Agriculture, Washington, DC.

_____. (2001b) *Higher Cropland Value from Farm Program Payments: Who Gains* Agricultural Outlook, U.S. Department of Agriculture, Washington, DC.

_____. (2002) Value Added Net Farm Income series, 1950-2000, various files. Internet Website: http://www.ers.usda.gov/Data/FarmIncome

USDA/NASS (U.S. Department of Agriculture, National Agriculture Statistics Service). (2001) *Agricultural Economics and Land Ownership Survey, (AELOS 1999)*. Internet Website: http://www.nass.usda.gov/census/census97/aelos/aelos.htm

Section II:
Government Policies and
Farmland Values

Chapter 4

The Economics and Politics of Farmland Values

Andrew Schmitz and Richard E. Just
University of Florida
University of Maryland

INTRODUCTION

Land is a major input in world agriculture. Historically, farmland prices have shifted significantly over time, and so has the wealth of those who own farmland. Fluctuations in farmland values and the associated variations in wealth are akin to the rise and fall of the stock market. A drastic reduction in stock-market prices represents a reduction of household wealth. In his Waugh lecture on "Boom-Bust Cycles and Ricardian Rent," Schmitz emphasizes the volatile nature of land markets and the significance of wealth changes. He distinguishes between changes in wealth from net realized farm income and cases in which farmland prices increase even though net farm income decreases. According to Schmitz (1995: 1110–11)

> Critical to understanding the boom-bust phenomenon are the dynamic changes in wealth as distinguished from net realized-farm income. During the North American agricultural boom that peaked in 1981/82, the value of farm real estate in the United States exceeded U.S. $800 billion—more than double the 1974 value—only to drop in 1987 to U.S. $597 billion. The changes in Saskatchewan, Canada, were even more dramatic. Between 1972 and 1982 the value

of land and buildings increased by a factor of roughly 7.0, from CDN \$3.6 billion to CDN \$27.1 billion. The value of implements and machinery increased six-fold between 1972 and 1986. Realized net-farm income, however, did not grow nearly as rapidly as did asset values. Interestingly, Kansas and Montana, two large wheat-growing areas, did not experience the same degree of escalation and depreciation in land values that Saskatchewan did, even though U.S. farm programs provided much more income insurance than did Canadian programs, especially during the boom phase of the cycle. The U.S. government wrote off proportionately more farm debt than did the Canadian government.

In some regions, land has many competing uses both within agriculture and from urban growth. The latter, in certain states, has caused land values to rise and wealth to landowners to skyrocket to the point at which farmland is bid out of agricultural use. This is best highlighted in the Gilbert and Akor (1988: 64–5) study on dairying in California that states:

> In a capitalist society with fee-simple ownership, few controls on property use, and a 'free market' in land, rapid urbanization creates real estate booms. ...In the 1950s, [Los Angeles] (L.A.) County added almost 2 million people to its total and in the next decade, [they added] another million. Between the late 1950s and late 1970s, it lost nearly 16,000 farm acres. Land prices were U.S. \$8,000 per acre to U.S. \$30,000 per acre, occasionally reaching U.S. \$90,000 per acre. Dairy owners with 15 acres to 20 acres sold their land for development and for windfall profits. ...Some of the former L.A. dairy owners moved to the rural southern San Joaquin Valley. But most went only to the western tips of neighboring San Bernardino and Riverside counties, to an area called the Chino Valley. ...The dairy operators purchased 10 acres to 80 acres of farmland, built state-of-the-art dairies, and constructed lavish new homes beside the facilities. Finally, with the remaining capital, they were

> still able to enlarge their herds, which often
> doubled or tripled their size. ...It was
> through these cycles of urbanization, reloca-
> tion, and expansion that industrial dairying
> was born. The windfall of urban prices for
> farmland, combined with the Dutch 'dairy
> culture' of Southern California, enabled the
> capitalization of large-scale dairying. A case
> study sums up the process: One farmer re-
> lates that he bought his first dairy farm in
> Artesia [southeast of L.A.] in 1950 at a price
> of 12 acres for U.S. $12,000. He sold the
> land for subdivision for U.S. $300,000 and
> bought 30 acres in Dairy Valley, now
> Cerritos. In 1970 he again sold to sub-
> dividers for U.S. $1.3 million and moved to
> Chino. (1988: 64–5)

Because of increased urbanization, there are many cases in which the price
of farmland is a result of its potential sale value for housing and for real-estate
development. On these types of land, agriculture eventually disappears, because
the income from agriculture cannot support the high price of land. This is especially true in parts of Florida such as Miami-Dade County and in Central Maryland around Washington D.C. where land has extremely high values because of
urban pressures.

A major factor that must be considered when understanding land prices is
agricultural policy. In the United States, agricultural policy provides a major
source of farmer income. For example, in 2000 and 2001 over 70 percent of net
farm income for North Dakota farmers came from government subsidies. The
new U.S. Farm Bill of 2002—the U.S. Food, Security, and Rural-Investment Act
(FSRI)—will provide record subsidies to farmers. Part of this book is devoted to
determining the extent to which farm payments are capitalized into land values
and whether or not farm payments distort production and trade. Researchers, like
Tweeten (1986), argue that the main beneficiary of farm programs is the landowner. However, this book points out that considerable controversy still exists
on this topic. Interestingly, in farm states like Illinois, land values and cash rents
both reached an all time high in 2001 in the presence of some of the lowest
commodity prices in history. This illustrates that subsidies are ultimately transferred, at least in part, to farmland owners through various mechanisms. These
mechanisms include adjustments in cash-rental rates as a response to changing
subsidy levels and adjustments in lease writing such that crop-share leases include a percentage of the subsidy payment (Gardner 2003). As an example of a
crop-share lease that includes a percentage of the subsidy payment is a statement
by one landowner who rents his land on a crop-share basis who said his lease
reads "...paying, therefore as rental for said term [one-fourth of fall crops pro-
duced] on said land and one-fourth of any payment for participation of this property in any government entitlement of subsidy program during said term...."

This chapter provides a broad perspective of the economics of land values. It provides some theoretical arguments that must be considered when studying land markets. We recognize the many studies that have been done on the factors that influence farmland markets, although we note that some land models are only modestly successful at tracking land values. For example, Schmitz (1995) finds that land values during boom-bust cycles are positively correlated with interest rates. However, more consistent with the model suggested by Just and Miranowski (1993) and models of earlier capitalization, we show in this chapter that land values over time bear a relationship to net farm income that is influenced by farm policy. We do so by comparing land values among selected Canadian provinces and by comparing Canadian land values with those of the Northern United States. We then discuss land markets in the context of farm policy and show the relationship between land markets and decoupled farm payments. Next, we discuss the impact of urban growth and trade on land values and show how comparative-advantage principles affect land markets. (For example, lowering import quotas can have a major impact on land values in an import-competing country, yet quotas can have little impact on a country's production and imports.) We conclude our chapter with a discussion of the impact of politics on land values. Also, we note that since the land-value collapse of the mid-1980s, land values in the United States have risen consistently over time. U.S. farmers will continue to lobby governments through rent-seeking activities to prevent a decline in land values and to avoid another land-market crash and its disastrous consequences.

THEORY

Land is an important input in agricultural production. As a fixed factor, it is assumed that land garners most of the producer surplus (Just and Hueth 1979). Theory suggests that land prices vary with future agricultural earnings. For example, a simple land-price model follows the capitalization framework

$$V_t^A = \sum_{i=0}^{\infty} \delta^i E_t\left(Y_{t+i}\right) \tag{1}$$

where V_t^A is the value of farmland at time t, $E_t(Y_{t+i})$ is the per-acre net rental rate for farmland at time $t+i$ expected at time t, and δ^i is a discount factor.

Farmland often has alternative uses that include recreation and urban development. The model in Equation (1) can be expanded to include alternative uses. Assuming conversion to recreation or to urban development is an irreversible decision, a model with alternative uses may follow

$$V_t^D = \max_n \left\{ \sum_{i=0}^{n} \delta^i E_t\left(Y_{t+i}\right) + \sum_{i=n+1}^{\infty} \delta^i E_t\left(R_{t+i}\right) \right\} \tag{2}$$

where V_t^D is the value of farmland considering its development potential, n is the year farmland is converted to recreation or urban development, and $E_t(R_{t+i})$ is the rental rate from land that has been converted to recreation or to urban development at time $t+i$, expected at time t. Clearly, the value of land considered to be of development potential in Equation (2) is higher than the value of land for agricultural purposes in Equation (1), $V_t^D \geq V_t^A$.

In Figure 4.1, we consider the relationship between product-market prices and land-rental rates. Equilibrium price and quantity (P_e, Q_e) in the product market corresponds to the equilibrium price and quantity (W_e, X_e) in the land and other input markets. Suppose the final output price decreases as a result of a change in agricultural policy. Production will then fall to Q^1 in the product market given supply $S_0(P, W_e)$, and land utilized for production will decrease to X^1. In turn, the associated decrease in the rental rate for land will lower the cost of production of the final good, and will shift the supply curve from $S_0(P, W_e)$ to $S_1(P, W^1)$. This increases production at P^1 to Q^2. The resulting supply curve $S^*(P)$ includes land-market adjustments (i.e., it does not consider the land market as fixed), and intersects (P_e, Q_e) and (P^1, Q^2) at a and c, respectively. The change in producer surplus in the product market, area $P_e P^1 ca$, will be exactly equal to the change in total economic surplus in the land market, area *dehg*. In cases in which farmers rent their land, the welfare loss for the farmer in the product market is area $P_e P^1 ca$, which is equivalently measured by area *defg* in the land market. The welfare loss for the landowner in the product market is area *abc*, which is equivalently measured by area *fhg* in the land market. For farmers

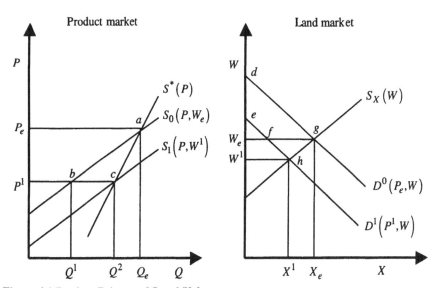

Figure 4.1 Product Prices and Land Values

who own their land, welfare effects include both owner and renter effects combined because the producer is both the consumer and the supplier in the market for land (Just et al. 1982). Note, however, that these impacts do not consider the wealth effects that reduced rental rates have on land prices.

Consider the extreme case in which all farmers produce on rented land (Figure 4.2). Assume that rents are paid on a cash basis and the supply of land for rent is perfectly inelastic. Initial equilibrium is again at (P_e, Q^1) in Figure 4.2a and (W_e, X_e) in Figure 4.2b. Suppose a support price of P^S is introduced in Figure 4.2a. Ignoring for simplicity the possible increased use of other production inputs, as in the case of fixed-proportions production, output does not increase because additional land is not brought into production. The demand for land increases from $D^0(P_e, W)$ to $D^1(P^S, W)$ in Figure 4.2b but this only causes the rental rate for land to increase because no more land is drawn into use. Production does not increase but rather the supply of output shifts from S to S^1 in Figure 4.2a due to the higher cash rents the farmers must pay. The equilibrium supply corresponding to $S^*(P)$ accounts for land-market adjustments that become perfectly inelastic at Q^1. Thus, the tenant pays the landlord the

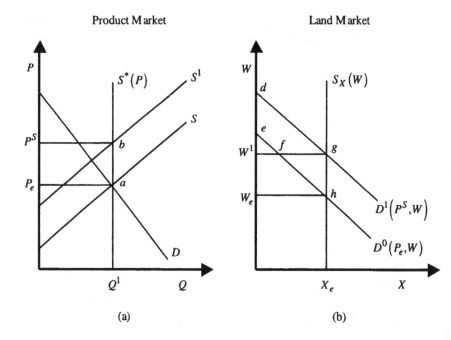

Product Market   Land Market

(a)   (b)

Figure 4.2 Inelastic Land Supply and Output Price Subsidies

increased revenue generated by price supports. In this extreme of inelastic-land supply and fixed-proportions production, landowners rather than farmers collect all the extra rent $P^S P_e ab$ due to price supports as landowners command a higher rent from their renters for their land.

Using this same analysis, we assume that inelastic-land supply and fixed-proportion agricultural production can be relaxed to show that a farmer can receive a proportion of the increased benefits from his subsidized-output prices. But the proportion of the benefits received by farmers is lower with more inelastic land supply and with more inelastic supply of other inputs. Even though land may have urban-development alternatives, the one-way and permanent nature of farmland conversion likely causes farmland supply to be highly inelastic in the short run. And while agricultural production is not restricted to fixed proportions, the fixed-proportions assumption has approximated reality sufficiently that linear programming models are widely used when modeling farm production. Thus, a large share of the benefits of an agricultural subsidy goes to the landowner. These benefits go to the farmer only if the farmer owns the land he uses for production.

The ownership of farmland by those currently farming has decreased over time (Table 4.1). In Iowa, for example, the percentage of farmland that was cash rented increased from 21.1 percent in 1982 to 34.9 percent in 1997, while the percentage under crop-share lease increased from 21.1 percent to 23.7 percent. At the same time, the percentage of land operated solely by the owner decreased from 54.1 percent to 30.8 percent. As a result, if land values were to increase due to farm programs, many of the benefits would not flow to the producer. Instead, landowners who do not farm would capture a large share of the program benefits through the rents they receive from their renters.

BACKGROUND

A number of studies have tested variations of the land-market model found in Equation (1) or (2). Although a correlation between farm income and land price has been found, it is not the only, or even the major, determinant of farm

Table 4.1 Tenure of Operators of Iowa Farmland, 1982 and 1997

	Percentage of farmland	
Tenure	1982	1997
Operated solely by owner	54.1	30.8
Operated by owner with help	0.9	7.8
Operated under cash-rent lease	21.1	34.9
Operated under crop-share lease	21.1	23.7
Operated under other lease agreement	1.0	2.8
All others	2.7	0.0

Source: Schmitz et al. (2002)

land values. For example, although farmland values and incomes in the United-States have been closely linked from 1910 to 1950, in the 1960s and 1970s farm-land prices increased while farm incomes decreased (Shalit and Schmitz 1982). In addition, Melichar (1979) questions the use of net farm income as a proxy for expected income and believes that only expected income from farmland should be considered. Since this finding, most studies have used net rent as a proxy for expected income from farmland. Alston (1986) tests the hypothesis of the association between farmland values and inflation rates. He also tests the hypothesis of whether or not land acts as a hedge against inflation. He finds almost no evidence to support the hypothesis. On the other hand, Schmitz (1995) finds that inflation was a major factor driving land prices up in the late 1970s and early 1980s.

Land models are not easy to estimate empirically. Many factors other than net farm income and interest rates affect land values. Studies by both Pope (1985) and Clark et al. (1993) find that farm income alone is incapable of explaining the level of land value. Robison et al. (1985) find that value of land for nonfarm use has a significant effect on farmland prices. For example, the price of farmland near large cities is driven more by urban expansion and the demand for open space than it is by net farm income. Also, capital gains taxes play a role in the value of farmland (Gilbert and Akor 1988).

Due to the large number of variables that affect farmland prices, simple regression models focusing on one or a few aspects of the problem can yield misleading results. For example, in Table 4.2 we present ordinary least-squared (OLS) regressions for Florida in which both pastureland prices and developed-land prices are regressed against interest rates and cattle prices. Note that the interest rate variable has the wrong sign. Models that omit part of the explanation for land prices may be misleading about the estimated impacts of other

Table 4.2 OLS Regressions of the Land Market in Florida[*]

Dependent variable	Price of unimproved land in Florida		Price of improved land in Florida	
	Model 1	Model 2	Model 3	Model 4
Constant term	−1577.06	960.97	−2802.00	1441.17
	(−3.10)**	(−1.76)*	(−3.58)**	(−1.94)*
Interest rate on farm debt	157.15	188.82	265.45	274.59
	(3.88)**	(3.70)**	(4.11)**	(3.79)**
Price of steers	24.78		43.15	
	(5.66)***		(6.16)***	
Price of calves		10.67		17.60
		(4.22)**		(5.04)**
R^2	.7267	.6077	.7565	.6807
Adjusted R^2	.6903	.5554	.7241	.6381
F-value	19.94	11.62	23.30	15.99
Probability $> F$	.0001***	.0009**	.0001***	.0002**

[*]Significance is indicated by * at the .05 level, by ** at the .01 level, and by *** at the .001 level. Values in parentheses are *t*-ratios.

variables. A common problem in econometric studies with many independent variables is the lack of identification of individual coefficients due to multicollinearity, particularly when most variables have a high positive correlation naturally as in the case of land prices. The Just and Miranowski (1993) solution to the identification problem is to use economic theory to determine the relative role of the various economic variables, in which case econometric estimation is required only to determine few absolute impact coefficients. Their model estimates not only the plausible impacts of income from farming but also it estimates the plausible effects of government payments; inflation; risk associated with farm income that includes government payments; taxation of current income, capital gains, and land; and credit availability and interest rates.

LAND VALUES: SELECTED CANADIAN PROVINCES

Cash available for investment influences land values. In part, this cash is dependent on net farm income. Without a doubt, net farm income is influenced by government policy. Supply-management commodities, including dairy and poultry, dominate agriculture in both Quebec and Ontario. And supply management has afforded a significant degree of protection for these commodities. On the other hand, government policies play a much smaller role in the Prairie agriculture sector and these government payouts have declined over time (Schmitz et al. 2002). Also, the Agricultural Income Disaster Assistance (AIDA) government program that was introduced in the late 1990s largely benefited the major hog producers in Quebec and Ontario at the expense of the farmers of the Prairie Provinces. Because of intensive livestock production in Quebec and Ontario, farmers are purchasing additional land for manure disposal. And this is driving up the price of farmland. Unlike the Prairie Provinces, Quebec and Ontario have had significant population and urban growth, which also drives up the price of farmland.

Canadian land values are given in Table 4.3. They are highest for Ontario and are closely followed by British Columbia and Quebec. Note that, except for Saskatchewan, where the price for farmland was stagnant from 1988 through 1995, land values elsewhere have been rising sharply since 1988.

As one might expect, land values for the provinces of Quebec and Ontario are positively correlated (Table 4.4). For the province of Saskatchewan, however, land values are negatively correlated to those land values in Ontario and Quebec. What explains the difference?

Cash that is available to farmers for the purchase of farmland influences land values. Available cash is influenced by many factors, including net farm income but also farmland collateral values and anticipated appreciation of land values for nonfarm uses. Notably, net farm income in Quebec and Ontario is less variable than in provinces such as Saskatchewan and has risen more rapidly over time. These relationships are heavily affected by the urban demand for land inclusive of the demand for open space, which has been rising steadily in Quebec and Ontario. In addition to demand for land for urban purposes and open

Table 4.3 Per-Acre Value of Farmland and Buildings

Year	PQ	ON	MB	SK	AB	BC	Canada
				Canadian dollars per acre			
1981	668	1,695	410	382	600	1,191	615
1982	690	1,659	380	413	592	1,083	614
1983	707	1,542	380	405	543	1,091	588
1984	690	1,509	387	393	493	1,035	558
1985	659	1,402	359	357	453	981	517
1986	652	1,288	344	332	407	884	478
1987	897	1,258	325	298	366	833	456
1988	727	1,488	304	286	374	873	454
1989	781	1,905	328	286	411	982	516
1990	850	2,147	359	284	432	1,083	555
1991	918	2,303	357	285	414	1,190	580
1992	943	2,184	360	256	405	1,242	547
1993	977	2,144	373	253	413	1,399	555
1994	1,031	2,134	388	271	450	1,589	584
1995	1,114	2,188	413	289	515	1,767	634
1996	1,220	2,384	443	314	553	1,890	689
1997	1,274	2,471	456	317	573	1,894	709
1998	1,342	2,507	458	316	580	1,882	717
1999	1,457	2,709	486	329	605	1,964	760
2000	1,501	2,709	491	329	617	1,888	763

Source: Authors' compilations from FCC (2001) data.

Table 4.4 Correlation of Land Values in Selected Provinces, 1981 to 2000[*]

Province	Quebec	Ontario	Saskatchewan
Quebec	1.00	0.89	−0.32
Ontario	0.89	1.00	−0.38
Saskatchewan	−0.32	−0.38	1.00

[*]Authors' estimations from the data in Table 4.3.

space, the demand for farmland to deal with the waste from intensive livestock operations has been steadily increasing because of environmental pressures.

U.S. LAND VALUES

We plot U.S. land-value data for selected states in Figure 4.3. Obviously, these land values are highly correlated. Also, in all these states, land values have increased steadily since 1986.

Land values are influenced by many factors, including urban growth. This is especially true in states like Florida. Recall from Table 4.2 that farmland values in Florida were not related to traditional economic variables such as cattle prices

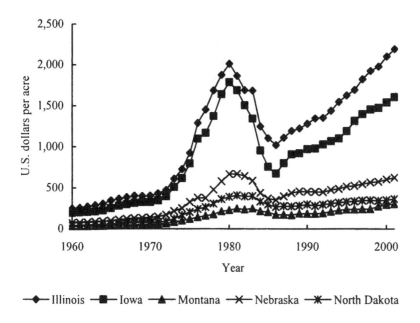

Figure 4.3 U.S. Land Values, Selected States, 1960 to 2000

and interest rates. We confirm these results in Table 4.5, in which Florida land values bear little relationship to land values in the Corn Belt states (e.g., Illinois, Indiana, and Iowa). Nor do Florida land values resemble states in the Great Plains (e.g., Kansas, Montana, and North Dakota) where prosperity in agriculture appears to be the major factor that influences land values. However, farmland values are highly correlated among states within the Corn Belt and the Great Plains, and particularly so between Illinois and Indiana, Illinois and Iowa, and Kansas and North Dakota.

We give the land-value and cash-rental data for Indiana in Table 4.6. Land values in Indiana reached near record highs in the year 2000 despite rock-bottom commodity prices. Even so, average cash rents reached their peak from 1998 to 2000. Obviously, cash rents and land values are highly correlated. The correlation between average rent and average land value is 0.91 (Table 4.6).

U.S.-CANADA LAND COMPARISONS

Total direct government payments to U.S. farmers reached a record U.S. $23.3 billion in the year 2000 and are expected to be even higher under the 2002 Farm Program. And indirect payments in 2000 were also significant. For some states, up to 85 percent of net cash farm income was represented by government subsidies (Table 4.7). Canadian farm programs for Western Prairie grain and oilseed

Table 4.5 Land Values: Pair-Wise Correlations, Selected States, 1981 to 2002

	Florida	Illinois	Indiana	Iowa	Kansas	Montana	North Dakota
Florida	1	0.62	0.63	0.46	0.28	0.58	0.12
Illinois	0.62	1	0.99	0.96	0.89	0.95	0.82
Indiana	0.63	0.99	1	0.93	0.85	0.97	0.79
Iowa	0.46	0.96	0.93	1	0.96	0.88	0.90
Kansas	0.28	0.89	0.85	0.96	1	0.84	0.97
Montana	0.58	0.95	0.97	0.88	0.84	1	0.80
North Dakota	0.12	0.82	0.79	0.90	0.97	0.80	1

Source: Authors' computations.

Table 4.6 Cash Rents and Land Values in Indiana, 1976 to 2000

	Cash rents			Land values		
Year	Top rent	Average rent	Low rent	Top value	Average value	Low value
	U.S. dollars per acre			U.S. dollars per acre		
1976	100	77	54	1,600	1,225	900
1977	116	90	64	2,136	1,585	1,120
1978	113	88	64	2,230	1,741	1,280
1979	117	92	68	2,419	1,898	1,401
1980	123	97	72	2,443	1,888	1,400
1981	137	106	78	2,679	2,100	1,528
1982	124	98	71	2,210	1,714	1,201
1983	122	97	73	1,989	1,577	1,142
1984	120	95	70	1,876	1,451	1,030
1985	112	89	65	1,570	1,195	836
1986	102	79	57	1,284	976	680
1987	91	72	52	1,196	913	643
1988	97	77	57	1,365	1,034	721
1989	106	84	63	1,518	1,154	818
1990	108	87	65	1,589	1,223	861
1991	110	88	67	1,633	1,245	893
1992	112	90	68	1,673	1,264	903
1993	114	89	69	1,727	1,304	936
1994	118	94	71	1,892	1,439	1,040
1995	122	98	73	2,029	1,545	1,099
1996	129	104	80	2,274	1,765	1,303
1997	135	110	84	2,549	1,997	1,493
1998	140	112	86	2,715	2,155	1,632
1999	138	110	84	2,643	2,092	1,546
2000	140	112	86	2,715	2,173	1,630

Source: Dobbins (2001).

Table 4.7 Government Payments as a Percentage of Net Cash Farm Income, Selected States and Provinces, 1995 to 1998

State	Percentage	Province	Percentage
Minnesota	38.7	Ontario	16.2
North Dakota	85.9	Manitoba	41.5
Montana	68.6	Saskatchewan	19.0
		Alberta	13.1

Source: Schmitz et al. (2002).

producers are very different from those programs in the United States. In addition, the level of support in Canada for grains and oilseeds is much lower (Figures 4.4 and 4.5). This is illustrated clearly through the calculation of producer subsidy equivalent (PSE) measures. The higher level of support in the United States is not only through high loan rates and target prices, but it is also through crop insurance payments (Tables 4.8, 4.9, and 4.10).

As a result of these differences in government support and program structure, land values in Saskatchewan follow a different pattern than they do in the neighboring U.S. states of Montana, North Dakota, and South Dakota (Figure 4.6). For example, unlike Saskatchewan, a large percentage of net cash farm income is made up of government payments in North Dakota (Table 4.7). Since 1997, Saskatchewan land values have fallen, but this is not the case with the northern U.S. states. The correlation between Montana and North Dakota land values is .81 (Table 4.11). The correlation between North Dakota land values and those in Saskatchewan is substantially lower at .68, and between Montana and Saskatchewan the correlation is even lower at .38. Saskatchewan land values

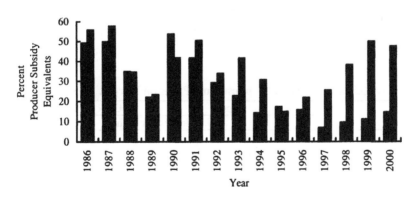

Source: OECD (1986 to 2000).

Figure 4.4 Producer Subsidy Equivalents (PSEs) for Wheat in the United States and Canada, 1986 to 2000

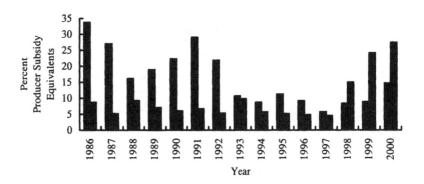

■ Canadian Oilseeds ■ U.S. Oilseeds

Source: OECD (1986 to 2000).
Figure 4.5 Producer Subsidy Equivalents (PSEs) for Oilseeds in the United States and Canada, 1986 to 2000

Table 4.8 Crop Insurance Coverage for an Average Wheat Producer in Montana and Saskatchewan, 1995 to 2000

	Price of coverage per bushel[a]		Yield guarantee	
	Montana	Sask.	Montana	Sask.
Year	U.S. $	CDN $	Bushels per acre	
1995	3.35	3.43	21.6	21.5
1996	3.55	4.35	26.4	20.6
1997	3.85	4.08	20.0	19.5
1998	3.65	3.37	25.6	20.6
1999	3.30	3.89	24.0	21.0
2000	3.15	2.94	28.8	22.6

[a]The price guarantee is calculated for No.2 hard red spring (HRS) wheat in Saskatchewan in Canadian dollars and No.2 wheat in Montana in U.S. dollars.
Source: SCIC (2001) and FCIC (2001).

Table 4.9 U.S. and Canadian Crop Insurance Prices for Durum, 1995 to 2000

	U.S. durum price	CDN durum price
Year	U.S. dollars per bushel	
1995	3.35	2.61
1996	3.55	3.43
1997	3.85	3.40
1998	4.25	2.86
1999	3.60	2.71
2000	3.40	2.43

Source: SCIC (1995 to 2000).

Table 4.10 Revenue Guarantee Features: Three U.S. Corn Risk-Management Plans

Plan feature	Income protection	Revenue assurance	Crop revenue coverage
Revenue guarantee calculation	Historical production multiplied by 100 percent of the Chicago Board of Trade's February price for the December contract	Historical production multiplied by 100 percent of a projected county price (the Chicago Board of Trade's February price for the December contract, adjusted by a county factor)	Historical production multiplied by the higher of (1) 95 percent or 100 percent of Chicago Board of Trade's February price for December delivery or (2) 95 percent or 100 percent of the Chicago Board of Trade's price in November for the December contract
Actual harvest revenue calculation	Actual production multiplied by 100 percent of the November price for the December contract	Actual production multiplied by USDA's posted county price	Actual production multiplied by 95 percent or 100 percent of the November price of the December contract

Source: Authors' compilation based on GAO (1998: 23) data.

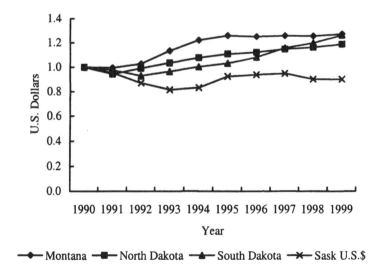

Figure 4.6 Index of Total Farm Real-Estate Values

Table 4.11 Land Values Correlations: North Dakota, Montana, and Saskatchewan, 1981 to 2000*

State/Province	North Dakota	Montana	Saskatchewan
North Dakota	1.00	0.81	0.68
Montana	0.81	1.00	0.38
Saskatchewan	0.68	0.38	1.00

*Estimated from the data in Table 4.3 and Figure 4.7 of this chapter.
Source: Authors' calculations.

dipped in the early 1990s. Simultaneously, the land values in the Northern Plains states increased with the expansion of federal support for crop insurance in the United States in the early 1990s.

Earlier we considered land-value comparisons for Montana, North Dakota, South Dakota, and Saskatchewan (Table 4.11). We give additional information on the relationship between land values in Indiana, Iowa, and Kansas as compared to Quebec and Saskatchewan (Table 4.12). As emphasized earlier, farm programs heavily influence land values. High levels of subsidies exist for farmers in states like Indiana and provinces like Quebec, but in Saskatchewan the subsidy levels are low. (Also, the levels of farm subsidies have moved together over time for Indiana and Quebec.) Note the high positive correlation of land values between Indiana and Quebec and the negative correlation of land values between Indiana and Saskatchewan. There is also a low correlation between land values in Iowa and Kansas as compared to those in Saskatchewan.

Table 4.12 Land Value Correlations in Selected States and Provinces, 1981 to 2001

	Quebec	Saskatchewan
Indiana		
With exchange rate adjusted	0.73	–0.14
Without exchange rate adjusted	0.82	0.21
Iowa		
With exchange rate adjusted	0.59	0.09
Without exchange rate adjusted	0.63	0.31
Kansas		
With exchange rate adjusted	0.43	0.27
Without exchange rate adjusted	0.48	0.47

Source: Authors' calculations.

U.S. LAND VALUES AND DECOUPLED FARM PROGRAMS

U.S. farmland prices have risen significantly since 1985, after dropping sharply for the previous five years. Coupled with rising land values, net farm income that includes government payments has risen steadily. Also, as farm income has been increasing, cash rents have been escalating in parts of the United States. In the spring of 2002, cash rents reached an all-time high. Interestingly, this increase in farm income and cash rents occurred as commodity prices reached record lows. These observations call into question whether or not farm programs have truly moved toward the decoupling that was intended to happen under the Federal Agricultural Improvement and Reform Act (FAIR) (Tweeten and Martin 1976; Shalit and Schmitz 1982; Clark et al. 1993).

In what follows, we argue that farm programs generally cannot be decoupled from production and that farm program payments, at least in substantial part, are continuing to be bid into the price of farmland. According to Schmitz and Gray (2001: 474):

> ...when comparing U.S. and Canadian real-estate values, we see that in the 1990s land prices increased in Montana and North Dakota but remained relatively constant in Saskatchewan (in some areas of Saskatchewan, land values have fallen sharply). Since input costs, crop yields, and crop prices are similar in the three regions, the remaining explanation for the difference in land values is government policy. The producer subsidies for grain and oilseed production are much higher in the United States. Many of these subsidies are not decoupled, which explains partly why U.S. land values remain high (the old saying still remains: Many of the benefits of farm programs get capitalized into land values).

Under the 1994 General Agreement on Tariffs and Trade (GATT), government-support programs must fall within the green-box category of the World Trade Organization (WTO), which are programs deemed not to affect production or trade. Policies that are truly decoupled are more efficient than traditional price-support policies. But where can these policies be found?

In Figure 4.7, domestic supply is given by S and total demand including both domestic demand and exports is given by D. The free-trade price is P^0 where output Q^0 is produced. For various reasons (including the use of production subsidies by competitors), suppose the world price falls to P^1. With no

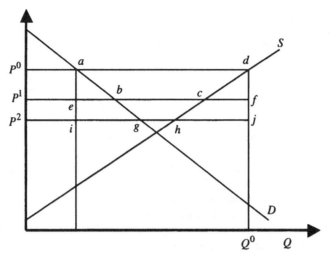

Figure 4.7 Effects of Price Supports in Large- and Small-Country Cases

retaliation by competitors; this country suffers a welfare loss greater than area *abcd*. But what happens if the government maintains domestic price at P^0 by means of a price support given a world price of P^1? The welfare cost of the price-support program is given by the sum of areas *aeb* and *dcf* in the small-country case. In the large-country case, the cost of the price-support program increases because increased production lowers the world price to P^2.

We contrast this model with the case of a decoupled farm program (Figure 4.8). Under free trade, price P^0 and quantity Q^0 prevail. Suppose that price

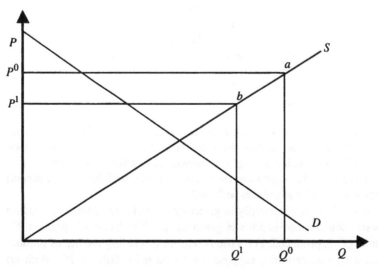

Figure 4.8 Trade with Decoupled Farm Programs

drops to P^1 possibly because of foreign-country subsidies. In the absence of government intervention, producers lose the entire area P^0P^1ba. What if the country retaliates by introducing a decoupled program and leaves producer welfare unchanged? Governments must provide a decoupled payment equal to area P^0P^1ba. Given a decoupled program, however, the world price will not be affected. Output will remain at Q^1 whether or not the government subsidizes its producers. To do this, the country must devise a way to provide subsidies so that farmers can adjust production and consumers can adjust consumption to levels that would prevail in the absence of subsidies. If this can be done, then no net welfare costs will be associated with such decoupled programs. Thus, such programs are more efficient than price-support policies. And they do not antagonize the farmers of export-competing countries.

We suggest, however, that a great deal of confusion exists about the decoupling of farm programs and the relationship of such programs to land values. A program can be decoupled in a production sense but it can still cause land values to increase. Consider Figure 4.9 in which S is the producer supply curve and producer price is P^1. Suppose a support price of P^2 is introduced. It is decoupled neither in production nor in rents because output increases from Q^1 to Q^2 and economic rent increases by an amount P^2P^1ba that gets bid into land prices. Suppose instead that a water subsidy is instituted that shifts supply to S^1, which is also not decoupled. Output then increases by $Q^2 - Q^1$ and producer rents go up by *edcb*. Whether these rents are bid into land prices depends on

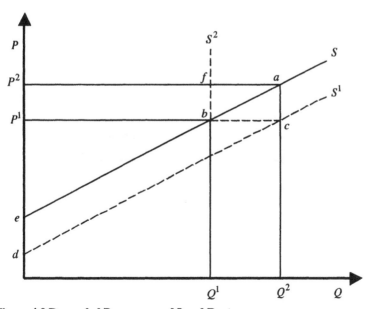

Figure 4.9 Decoupled Programs and Land Rents

whether water rights are tied to the land. For example, if water rights are assigned to individuals, they will not affect land prices; if water rights are tradable, the increase in rents will be bid into the price of water rights.

Consider further a different case in which the supply of output is limited to S^2 (Figure 4.9). In this case, a price support of P^2 is decoupled from production because output does not change in response to the institution of a price support. But rents go up by an amount P^2P^1bf due to the price support. Thus, even though a program is decoupled with respect to production, it may generate rents. In this case, the increased rents would be bid into land prices just as in the case of Figure 4.2, in which land is available in fixed supply because acquiring land is the only way to get access to the increased rent generated by the program. For a program to be decoupled it would have to be one that affects neither production nor the rents earned by assets.

In reality there is a mixture of programs that affect producer welfare. For example, under the 2002 U.S. Farm Bill, there are both target prices and loan rates for wheat and corn along with drought-relief assistance and crop-insurance subsidies. It is possible that one of the elements of the farm program is not decoupled while others are.

We question the extent to which farm programs are generally decoupled. The supply and demand schedules are S and D in the absence of farm programs (Figure 4.10). The free-trade price is P^1 and exports total Q^1-Q^2. Suppose that, due to a change in international conditions, price falls to P^2. Without government involvement, the price decline causes output to fall to Q^3 and exports also fall. Total producer revenue declines by $(P^1-P^2)(Q^1-Q^3)$.

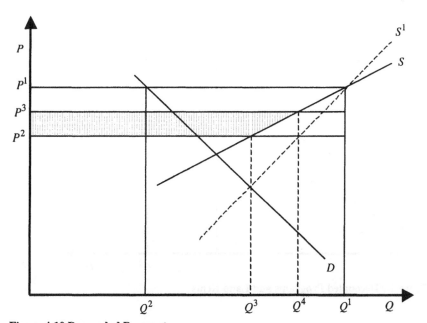

Figure 4.10 Decoupled Payments

What happens if the government responds to a price decline by subsidizing farmers' income with a supposedly decoupled payment to make up for the lost revenue? If the program is truly decoupled, then output responds only to price P^2 and remains at Q^3. If output increases beyond Q^3 in response to government payments, then the program is not decoupled. Suppose, for example, that production increases from Q^3 to Q^4 due to government payments. Then the program is only partially decoupled. In essence, government payments create a new supply curve S^1. (Note that the supply curve S^1 is drawn with reference to price P^2.) Alternatively, output Q^4 corresponds to price P^3 on the original supply curve S. Accordingly, trade increases by the same amount as if output price were raised to P^3. The corresponding Ricardian rents from the farm program are given by the shaded area in Figure 4.10. So the effects are the same as if a non-decoupled program were imposed to restore only part of the lost revenues. Thus, to the extent that programs are not truly decoupled, farm payments will be bid into rents and, in turn, into the price of land. The fact that land prices seem to have been boosted by rising farm payments in recent years suggests that farm programs have not been truly decoupled.

LAND VALUES, URBAN DEVELOPMENT, AND TRADE

In land-market analyses, the effects of urban development on land prices must be considered. In many areas of the United States, land prices appear to be positively impacted by urban growth. As a result, the price for agricultural land would be much higher if the land were sold for urban real estate than it would be if the land were considered for agricultural purposes exclusively.

Consider tomato production in Miami-Dade County, Florida. According to the data in Table 4.13, the total per-acre cost of land used for producing tomatoes is roughly U.S. $11,507. Suppose, for purposes of discussion, that per-acre land rent has been stable at U.S. $450. Using these numbers, the net per-acre profit has been both positive and negative since 1995 (Figure 4.11).

Now consider farmland prices for tomato land in Miami-Dade County. According to the data in Table 4.14, land that is situated less than five miles from a major town was priced at U.S. $40,000 per acre in 2001. Even at 5 percent interest, the per-acre land-rental payment would be U.S. $2,000 annually, which is far above the rental rate assumed in Figure 4.11. In other words, the per-acre income generated from tomatoes cannot support farmland at U.S. $40,000. Of course, the price of farmland remains high in Miami-Dade County because of its potential for urban development.

Cases such as these, which can be found around most major cities, raise questions about what role land values play in comparative-advantage arguments in trade. In the tomato case over the long term, the Miami-Dade County area cannot compete with Mexico, for example, if new or existing farmers might find continued production advantageous depending on the expected appreciation of their land prices. Thus, continued production may depend largely on the growth

Table 4.13 1999 to 2000 Florida Staked Tomato Production Costs, Dade County

Category	Per-acre average (U.S. Dollars)
Operating costs	
Fertilizer and lime	350.50
Fumigants	625.50
Fungicide and insecticide	751.97
Labor and machinery	797.21
Other	1,558.75
Total operating cost	4,083.91
Fixed costs	
Land rent	450.00
Supervision and machinery	1,029.68
Overhead	1,113.52
Total fixed costs	2,593.20
Harvest and marketing costs	
Pick and haul	1,190.00
Pack	2,240.00
Containers	1,120.00
Other	280.00
Total harvest and marketing costs	4,830.00
Total cost per acre	11,507.12

Source: Smith and Taylor .

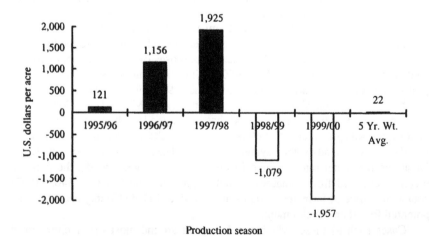

Figure 4.11 Staked Tomatoes: Estimated Net Profit per Acre, 1995/96 to 1999/2000

Table 4.14 Transition Land Values, Southeast Miami-Dade County, FL, 1994 to 2001

Year	Less than 5 miles to a major town Dollars per acre	More than 5 miles to a major town Dollars per acre
1994	25,166	15,000
1995	28,500	16,200
1996	30,167	17,375
1997	28,400	19,000
1998	28,000	20,600
1999	32,063	21,953
2000	34,000	22,917
2001	40,000	26,250

Source: Reynolds and Dorbecker (2002).

rate of nonagricultural factors rather than on comparative advantage principles applied to agricultural production or even the current relative values of land in urban as opposed to agricultural use.

Often, trade barriers affect land prices in the same manner as increased urbanization, and at times both forces work together to prop up land values. We consider first the impact of trade on land values. The U.S. supply curve for tomatoes is S_Q while domestic demand is D_Q (Figure 4.12a). Under a free-trade scenario, price is P^0 and imports are $Q^1 - Q^0$. The land market is modeled in Figure 4.12b. If the land on which tomatoes are produced can be used only for

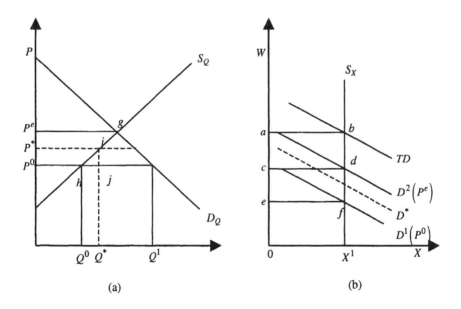

(a)   (b)

Figure 4.12 Land Rent, Urban Development, and Trade Barriers

agriculture, the rent is area eOX^1f in Figure 4.12(b) in which S_X is the supply curve of farmland and $D^1(P^0)$ is the derived demand curve for farmland, given the tomato price P^0 in Figure 4.12a.

Now, suppose producers lobby the government to impose an import quota with the effect of raising the commodity price to P^* causing output to increase to Q^* (Figure 4.12a). Producer rents increase by P^*P^0hi as the demand curve for land shifts to D^* in Figure 4.12b. Compare this situation to one that, instead of imposing an import quota, the government imposes a deficiency-payment scheme supporting producer price at P^*. The deficiency payment to producers would be equal to area P^*P^0ji. (Note that producer rents are the same under either an import quota or a deficiency-payment scheme because area ihj is lost to the increasing marginal cost of production.) If producers lobby for the complete elimination of imports, their rents will increase by area P^*P^0hg as in Figure 4.12a, which is equal to rent area *cefd* in Figure 4.12b. The derived demand for farmland in Figure 4.12b thereby shifts to $D^2(P^e)$.

Studies often correlate land values with producer deficiency-payment levels but ignore the effects of hidden subsidies (e.g., import quotas). This is especially a problem in cross-country comparisons in which one country may use hidden subsidies while another may use direct subsidies to protect its producers. As shown above, either type of subsidy can have a similar effect on producer rents and land values.

Now consider the case in which urban development has pushed the derived demand for farmland outward to total demand *TD* as in Figure 4.12b. The so-called urban demand rent is area *acdb*. If the farmer continues to use this land for tomato production, then this rent is a lost opportunity for increased income from the sale of the property. Alternatively, if the farmer chooses to continue farming in the anticipation that the appreciation of his land will be equal to or greater than the additional urban-demand rent, then these economic rents will not be realized until the land is sold for development purposes. Under these circumstances, new entrants will not be able to afford to buy the land for growing tomatoes unless the land is zoned solely for agricultural use. If the agricultural land is zoned residential, new entrants will probably not enter into tomato production. And eventually the existing viable farmers may be able sell their land to developers.

It is difficult to assess the meaning of cost of production when the demand from urban growth is present. If governments subsidize producers to cover the full costs of production, then huge subsidies will be needed where land is valued at much higher urban opportunity cost. Furthermore, subsidizing farmers for foregone urban rents may offer them a windfall if the land continues to appreciate at a high rate due to increasing urban-development opportunities. Policymakers are having a very difficult time coming to grips with these urban-rural interface issues. At one extreme, some groups argue that prime agricultural land should be zoned such that it cannot be sold for urban development. But if this were done, should the producers be compensated for their loss in land values due

to lost urban-development opportunities? If they are reimbursed, how much re-
muneration should they receive? The answer to this question involves legal arguments surrounding both property rights and economic compensation.

SHUT DOWN POINT, LAND VALUES, AND PRODUCTION

Consider the case in which output ceases in the absence of subsidies. In Figure
4.13, the free-market price is P^1. Here a subsidy that raises price to P^0 causes
output to increase from 0 to Q^1. Of this subsidy in the amount of $(P^0 - P^1)Q^1$,
the rent component is only $P^0 cb$. Compare this to a subsidy in which the market
price is P^0 and the subsidy raises the free-market price to P^2. Output increases
from Q^1 to Q^2. In this case the rent component of the subsidy is $P^2 P^0 ba$,
which accounts for a much greater share of the total subsidy $(P^2 - P^0)Q^2$. That
is, the rent component of the subsidy is smaller for the larger subsidy than it is
for the smaller subsidy. Only the rent component of the subsidy causes land values to increase. Therefore, the impact of policy on land values depends critically
on, among other factors, the profitability of the agriculture sector.

In the case for risk-averse producers, programs such as crop insurance can
also reduce risk and cause output supply to shift outward. A variety of circumstances can occur in this case as well. For example, if the demand schedule for
the product is highly inelastic, then an outward shift in supply can cause output
prices to fall. Depending on the form of the supply shift (e.g., vertically parallel

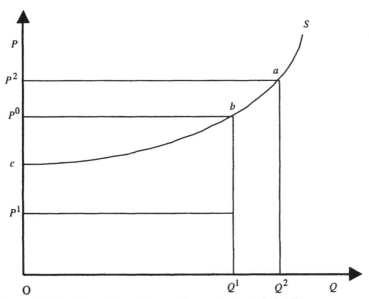

Figure 4.13 The Case Where Output Ceases Without Subsidies

as opposed to horizontally multiplicative), rents for producers may either change little or decrease. On the other hand, if land supply is highly inelastic and technology approximates fixed-proportions production, then an outward shift in supply can cause rents to increase without substantial effects on output price.

As these various cases imply, the structure of technology, supply, and demand as well as the form of government intervention influence the extent to which subsidies affect rents and get bid into land values.

CONCLUSIONS

U.S. farm policy has significantly enhanced farm income. Whether these transfers are merely redistributive in nature or whether they create inefficiencies in resource use will continue to be debated. In theory, if farm programs are totally decoupled, they will not distort production or trade and, hence, they will fit within a truly theoretical definition of the WTO green-box policies. However, under the WTO definition of green-box policies, programs can be trade distorting and still can fit the WTO definition. We argue that designing a farm program that is totally decoupled is virtually impossible in most common circumstances. As our results show, land values have been correlated positively with the size of government transfers regardless of whether these transfers were intended to be direct or indirect. Correspondingly, output is greater, trade volumes are higher, and land values are higher than would otherwise be the case.

Because of the wealth that land represents, farmers obviously have an interest in lobbying governments for farm policies. Considering the effects of subsidies on land values, the incentive to lobby for revenue-replacing subsidies is even greater. In this regard, farmers have been very successful with the passing of the 2002 Farm Bill. Landowners have no interest in having land values drop with the accompanying decrease in wealth that occurred in the mid 1980s. However, as we point out, because of the increasing separation between farm operators and landowners, more and more farm subsidies will end up in the hands of landowners than they will in the hands of farmers.

REFERENCES

Alston, J.M. (1986) "Analysis of Growth of U.S. Farmland Prices, 1963–82." *American Journal of Agricultural Economics* 68 (1): 1–9.
Clark, J.S., K.K. Klein, and S.J. Thompson. (1993) "Are Subsidies Capitalized into Land Values? Some Time Series Evidence from Saskatchewan." *Canadian Journal of Agricultural Economics* 41: 155–63.
Dobbins, C.L. (2001) Personal communications with Craig L. Dobbins, Department of Agricultural Economics, Purdue University, W. Lafayette, Indiana.
FCC (Farmland Values Report). (2001) Internet website http://www.fcc.fac.ca.
FCIC (Federal Crop Insurance Corporation) (2001) Personal communication with Terry Katzer of the U.S. Federal Crop Insurance Corporation (FCIC)
Gardner, B. (2003) See Chapter 3, this volume.

GAO/RCED (U.S. General Accounting Office, Resources, Community and Economic Development Division). (1998) *Crop Revenue Insurance: Problems with New Plans Need to be Addressed.* GAO/RCED-98-111. Washington DC.

Gilbert, J. and R. Akor. (1988) "Increasing structural divergence in U.S. dairying: California and Wisconsin since 1950." *Rural Sociology* 55 (1): 56–72.

Just, R.E and J.A. Miranowski. (1993) "Understanding Farmland Price Changes." *American Journal of Agricultural Economics* 75 (1): 156–68.

Just, R.E. and D.L. Hueth. (1979) "Welfare Measures in a Multimarket Framework," *American Economic Review* 69 (December): 947–54

Just, R.E., D.L. Hueth, and A. Schmitz. (1982) *Applied Welfare Economics and Public Policy.* Upper Saddle River: Prentice Hall.

Melichar. 1979. "Capital Gains Versus Current Income in the Farming Sector." *American Journal of Agricultural Economics 61* (5): 1085-92

OECD (Organization of Economic Community Development). (1986–2000) (13)

Pieper, C. and N. Harl. (1999) "Iowa Farmland Ownership and Tenure, 1993 to 1997: Analysis and Comparison." Department of Economics Working Paper. Ames: Iowa State University.

Pope, C.A. III. (1985) "Agricultural Productive and Consumptive Use Components of Rural Land Values in Texas." *American Journal of Agricultural Economics* 67 (February): 81–6.

Reynolds and Dorbecker. (2002) Internet Website:
http://www.agmarketing.ifas.ufl.edu/difiles/AppB_Land_Values.pdf

Robison, L.J., D.A. Lins, and R. VenKataraman. (1985) "Cash Rents and Land Values in U.S. Agriculture." *American Journal of Agricultural Economics* 67 (November): 794–805.

Schmitz, A. (1995) "Boom/Bust Cycles and Ricardian Rents." *American Journal of Agricultural Economics* 77 (5): 1110–25.

Schmitz, A. and R. Gray. (2001) "The Divergence in Canada-US Grain and Oilseed Policies." *Canadian Journal of Agricultural Economics* 49 (December): 459–78.

Schmitz, A., H. Furtan, and K. Baylis. 2002. *Agricultural Policy, Agribusiness, and Rent Seeking Behaviour.* Toronto: University of Toronto Press.

SCIC (Saskatchewan Crop Insurance Corporation). (2001) Personal Communication with Keith Hayward of the Saskatchewan Crop Insurance Corporation.

Shalit, H., and A. Schmitz. (1982) "Farmland Accumulation and Prices." *American Journal of Agricultural Economics* 81 (November): 710–19.

Smith, S.A. and T. T. Taylor. Internet Website: http://www.agbuscenter.ifas.ufl.edu/cost.

Tweeten, L. (1986) "A Note Explaining Farmland Price Changes in the Seventies and Eighties." *Agricultural Economics Research* 38 (Fall): 25–30.

Tweeten, L. and J.E. Martin. (1976) "A Methodology for Predicting U.S. Farm Real Estate Price Variations." *Journal of Farm Economics* 42 (May): 378–93.

Chapter 5

U.S. Commodity Policies and Land Values

Bruce Gardner
University of Maryland

INTRODUCTION

Many economists have presumed that the benefits of commodity programs accrue entirely, or almost entirely, to landowners. Others have questioned that presumption. Alston and James (2002: 17–18) state "The conditions under which all of the benefits from an output subsidy accrue to landowners are extreme, but may be appropriate at some levels of aggregation." They argue that analysts err by assuming too readily that land is fixed in supply while other inputs have fixed prices (i.e., are perfectly elastic in supply), an especially egregious error for individual commodities that use far less than the aggregate supply of cropland.

The classical model for the analysis of income-distributional consequences of commodity policies is that developed by Floyd (1965). In his model of a single output and two inputs, the input-price effect resulting from producer-price support via a production subsidy is

$$\frac{EP_a}{EP_x} = \frac{\sigma + e}{\sigma + K_a e_b + K_b e_a},\tag{1}$$

where P_a and P_x are input and output prices, E is the elasticity (percentage change) operator, σ is the elasticity of substitution between the two inputs a and b, e_a and e_b are own-price elasticities of supply for each input, and K_a and K_b are their shares in total costs. The changes in the relative price of the two inputs are related according to

$$\frac{EP_a}{EP_b} = \frac{\sigma + e_b}{\sigma + e_a}.$$

(2)

From Equation (1), letting land be factor a and non-land inputs be factor b, if non-land inputs are perfectly elastic in supply (available at prices determined outside the agricultural sector at given prices), we have

$$EP_a / EP_x = 1 / K_a.$$

(3)

If P_x is increased 10 percent by a policy and the land share is one-third, then the (rental) price of land rises 30 percent (i.e., exactly enough to capture all the benefits of the higher commodity price).[i]

It is typically expected that some non-land inputs will not be perfectly elastic in supply, and that land and non-land input supplies will divide the benefits in the proportions given by Equation (2). Still, with land relatively inelastic in supply, we may expect landowners to reap the benefits disproportionately.

A tricky and too often neglected aspect of the economic analysis of policies is the appropriate specification of how actual policies apply. Several types of policies are relevant. For production subsidies, if payments are tied to current production, we get classical incidence as in Equation (1). For decoupled payments, Barnard et al. (2001: 27) discuss the difference between loan deficiency payments (LDPs) and production flexibility contract payments (PFCPs), as introduced under the 1996 Federal Agriculture Improvement and Reform Act (FAIR). They write "Because LDPs are paid on each unit produced, farm operators have an incentive to increase production through greater use of fertilizer, herbicides, and other inputs. As a result, input suppliers capture a share of LDP benefits, and consequently, LDPs may have a lesser effect on cropland values than PFCPs and other decoupled, lump-sum payments" (2001: 27). The LDPs are classical production subsidies, so Equation (1) applies. But that is not the only reason why landowners do not get all the program benefits. In addition, the use of land and non-land inputs induced by LDPs increases commodity output, and this causes market prices of the supported commodities to be lower than would be the case without the programs (as foreign producers of these commodities like to remind U.S. policymakers). The practical point for gainers from payments is that buyers of the commodities share in the gains. For example, if the elasticity of product supply equals (in absolute value) the elasticity of product demand, then gains from the subsidy will be shared equally between buyers and input suppliers, and the gains to landowners and other input suppliers would be just half of what Equation (1) indicated.

For PFCPs the analysis is different because, if truly decoupled, the policy has no effect on output, or on product or input prices. Whoever gets the payment gets an income increase equal to the payment, as a first-order approximation. Second-order effects involve the price effects of choices by the recipient on how to use the increase in income that the payments provide. A farmer might invest in the farm, or instead might invest in financial assets or spend more on consumer goods. The key point is that the payment itself provides no more incentive to use PFCPs to increase production on the farm than would be the case for any other lump-sum cash infusion.[ii]

With respect to landowner gains, the issues are: What determines who gets the payments (i.e., eligibility criteria for participation and the amount for each participant)? How do contractual arrangements change once payment eligibility is determined? For FAIR Act PFCPs, eligibility is tied to land at a particular location that has grown program crops from 1991 to 1995, and the amount of payment is tied to a quantity that was to have been produced for program purposes on that land (typically different from the quantity actually produced). Does this mean that payment has to end up increasing the income only of the owner of payment-receiving plots?

A complicating issue is the requirement that payments must be shared between landowner and tenant (e.g., the case of a crop-sharing contract), or be paid to a cash renter if he or she is the active farmer. Ryan et al. (2001: 23) report testimony of a panel comprised of farm managers with cash rentals, in which terms of leases were negotiated with "lease rates being bid up until the landowner had captured most of the tenant share of the PFCP." With crop-sharing contracts the issue is more complicated in that PFCPs are supposed to be shared in proportion to crop shares. If the terms of such leases are not adjusted, the landowner will not reap his or her full benefits (which is the apparent legislative intent).

Might payment limits make a difference? If actually enforced, they could result in some otherwise eligible land not receiving payments. But even if this happens, it is not the case that some other economic interests in production would get the benefits. The benefits would still go to the land only.

The debate on payment limitations shows the inevitability of landowners getting the benefits as Congress sees the issue. In the 20 April 2002 audio webcast discussion of the Congressional Conference Committee on the farm bills it was asserted, and no one disputed, that owners of cropland who are not active farmers (identified as farmers' widows in that debate) would be forced to cash rent if legislation impaired their ability to capture payments through either payment limits or restrictions on crop-sharing contracts. Under a cash-rental scheme, the analytical point of the debate requires the acceptance of the proposition that the rent paid to the landlord would inevitably contain the program benefits as long as there would be competition among renters for the leased land.

There is an alternative under which PFCPs would not accrue to landowners. It is that eligibility for payments could be tied in the base period to the number of bushels grown by a person who would be identified by name and who, in the case of cash rental, would be legislated to be the renter. If that renter were to choose to move to another landlord, the payment would move as well. Occupational licensing is the closest example of this approach.

EMPIRICAL EVIDENCE

There are several ways to marshal evidence regarding the effects of commodity programs on land values. The central question is how much lower would the

value of U.S. farmland be in the absence of U.S. commodity programs, everything else remaining the same? Two important variants of the question are:
(1) What would the effects be on land values in 2002 if the programs were removed in 2002 (as some once hoped the 1996 FAIR Act might do)? (2) How
much lower would land values be in 2002 if the commodity-program scheme
had never been instigated? Question (2) looks impossible to answer because to
answer it properly we would have to have an accurate econometric history of
U.S. agriculture since the inception of LDPs and PFCPs. Question (1) appears to
be more easily answered, but it raises many questions including expectations and
time to adjust. The researchers of substantial literature on the econometrics of
land values have had considerable difficulty sorting out the dynamics of land-
price determination. This situation has occurred especially when land values
have been determined not only by its present and future expected rental returns
in agriculture, but also by its value if it were converted to nonagricultural use
and to the range of macroeconomic factors (mainly inflation and interest rates)
that might influence the valuation of land as a financial asset.

Average Land Values in the United States

The average value of farm real estate fell in real terms after 1910 until the mid-
1930s. It then was essentially flat for two decades, after which time real prices
generally increased (Figure 5.1).[iii] What plausible linkages could there be between this story and that of U.S. commodity programs?

The New Deal was introduced in 1933, but if it resulted in any significant
capitalization of rents into land values during its first decade, such capitalization
is not apparent. Some idea of a relatively minor role of the programs is suggested by the existence of a market-driven event—the commodities boom and
bust periods of the 1970s and 1980s. One does not see land-value changes any-
where near this substantial that can be attributed plausibly to commodity programs. But maybe the turnaround from a long-term trend of declining real farmland values to a trend of rising farmland values in the post-World War II period
is in part attributable to the existence of commodity support (e.g., if the commodity support had been absent, land values might have continued to fall).
Econometric studies of time-series data of farmland prices have not found solid
empirical evidence of farm-program effects on land prices. Indeed, the general
tenor of these studies is that macroeconomic variables are the dominant explanatory factors to changes in U.S. farmland values (e.g., Just and Miranowski
1993). A particular difficulty when estimating program effects in time-series
data is that government payments fall during periods when market prices are
high (e.g., the 1970s). These periods tend to be the time when farmland prices
increase the most. In this situation, one needs to hold the relevant underlying
market conditions constant in order to identify program effects, which is a task
not yet accomplished in empirical work.

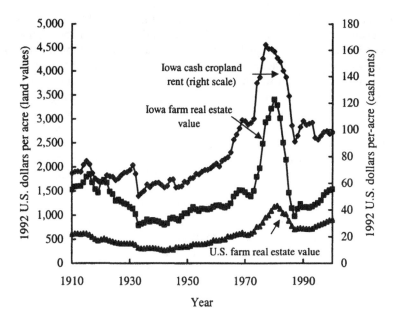

Figure 5.1 Real Farmland Value and Rent

Disaggregated Land Values

We could look cross-sectionally for the effect commodity policies have on land values by using data on land prices for different farms in which policies affecting those farms are different. The analysis would be similar to estimating the land-price effects of irrigation by observing the values of irrigated and nonirrigated acreage. However, the situation is more complicated for commodity policies. First, for policies that support market prices (e.g., the sugar program), all land that grows the supported commodity (which is likely to be all the comparable land in any particular region) will be affected in the same way, at least to a first approximation. In this case, we do not have the necessary contrast between the policy and its absence. Of course, we will always find some landowners who are not enrolled in a program, but if the market price were supported, that land would reap benefits anyway. Moreover, even if a particular farm could not grow the supported commodity, but could if the owner were to choose to do so, the market value of the land would also reflect that option and would be affected by the support program.

Second, if the program is one that makes payments but does not support market prices, [e.g., current PFCPs, market loss assistance payments (MLAPs) and LDPs], land values in an area would be affected even for nonparticipants if they were eligible to participate. Participation-tied payments have been the situation for most of the U.S. commodity programs since the 1960s.

Nonetheless, there is a real contrast between farmland that is eligible to participate in a commodity program and farmland that is not eligible to participate. In part, it can be reasonably hypothesized that eligibility is attributable to characteristics of location (e.g., soil quality, climate, and distance from the market). If the differences were only those of comparative advantage (e.g., the situation of no commodity programs in which land at different locations would grow different crops but on average would have the same value), then comparing land values at locations with and without program coverage would provide an estimate of the effects of commodity programs on land values. But, if land at locations without commodity programs had lower values even if there were no programs (the counterfactual situation we cannot observe), then we would not be able to use the difference in land values between the two locations to estimate the effects of programs.

An econometric approach to these problems is to attempt to hold constant non-program factors that make the value of land at two locations different using a standard-regression model, and then to observe how many of the residual differences can be explained by variables that pertain to commodity programs. As usual, the difficulty to this approach includes getting appropriate observations, obtaining data that measure both policies and relevant non policy variables that influence land values, and calculating the effects of these policies on land values rather than calculating the effects of other variables on policies (e.g., omitted variables correlated with land values).

A notable effort to estimate program effects cross-sectionally is that of Barnard et al. (2001). They provide insights into the per-acre value of farmland at the county level, as reported by farm operators in the 2000 Agricultural Resource Management Survey conducted by the U.S. Department of Agriculture, National Agricultural Statistics Service (USDA/NASS/ARMS). The approach utilizes regressions that explain per-acre land value as a function of commodity-program payments received, soil quality, availability of irrigation, urban influence, and other factors not specified in the report. They estimate that U.S. $61.6 billion of the U.S. $312.3 billion value of land harvested for eight program crops (i.e., wheat, corn, soybeans, sorghum, cotton, rice, barley, and oats) is attributable to program payments. Since payments received in the year 2000 for these programs amounted to about U.S. $21 billion, it appears that each dollar of payment generated about U.S. $3 of land value.[iv] Of course, we do not expect the 2000 level of payments to be a precise indicator of program effects on land values. For one thing, in 2000 a larger than usual proportion of payments was made up of LDPs, which were expected to have a smaller effect on land values than were PFCPs because LDPs had the additional effect of lowering commodity prices (Barnard et al. 2001). More broadly, land values were not expected to reflect payment levels of a single year, but were to discount expected future benefits. The observational basis for farmers' expectations about these benefits is not only generally current payments but also recent past payments and commodity-market conditions underlying forecasts of future payments.[v]

To take a related but independent approach to empirical evidence, I used a sample of 315 counties observed over the 1950 to 1992 period. The counties are noncontiguous and far enough apart to represent arguably distinct land markets. Each county is near the center of a state-economic area—a classification developed by the Bureau of the Census in the 1940s. In nonmetropolitan areas, these areas coincide with type-of-farming areas as defined by the USDA that were used in the 1950 Census of Agriculture (USDA/NASS 2001). Their numbers vary widely by state and depend on the size of the state and the variety of agriculture within it.

Estimating an ordinary least squares (OLS) model similar to that of Barnard et al. (2001) on these data provides some evidence of the effect of per-acre payments on a county's average per-acre farmland value, but it is not a quantitative estimate in which one can be confident. The scatter diagram of per-acre payments and average county per-acre farmland value is shown in Figure 5.2a. The simple OLS-regression line shows a positive effect, but the fit is poor, and one may reasonably question whether or not this association between payments and land prices indicates the results of commodity-program support. The very high land-value counties are in urbanized areas where their prices are unlikely to be caused by commodity programs. In their study, Barnard et al. (2001) skirt this problem by estimating separate regressions for different regions of the county. They also include additional right-hand-side variables to account for the effects of urbanization.

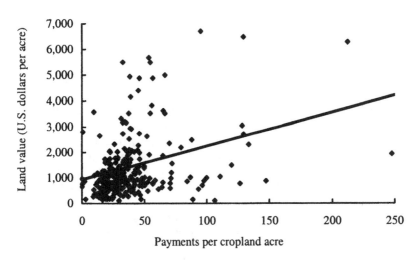

Figure 5.2a Per-Acre 1992 U.S. Government Payments and Per-Acre Land Values in 315 U.S. Counties

In Figure 5.2b, the sample of counties is restricted to the 92 in my sample in which the farm population was 30 percent or more of the total population of the county. This excludes all counties whose 1992 average farmland values were more than U.S. $2,000 per acre, but the relationship between government payments and land values is still quite loose at $R^2 = 0.24$.

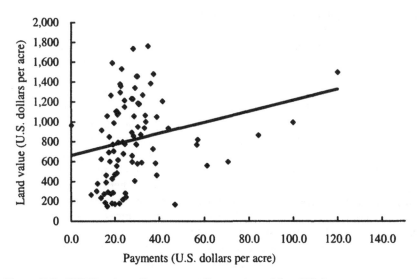

**Figure 5.2b 1992 Per-Acre Government Payments and Land Values
in 92 Rural U.S. Counties**

In regressions that include measures of irrigated land, size of farms, per-centage of rural and urban population, and rate of population growth in a county, three measures of commodity-program impacts are used to estimate their effects on county land values. These measures are per-acre government payments of land dedicated to farms, per-acre payments of cropland, and percentage of county land that is devoted to program-supported crops. As expected, the three measures have positive effects on land values in both regressions on the whole set of 315 counties and on the restricted set of 92 highly rural counties. But the magnitude of the estimated effect varies widely depending on the exact specifi-cation of the regression equation, indicating that an additional dollar of per-acre payment support increases the value of land devoted to farms by U.S. \$3 to U.S. \$15 an acre.

Two underlying problems in this and other cross-sectional attempts to measure commodity-program effects pervade these regressions, as is also the case in the cross-sectional studies by Barnard et al. (1997 and 2001). First, coun-ties with lower commodity-program benefits may have lower land values for reasons other than commodity programs or other variables included in the re-gressions. If commodity programs were to end, land prices in the counties that are now heavily supported would not fall to the levels of the less-supported counties (as these regressions should predict). Second, and perhaps more sys-tematically important in biasing the results toward overestimating program effects using cross-sectional data, these regressions hold national-level market conditions constant by construction, since all observations are for a single year.

Commodity programs encourage more input use and production of sup-ported commodities than would be the case in the absence of the programs. If the programs were removed, market prices of the supported commodities would rise. In cross-sectional observations, prices received by farmers would increase more in heavily supported counties than they would in those counties that rely on the programs to a lesser extent. Therefore, the cross-sectional regressions that hold commodity prices constant overstate the program effects.

To improve our ability to estimate U.S. commodity program effects on land values in the presence of both of these problems, I make use of cross-sectional U.S. county data for 1950 and 1992. The idea is to account for the growth of farmland values between 1950 and 1992 as a function of commodity support provided during that period. The first problem above is dealt with by including the 1950 value of each county's farmland as a right-hand-side variable. The characteristics of a county's farmland, which affect its land value but are not captured by the other variables in the regression, are held constant by including the 1950 value as an additional variable. The second problem is dealt with through the use of market-value changes between 1950 and 1992. If programs were to reduce market prices, the relevant effects of that decline would show up as a corresponding reduction in the 1950 to 1992 increase in farmland value.

The case for using 1950 as a base year after World War II is that the U.S. agricultural economy and U.S. agricultural policy (as well as the macroeconomic picture) were unsettled. A general worry about a recurrence of the Great Depression kept asset prices low (e.g., stock-market indices as well as land prices). Commodity programs provided support, albeit relatively little by later standards. Therefore, we do not expect to see a large capitalized benefit of farm programs in the 1950 farmland prices, and indeed as the data in Figure 5.1 reveal, real farmland prices were still below pre-1930 levels. During the 1950s, new government programs were introduced to support commodity markets, including the PL480 program (grain used for subsidized foreign-food assistance) and the Soil Bank program (supply control). After a landmark wheat referendum in 1963, farm policy moved decisively toward direct payments to producers as a means of support, and that approach continued in the form of LDPs throughout the years of the 1990 Farm Act.

An indicator of the effects of this four-decade program of support on land values is how land values changed in counties that varied in their reliance on these support programs. The data in Table 5.1 reveals what happened to land values in two subsets of counties. These county subsets include those with more than 80 percent of their farmland in program-supported crops (e.g., grains, cotton, sugar beets, sugarcane, tobacco, or peanuts), and those with fewer than 20 percent of their cropland allocated to those crops.

The former are called program-intensive counties. There are 83 of these counties included within the sample of 315. They are located almost entirely in the Southern and Midwestern United States. There are 45 nonprogram-intensive counties, located almost entirely in the Northeastern and Western United States.

The non-program counties had initially somewhat higher land values in 1949, and notably for our purposes, the rate of increase in land values in those counties was substantially larger than it was in the program-intensive counties. So we have no evidence of commodity programs boosting land values more than in those counties that were less reliant on commodity programs. The data in Figure 5.3 point out the relationship between a county's percentage of cropland in

Table 5.1 Per-Acre County Land Values in 1949, 1992, and 1997

		1949	1992	1997
83 program-intensive counties[a]	Per-acre value	U.S.$108	U.S.$1,035	U.S.$1,365
	Rate of increase[c]		5.3 percent	5.3 percent
43 non-program-intensive counties[b]	Per-acre value	U.S.$120	U.S.$2,975	U.S.$3,129
	Rate of increase[c]		7.5 percent	6.8 percent

[a]Program-intensive = 80 percent or more of the county's cropland is in program crops.
[b]Non-program-intensive = fewer than 20 percent of the county's cropland is in program crops.
[c]Rate of increase is the annual percentage rate of increase since 1949.
Source: USDA/NASS (2001).

program crops in 1950 and the subsequent increase in farmland value in that county.[vi] The relationship appears weakly negative (the more of a county's land that was in program crops in 1950, the less rapidly the value of its farmland grew over the next four decades, despite all the commodity programs of that period).

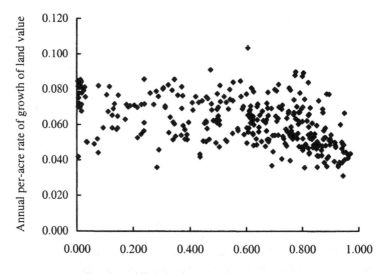

Fraction of U.S. land in commodity-program crops

Figure 5.3 Program Crops and Per-Acre Land-Value Growth from 1950 to 1992

It may be said that comparisons of the data from Figure 5.3 and Table 5.1 tell us little about farmland values because the non-program counties happen to have been located in the more urbanized, faster-growing parts of the county. These farmland values were determined largely by expected opportunities for conversion to nonagricultural uses, the demand for which grew rapidly over the period under consideration. Indeed, studies of land-value determination have concluded that returns from agricultural production are not key factors when determining land prices in many such areas (Hardie et al. 2001), and this again is the rationale, as discussed earlier, for analyzing different regions of the county and for carrying out a multiple-regression analysis.

Regression results for models, including non-program variables are shown in Table 5.2. The non-program variables are the size of farms in 1949 (indicated by per-farm crop sales), the rurality of a county (measured as the fraction of the population living on farms), the rate of growth of the county's population, per-acre county property taxes in 1949, the trend rate of total factor-productivity growth in the state's agriculture between 1949 and 1992 (using estimates

Table 5.2 Regression Results Explaining the Annual Rate of Increase in Farmland-Values for 315 U.S. Counties from 1950 to 1992[a,b]

Independent variables	Estimated coefficient		
	Model 1	Model 2	Model 3[a]
Intercept	5.16	5.01	3.96
	(13.0)[c]	(12.5)	(4.7)
Fraction of cropland	−0.62		
in program crops, 1949	(−2.8)		
Government payments		−0.00035	0.0078
per acre, 1992		(−0.1)	(2.2)
Value of farmland	−1.22	−1.29	−0.83
per acre (log), 1950	(−12.2)	(−13.1)	(−4.2)
Farm sales per acre, 1949	0.0092	0.0091	−0.0005
	(2.8)	(2.7)	(−0.1)
Rurality (percent of county population	−0.91	−1.13	−0.15
that is rural-farm, 1959)	(−2.2)	(−2.8)	(−0.2)
County population growth	1.67	1.78	1.7
rate, 1960–90	(11.3)	(12.1)	(6.8)
Property taxes per acre, 1949	0.56	0.68	0.17
	(5.3)	(7.1)	(0.8)
State-level multifactor-productivity	2.21	2.24	0.96
growth, 1949 to 1993	(9.3)	(9.3)	(2.5)
R^2	0.611	0.6	0.554

[a]Estimated for 92 most rural U.S. counties only, as defined in the text in this chapter.
[b]Mean per-acre value increase for counties = 2.4 percent.
[c]Numbers in parenthesis denote t-statistics
Source: Author's calculations.

from Ball et al. 1997, and Deininger 1995), and initial crop prices in 1950 (to hold constant value-determining characteristics not otherwise accounted for). These factors are all significant determinants of the rate of growth of a county's land value. Nonfarm influences are important. The data in Model 1 of Table 5.2 indicates that a 1 percent increase in the rate of growth of the county's population increases the rate of growth of the per-acre farmland value by 1.67 percent. The coefficient of rurality in Model 1 of Table 5.2 indicates that, given the county's population growth rate, a 1 percentage point increase in the fraction of a county's population living on farms reduces the annual rate of increase of farmland value in a particular county by 0.009 percent, showing the gains in farmland values that result from a county being in an area in which the nonfarm population is larger initially. The positive effects of total factor-productivity growth and the value of per-acre farm sales indicate that the effects of agricultural variables on farmland value are also significant.

The Model 1 data results of Table 5.2 say that, holding the other variables constant, having your county's acreage heavily in program commodities at the beginning of the support period makes no contribution to increasing land value per acre. Indeed, according to these estimates, farmland increases significantly less in value when the greater share of its acreage is in U.S. commodity-program crops.

The Model 2 data results of Table 5.2 have the same nonprogram variables but use a farm-program indicator more closely related to that used by Barnard et al. (2001) for government per-acre payments in 1992. Two problems with the variables used by Barnard et al. (2001) are that the single year, 1992, is used as representative of post-1950 assistance to the county's commodities, and that it is likely to be endogenous in a way leading to specification error and bias. The story for the latter problem is that counties whose commodities faced weak market conditions (not captured in the per-acre sales variable) tend to have both lower farmland prices and higher commodity-program payments than do counties that sell farmland into stronger markets. Therefore, the government-payment variable tends to be biased downward, and the insignificant coefficient estimated could well be positive. But the evidence as we have it is at best weak in uncovering commodity-program effects on land values in this set of counties.

Model 3 of Table 5.2 focuses on the 90 most rural counties as defined earlier, on the expectation that commodity policy is likely to make a more significant difference in the farmland market in those locations. Using the same specifications as in Model 2 of Table 5.2, we find a significantly positive effect of government payments and a weaker effect of all the other variables, although the rate of county population growth remains the most significant variable in the equation. The coefficient of 0.017 implies that increasing per-acre payments by one U.S. dollar would increase the rate of growth of farmland value by 0.017 percent. Since the mean value of the per-acre 1992 payments in these counties is U.S. $15, it implies that the elimination of the commodity programs could have caused the rate of increase of farmland values to decline by 0.26 percent instead of growing by 1.76 percent annually from 1950 to 1992. (If the commodity programs were not in place, the rate of growth of farmland value would have been 1.5 percent). However, the coefficient of government payments is not robust to alternative specifications of the Table 5.2 model (not shown) that try other right-hand-side variables[vii].

DISCUSSION OF FINDINGS AND CONCLUSIONS

Overall, the evidence from the sample of 315 countries during 1950 to 1992 provides only weak evidence that farm programs have increased farmland values. One may question how this can be given (a) strong a priori expectations that U.S. commodity programs were likely to have produced major effects on land values; (b) hedonic pricing models such as that of Barnard et al. (2001) find substantial effects of government payments on land values; and (3) direct observation that acreage, which carries the right to receive payments, rents and sells for more than similar land does that does not have the right to receive payments.

One likely reason that farm programs have had little effect on farmland value differences among counties is that even though the programs are commodity specific, commodity-program crops account for a large fraction of total cropland (i.e., 50 percent in 1950 and 42 percent in 1992 in our sample of 315 counties), so we are not looking at a specialized type of land use. In the long term, owners of land in almost every county of the United States can grow program crops and can benefit from the programs. Therefore, the relevant farmland market is a more integrated national market, and land in a given county is less a specific factor of production than one might expect. Thus farmland in all counties gains more uniformly from the existence of commodity programs than would be indicated by their initial percentages of acreage in program crops or by the level of per-acre payments in the county.

Also, when moving from a short-run to a long-run context, commodity supply may increase the supply elasticity more than the elasticity of demand increases (especially because U.S. commodity programs have often constrained supply response in the short run in ways that have not been sustained over the long run). Thus, program benefits from subsidy programs that accrue largely to landowners in the short run end up with a larger share of their benefits going to commodity buyers in the long run.

In summary, even though it is true that eliminating commodity programs today would cause substantial declines in cropland values, the evidence I present in this chapter suggests that it may not be true that U.S. farmland prices are significantly higher today than they would have been if commodity programs had never been implemented from 1950 to 1992. Moreover, if farmland values were increased by these programs, the increase would be for all farmland and would not be specific to land-growing, commodity-supported crops.

REFERENCES

Alston, J. and J. James. (2002) "The Incidence of Agricultural Policy." In B. Gardner and G. Rausser, eds., (1689–1749) *Handbook of Agricultural Economics* (Vol. 2B). Amsterdam: North-Holland.

Ball, E., J. Bureau, R. Nehring, and A. Somwaru. (1997) "Agricultural Productivity." *American Journal of Agricultural Economics* 79 (5): 1045–63.

Barnard, C., G. Whittaker, D. Westenbarger, and M. Ahearn. (1997) "Evidence of Capitalization of Direct Government Payments into U.S. Cropland Values." *American Journal of Agricultural Economics* 79 (5): 1642–50.

Barnard, C., R. Nehring, J. Ryan, and R. Collender. (2001) "Higher Cropland Value from Farm Program Payments: Who Gains?" *Agricultural Outlook,* USDA/ERS/AGO-286, 26-30. Washington, DC: USDA/ERS (November).

Deininger, K. (1995) *Technical Change, Human Capital, and Spillovers in U.S. Agriculture.* Minneapolis: University of Minnesota Press.

Floyd, J.E. 1965. "The Effects of Farm Price Supports on the Return to Land and Labor in Agriculture." *Journal of Political Economy* 73: 148–53.

Hardie, I., T. Narayan, and B. Gardner. (2001) "The Joint Influence of Agricultural and Nonfarm Factors on Real Estate Values." *American Journal of Agricultural Economics* 83(1): 120–32.

Just, R. and J. Miranowski. (1993) "Understanding Farmland Price Changes." *American Journal of Agricultural Economics* 75 (1): 156–68.

Ryan, J., C. Barnard and R. Collender. (2001) "Government Payments to Farmers Contribute to Rising Land Value." *Agricultural Outlook,* USDA/ERS/AGO-282. Washington, DC: USDA/ERS (June–July): 22–26.

USDA/NASS (U.S. Department of Agriculture, National Agriculture Statistics Service). (2001) *Agricultural Economics and Land Ownership Survey, (AELOS 1999).* Internet Website: http://www.nass.usda.gov/census/census97/aelos/aelos.htm

ENDNOTES

[i] Note, however, that land being perfectly inelastic in supply $(e_a = 0)$ is not sufficient for land to capture all the benefits, except in the special case of fixed proportions $\sigma = 0$. The reason for this is that land, even if fixed in supply, is not a binding constraint on production when other inputs can substitute for land.

[ii] This statement pertains to PFCPs and market loss assistance payments (MLAPs) under the Federal Agricultural Improvement and Reform Act (FAIR). The 2002 Farm Bill, called the Food, Security and Rural Investment Act (FSRI), creates different incentives by regularizing these payments and updating payment bases. The updating generates a production incentive by signaling that a farmer's future legislated payment is likely to be a function of that farmer's production in the years leading up to the legislation.

[iii] Throughout this chapter, I refer to land or real-estate values rather than to land prices. The data used are from the U.S. Agriculture Census periodically from 1910 to 1997 and from corresponding USDA surveys. These data are derived from surveys in which farmers are asked their estimate of what their farm, including land and buildings, would sell for. Thus, the data are not market prices. Farmers may over- or underestimate average market prices of their farm real estate. If this occurs, I assume they do it in a consistent way.

[iv] This calculation is crude, because of me, not because of Barnard et al. (2001). The USDA/NASS (2001) reports that all government payments received by farmers in 2000 totalled U.S. $22.9 billion. About U.S. $1.8 billion is received through conservation reserve payments (CRPs) and other environmental programs, and some payments are made for programs other than those for the eight major crops. CRP payments ar-

guably do not increase land values nearly as much as PFCPs, because the farmer is essentially renting the land to the government and foregoes the returns from growing crops on that land. Thus, a farmer may increase the farm's value only marginally, or perhaps not at all, by enrolling it in the CRP.

[v] For example, 2002 USDA and outside forecasts are for commodity prices to gradually increase from 2003 to 2012, a trend that would generate smaller program benefits in the later years.

[vi] Citations of data for 1949 and 1950 both refer to the 1950 Census of Agriculture. The Census was carried out in the spring of 1950, but many questions (e.g., yields and acreages) asked for 1949 information. When answering the question 'what would your farm sell for?' presumably the respondent was thinking of 1950. The 1992 Census was carried out in January 1993, and all its questions referred to 1992 data, using farm machinery, equipment, and other inventories from 31 December 1992 (even though the question referring to land value asked for its 'current value.')

[vii] The variable measuring the initial importance of program commodities in the county, as used in Model 1 of Table 5.2, gives more robust results and gives no indication that programs boosted the rate of growth of land values from 1949 to 1992.

Chapter 6

Explaining Regional Differences in the Capitalization of Policy Benefits into Agricultural Land Values

Barry K. Goodwin, Ashok K. Mishra,
and François N. Ortalo-Magné[*]
Ohio State University
U.S. Department of Agriculture, Economic Research Service
London School of Economics

INTRODUCTION

An extensive range of programs intended to raise and stabilize farm incomes for at least the last seventy years supports the U.S. agricultural sector. These programs serve to increase the profitability and decrease the risks of agricultural production; hence they raise the value of productive assets. The effects of program benefits vary substantially, not only with the type of agricultural activity and the location of production, but also according to the policy instruments used. To the extent that policy instruments evolve over time, and to the extent that policy benefits vary widely across crops and regions, statements regarding the overall effects of program benefits on land values may be too general and may be misleading. Furthermore, as policy instruments have evolved over the years with the declared objective to minimize market distortions (e.g., decoupled payments), it is worth finding out whether the new policy instruments have made a difference.

The contribution of this paper is twofold. First, we investigate the extent to which decoupled payments (farm program payments not directly related to current production) are capitalized differently into farmland values than market income or government payments that are based on current production levels. We are in the unique position to examine the different effects of various farm programs on farmland values thanks to the enactment of the Federal Agricultural Improvement and Reform (FAIR) Act of 1996 which provided for market transition payments that were decoupled (i.e., payments to farmers that are unrelated to current production). Second, the U.S. Department of Agriculture, National Agriculture Statistics Service, Agricultural Resource Management Survey (USDA/NASS/ARMS) provides us with farm-level information on land values and sources of cash flows (USDA/ERS 2002). This enables us to study the impacts of farm payments at the farm level. Furthermore, the information on the location of the farm enables us to take into account the local nonagricultural demands on agricultural land values, a key determinant in areas near expanding urban centers.

Clearly, any factor that raises the net income stream of agricultural production increases the value of agricultural land. However, three important caveats apply to this capitalization process.

The first caveat is that there will be uncertainty about the stream of policy benefits in the future, which varies considerably with different farm programs. The future benefits of some payment programs may be substantially discounted. For example, Vantreese et al. (1989) find that the value of land containing a tobacco quota fell substantially between 1976 and 1989, largely because of uncertainties regarding the long-run future of the U.S. tobacco-quota program.

The second caveat is that the extent to which policy-instrument benefits are shared among landowners and farm operators may differ greatly. According to Ryan et al. (2001), about 45 percent of total land operated by the average U.S. farmer is rented. Further, about 18 percent of farm operators rent more than 75 percent of their total land and 7 percent of farm operators rent their entire farm. The extent to which lease arrangements actually provide for the sharing of farm-program benefits may not always be clear. Leasing arrangements may be long-term and slow to adjust to changing policies. In addition, a number of intangible factors and informal agreements may be inherent in land-rental and land-sharing agreements. For some policies, the benefactor (whether owner or tenant) of the policy gain is clear. For instance, some benefits paid through market price supports go immediately to the individual who markets the crop. If the crop seller is also the renter of the land, we expect that the landowner will adjust land-rental rates to capture the program benefits from the renter. In other cases, benefits accrue directly to landowners rather than to tenants. The FAIR Act, for example, required that Agricultural Market Transition Act payments (AMTAPs) (also known as production flexibility contract payments or PFCPs) be shared between the tenants and owners on a fair and equitable basis. Thus, if land were rented on a cash basis under FAIR, program payments would be directed entirely to the renter. To enforce this, FAIR required that payment-division aspects of leases be reviewed and approved by the County Farm Services Agency (FSA) committee.

The third caveat is that the effects of nonagricultural demands for agricultural landholdings have become increasingly important because the rural-urban fringe has expanded (Barnard et al. 2003). In many of these areas, nonagricultural demands are more important factors affecting agricultural land values than are issues relating to agricultural profitability. Separating these effects can be difficult, however, especially in more aggregated analyses.

The extensive empirical literature addressing the effects of government payments on agricultural land values has generally involved aggregate analyses of average land values over a large area (e.g., county, state, region, or even country) that lumps all farm programs together. For example, Featherstone and Baker (1987) consider aggregate farm returns and land values for the entire United States. Goodwin and Ortalo-Magné (1992) focus on wheat-growing regions in the United States, Canada, and France. Clark et al. (1993) evaluate aggregate land values for Illinois alone and for the United States as a whole. Likewise, Burt (1986) considers land values for the entire state of Illinois. Not one of these studies has benefited from sufficiently disaggregated data to conduct an analysis of the relative contributions of the various farm programs in relation to the local economic environment of each farm (Raup 2003; Reynolds and Johnston 2003).

There is no doubt that farm policy affects land values in a substantial way. Shoemaker et al. (1990) suggest that U.S. farmland values would be 15 percent to 20 percent lower in the absence of farm-program benefits. A study by Shertz and Johnston (1997), at the time of the implementation of the 1996 FAIR Act, suggests that the effects of program benefits on land values vary widely across the United States. Also, they predict that eliminating government farm programs would decrease land values by amounts ranging from about 30 percent in the Corn Belt region to 69 percent in the Northern Great Plains region. Barnard et al. (1997) bracket the effects of government programs on land values to increases ranging between 7 percent and 38 percent.

These estimated effects of farm programs on farmland values are often quoted in farm-policy debates, particularly when discussions turn to the distributional effects farm-program subsidies have on the cost of buying or renting farmland (Belsie 2002). In addition, the skewed distribution of farm-program payments toward wealthier farmers often provides grounds for arguments in favor of limiting total payments to individual farms. Such debates were especially important during the 2002 U.S. Food, Security, and Rural Investment Act (FSRI) deliberations. A major point of contention in the Senate version of the FSRI ponderings involved an amendment that would substantially limit farm-program payments.[i] Improving our estimates of the effects of farm-program payments on farmland values is therefore of direct policy relevance.

This chapter takes advantage of a unique set of farm-level data covering the post-FAIR years from 1998 to 2000. This post-FAIR period was characterized by an interesting mix of policies including decoupled direct payments, payments due to be phased out (in principle at least), and regular market price supports. Farms with historically based acreage were entitled to decoupled AMTAPs that were fixed and intended to decline each year until the end of the FAIR in 2002. Although the extent to which farm policy actually changed is debatable, FAIR was intended to signal a transition toward a policy environment characterized by less government involvement in agriculture. However, FAIR did include price supports in the form of loan deficiency payments (LDPs). Because of relatively low market prices during the FAIR years, LDPs to producers over the period of our study resulted in large payments. In addition, legislators enacted a number of ad hoc support measures in the form of disaster-relief assistance and market loss assistance payments (MLAPs).[ii]

We examine the impact of farm-program payments and urbanization on farmland values using two empirical specifications. In the first specification (Model 1) we examine how differences in the degree and type of agricultural support (i.e., decoupled farm program payments, payments directly related to production, and disaster relief payments) affect farmland values. In the second model (Model 2) we examine whether the effect of the decoupled payments declines in anticipation of the phase out of the farm program specified in the FAIR Act of 1996. In contrast to much of the existing research, both model specifications include a careful consideration of the influence of nonagricultural demands on agricultural land values. In particular, we include a number of locally defined variables intended to reflect the value of the option to convert farmland to commercial or to residential use.

In Model 1, we examine how the degree and type of support paid to farmers vary across crops produced, regions of the country, and other policy parameters (e.g., base acreage). The base acreage is one policy parameter that directly affects the farm-program payments under the FAIR Act of 1996. The base acreage is the area on each farm that was previously allocated to a particular program crop under previous farm programs, including the deficiency-payment program. Base acreage is the sole determinant of the level of AMTAPs. Base acreage also plays an important role in determining the level of subsequent ad hoc disaster-relief payments made to individual producers.

Model 2 separates the aggregate AMTAPs used in Model 1 into annual payments for each year. If farmers viewed AMTAPs as truly decoupled, the effect of the decoupled payments on farmland values would decline over time and would become zero after farmers received the last program payment. However, if farmers anticipate that farm programs would continue (as actually observed under FSRI 2002) the effect of AMTAPs would remain relatively constant throughout the sample period.

In this chapter, we first present a brief conceptual framework for the consideration of the determinants of agricultural land values. In the following section, we discuss our data. Next, we present our two empirical models that utilize farm and county-specific variables to evaluate the effects of policy benefits on land values. We then use our parameter estimates to attempt to quantify the contributions of various government programs to the value of farmland in the main U.S. agricultural regions[iii]. Finally, we present a summary and offer concluding remarks on policy and land markets.

CONCEPTUAL FRAMEWORK

Most empirical evaluations of the determinants of land values adopt a present-value model, in which the value of a productive asset is given by the capitalized values of expected future streams of net income generated by the asset.[iv] If agents were risk-neutral and the discount rates were constant, land values L_t would be given by

$$L_t = \sum_{i=1}^{\infty} \frac{E_t(R_{t+i})}{(1+r)^i},$$ (1)

where R_t represents net returns to the asset in period t, r denotes the constant discount rate, and $E_t(.)$ denotes the expectation operator given information available in period t. This abstracts from differences in tax rates over alternative sources of rents and capital gains. The case in which net returns R_t are constant over time, implies the standard capitalization formula

$$L_t = \frac{R^*}{r} = bR^* \tag{2}$$

where R^* is the net return and b is the implied discount factor.[v] Melichar (1979) notes that this approach does not fit the experience of U.S. agriculture, since returns to farm assets have grown substantially over time. Thus he modifies the capitalization formula to account for growth in net returns, which occurs at a rate of g per year

$$L_t = \frac{(1+g)R^*}{(r-g)}. \tag{3}$$

In the case of risk-averse agents, the uncertainty associated with future net returns may also play an important role in the valuation of farmland. In particular, risk-averse agents may discount future returns that are subject to uncertainty. Weersink et al. (1999) note that different sources of net returns, such as returns from the market as opposed to those generated by government programs, may have differential effects on land values. This could result from differences in the discount rate for individual sources of returns that result from risk premia. Following this line of reasoning, Weersink et al. (1999) break net returns into the individual components that result from market returns P_t and from the government G_t

$$L_t = \sum_{i=1}^{\infty} (b_1{}^i E_t P_{t+i} + b_2{}^i E_t G_{t+i}), \tag{4}$$

where b_j represents the discount rate for the jth source of income. The discount rates differ due to differences in the risk premia implicit in the rates, which reflect the uncertainty associated with different sources of future net returns.

Taking this approach further, in this chapter we decompose the various components of net farm income into those resulting from agricultural earnings, government farm programs, and nonagricultural returns to land. All of these sources are likely to exhibit substantial differences across individual farms, crops, and regions. In addition, one may also wish to further decompose various sources of government-program payments. It is certainly the case that the expected lifetimes and uncertainties of individual program benefits differ substantially across different types of programs. For example, AMTAPs were, in principle at least, set to expire with the 2002 crop under the FAIR

Act of 1996 (although FSRI 2002 signaled a return to a less market-oriented agricultural policy).

Following the Weersink et al. (1999) extension presented in Equation (4), we decompose overall expected returns to owning farmland into those components resulting from agricultural market earnings, nonagricultural uses, and various farm-program payments. This suggests a basic model of the form

$$L_t = \sum_{i=1}^{\infty} (b_1^i E_t P_{t+i} + b_2^i E_t H_{t+i} + \sum_{j=1}^{g} b_{3j}^i E_t G_{j,t+i}), \qquad (5)$$

where G_j represents farm-program benefits from the jth farm program ($j = 1,\ldots,g$) and H_t represents net expected returns from nonagricultural land uses. (Note that each government-program benefit is allowed to have a distinct discount rate that represents differences in the uncertainty of future benefits.) This general framework suggests an empirical model that relates land values to observable indicators of farm-program benefits, as well as to nonagricultural uses for land and market returns from agricultural production.

EMPIRICAL ANALYSIS

Our empirical analysis utilizes farmland values that are drawn from the 1998 to 2000 data of the USDA/NASS/ARMS survey, which is a probability-weighted stratified sample project of more than 10,000 individual U.S. farms. Questions in the USDA/NASS/ARMS surveys ask farmers on 31 December of each year to estimate the market value of their land, dwellings, and other farm buildings and structures. The survey also collects detailed information for the same period for individual farm-program benefits.

In our analysis, we consider only the reported value of land, excluding trees and orchards. Of course, our data suffers from the caveat that land values are self-reported and do not represent actual market transactions. Having acknowledged this, we also argue that no apparent reason exists for those farmers surveyed to misrepresent land values. Further, the survey is administered by trained personnel from USDA/NASS, and is subject to checks for validity.

We are primarily interested in the effects of farm programs that pertain to crop production on farmland values. We therefore exclude any farms from our analysis for which livestock-product sales account for more than 50 percent of overall farm sales. We also exclude farmland with reported per-acre values in excess of U.S. $10,000. In addition, we also exclude farms for which there was incomplete data. Thus, we are left with a dataset of 11,207 individual farm-level observations.

In Model 1, we include three separate facets of the economic returns to agricultural land ownership including long-run expected-market earnings from farmland (measured by the county-average return on farmland), the relative return for the sample observation compared to the county average (which is used to adjust for relative productivity of the land in the sample), and differences in returns based on differences in government-program payments.

We first represent the long-run expected-market earnings from farmland by including a variable that measures net per-acre county returns (excluding government payments) using data reported in the 1997 Census of Agriculture (USDA/NASS 1999). This variable captures differences in earnings that may be due to variations in crops and land types across different regions.

Second, we represent differences in farm-level productivity by including land characteristics specific to an individual farm that may influence the farm's profitability relative to the county average. Of course, it is impossible to separate farm characteristics from operator characteristics. With such a limitation in mind, we consider normalized crop yields, given by the ratio of the yield for crop i on farm j to the county average yield for the same crop in that year. This normalization removes any systemic-yield effect that influences county-average yields. We also consider the average of this normalized yield over all crops produced on the farms. Our normalized-yield measure for a farm producing M crops is given by

$$\tilde{y}_{it} = \sum_{j=1}^{M} (y_{ijt} / \overline{Y}_{jt}) / M \tag{6}$$

where $\tilde{y}_{it}$ is the relative yield for farm i compared with the county average for across all crops in year t, y_{ijt} is the per acre yield of crop j for farm i in year t, and $\overline{Y}_{jt}$ represents the county-wide average yield for crop j in year t. Note that the normalized-yield indicator represents the inherent profitability of one farm relative to all others in the county while the per-acre, county-average earnings variable represents the profitability of land in one county relative to all others. Finally, the per-acre government payments are decomposed into the various farm-program components (AMTAP, LDP, and disaster payments) using the survey data.

Nonagricultural demands for agricultural land at the rural-urban interface have become increasingly important in many areas. These demands usually reflect housing pressures that result from suburban sprawl and expansion. We include three factors intended to represent the expected future capital gains inherent in farmland in areas facing nonagricultural pressures. First, we include the total value of housing permits in the year of observation that were issued in the county in which each farm is located. The permit data, collected from the U.S. Census Bureau, apply to all forms of residential housing, both single- and multi-family dwellings, and provide figures that reveal the total value of construction costs. Second, we include a series of discrete indicator variables, collected from Barnard (2000), that represent the extent of urbanization for each county. The ordinal ranking for each county ranges from 1 = rural to 5 = urban, where higher numbers indicate a greater degree of urbanization. In that less than 0.3 percent of farms were located in the most urban areas, we combine urban categories 4 and 5 into a single category, Urban 4 & 5. Third, we consider population-growth rates in each county in the preceding year of observation of the farm.

In Model 2, we consider the effects of differentiated farm-program benefits on 1998 to 2000 U.S. farmland values that were reported in the survey. In particular, we consider AMTAPs received by each farm under the production flexibility contract provisions of FAIR, plus LDPs, disaster-relief payments (which include both MLAPs and other ad hoc disaster-relief payments), and an aggregate of all other farm-program payments. (It should be noted that disaster-relief payments could represent different aspects of government support.)

Although MLAPs are not tied to any particular commodity, they have been particularly contentious in policy debates, since they have been driven by legislative concerns over market conditions for specific crops (especially wheat and corn). Thus, although decoupled in theory, MLAPs may have an indirect tie to market conditions and thus may represent a form of market price support. Ad hoc disaster-relief payments for yield shortfalls may have a different effect on agricultural land values. In an attempt to distinguish these two effects, we utilize unpublished county-level farm-program payment data to break disaster-relief payments into those that come through MLAPs and those that come from all other types of disaster-relief payments. These are expressed on a per-acre basis by using total farmland acreage in each county as reported in the 1997 Census of Agriculture (USDA/NASS 1999). All USDA/NASS/ARMS-reported farm-program payments were converted to a per-acre basis using the total acreage operated by each farm.

The data in Table 6.1 give the variable definitions and summary statistics for the specific variables included in our two empirical models. The average per-acre 1999-dollar value farmland for the entire United States is U.S. $1,427. Our data are concentrated in rural areas, with farms in the most rural counties comprising about 64 percent of the sample. The program-payment data of Table 6.1 indicate that, on average, the highest dollar-per-acre payment rates were generated by AMTAPs, which delivered an average of U.S. $13.05 per acre. LDPs accounted for the next highest payment rate, delivering a per-acre average of U.S. $12.80. In our study, disaster-relief payments include ad hoc disaster-relief-assistance payments delivered under various assistance measures in MLAPs in addition to other disaster-relief payments that were often delivered on the basis of AMTAP eligibility and accounted for an average of U.S. $5.84 per acre. We also utilize county-level averages of MLAPs and other disaster-relief-assistance payments data that were collected by the U.S. Department of Agriculture—unpublished U.S. Department of Agriculture, Farm Services Administration (USDA/FSA) data (USDA/ERS 2002)—as instrumental variables to segment the analysis below. A consideration of these variables shows that the vast majority of per-acre disaster-relief assistance is accounted for by MLAPs that average U.S. $6.09 and by other forms of disaster-relief assistance that average U.S. $0.69.

A number of important econometric issues underlie our empirical analysis. The dominant characteristic of the USDA/NASS/ARMS data is the stratified nature of the sampling used to collect the data. Two estimation approaches have been suggested to deal with this stratification issue—the jackknife procedure and

Table 6.1 Variable Definitions and Summary Statistics[a]

Variable	Definition	Mean	Std. dev.
Value	U.S. $/acre reported value[a]	1427.91	11367.20
LDP receipts	Loan deficiency payment receipts (U.S. $/acre)	12.7990	153.6777
Disaster	Disaster-relief payment receipts (U.S. $/acre)	5.8384	101.2351
AMTAP receipts	Agricultural Market Transition Act payments (U.S. $/acre)	13.0504	134.4673
Other GP	Other government payment receipts (U.S. $/acre)	3.3676	82.1149
Urban	Urban indicator (1 = most rural,..., 5 = most urban)	1.7339	10.4334
Urban 1	1 if urban = 1, 0 otherwise[b]	0.6429	4.5404
Urban 2	1 if urban = 2, 0 otherwise	0.1121	2.9792
Urban 3	1 if urban = 3, 0 otherwise	0.1155	3.0290
Urban 4&5	1 if urban = 4 or 5, 0 otherwise	0.1303	3.1904
Yield	Mean normalized yield	0.9719	2.3919
MLA (t-1)	County average market-loss assistance payments (t-1) (U.S. $/acre)	6.0850	61.4563
Disaster (t-1)	County average disaster-relief assistance payments (t-1) (U.S. $/acre)	0.6905	19.8385
Population	Population growth rate (percent)	0.1756	12.0782
Net returns	Net returns exclusive of government payments	62.3712	526.8973
Housing permits	Total value of housing permits (U.S. $ 10 million)	3.3292	109.9980

[a]Numbers of observations are 14,854 for all variables.
[b]Urban 1, Urban 2, Urban 3, and Urban 4&5 are dummy variables that take on the value one if the observation falls in each category and zero otherwise.
Source: Authors' computations based on USDA/NASS/ARMS (USDA/ERS 2002) data.

the bootstrap procedure. The simplest estimation approach involves a jackknife procedure, in which the data are split into a fixed number of subsamples and the estimation is repeated with a different subsample omitted each time the regression is run (Jones 1968). Under the jackknife approach, a sample is divided into m subsamples.[vi] The model of interest is then estimated m times using weighted-regression procedures with one of the respective subsamples omitted from the estimation data each time the regression is run. A simple expression for the variance is then taken by considering the variability of the estimates across each of the replicated estimates. Although this approach has been shown clearly to be appropriate in simple regression applications, we did not use this method in the present study because its suitability for complex estimation problems such as this one is unclear. In addition, it is unclear how this study's focus on one subsample of the overall USDA/NASS/ARMS sample would be affected by using the predefined jackknife groupings, since one subsample could be left with unbalanced jackknifed groups.

The alternative approach that we adopt here involves repeated sampling from the estimation data using a bootstrapping scheme in which the residuals of the first regression are randomly assigned back into the estimated values and the regression is run again (DiCiccio and Efron 1996). Ideally, rather than random sampling from the entire estimation sample (such as the jackknife procedure), an appropriate approach to obtaining unbiased and efficient estimation results involves random sampling from individual strata (e.g., Deaton 1997). In the USDA/NASS/ARMS data, however, this is not possible since each stratum is not identified. The USDA/NASS/ARMS database does, however, contain a population-weighted factor, which represents the number of farms in the entire U.S. farm population that is represented by each individual observation. Therefore we use a probability-weighted bootstrapping scheme in this chapter whereby the likelihood of the same farm being selected in any given replication is proportional to the number of observations in the population represented by each individual USDA/NASS/ARMS observation[vii]. The specific estimation approach involves selecting N observations (in which N is the size of the survey sample) from the sample data. The data are sampled with replacement according to the probability rule described above.[viii] The models are then estimated using a pseudo sample of data. This process is repeated a large number of times and estimates of the parameters and their variances are given by the mean and variance of the replicated estimates.[ix]

RESULTS

Table 6.2 contains parameter estimates and summary statistics for the 1998 to 2000 U.S. farmland-value determinants in which we consider two alternatives. In Model 1 we consider each source of program benefits individually. In Model 2 we allow AMTAPs to have different effects in each of the three years from 1998 to 2000. Recall that, under the provisions of FAIR, AMTAPs were set to decline in each subsequent year and, in theory at least, they were to expire in 2002. Ex ante we anticipate that the effects of AMTAPs on land values should diminish over time. However, the legislative debate leading up to the passage of FSRI suggested that AMTAP policies would end. FSRI extended FAIR's decoupled benefits, even adding a modest increase in their level of support and making soybeans eligible as base acreage from which program payments could be earned. If such legislative actions were anticipated or were to surprise producers, these actions should have been reflected in Model 2 when we examined the effects of AMTAPs on land values. If AMTAPs were expected to end, modest effects on land values should have been realized, in particular for 2000.

The results for Model 1 and Model 2 indicate that price supports, delivered through LDPs, have a significant impact on agricultural land values. In particular, an additional LDP dollar raises per-acre agricultural land values by U.S. $7.65 to U.S. $7.78, which suggests that this market support signals more than just the benefits conveyed in a given year. Price supports through LDPs likely convey an expectation of future support through government actions. Indeed, FSRI

Table 6.2 Model of Land Value Determinants: Parameter Estimates and Summary Statistics[a]

Variable	Model 1	Model 2
Intercept	345.9118	340.4701
	(68.8789)*[b]	(68.6303)*
LDP[a]	7.7829	7.6466
	(0.7307)*	(0.7773)*
Disaster-relief payments	5.5088	5.5326
	(1.2197)*	(1.2296)*
AMTAP (Combined across all years)	4.0640	
	(0.7894)*	
AMTAP (1998)		4.2904
		(1.1112)*
AMTAP (1999)		2.5971
		(0.9233)*
AMTAP (2000)		6.5682
		(1.2704)*
Other GP	−1.8819	−1.8591
	(1.0724)*	(1.0726)*
Urban area 2	351.1932	347.7551
	(31.8646)*	(31.7746)*
Urban area 3	579.5412	578.0024
	(42.8927)*	(42.8446)*
Urban area 4&5	647.8030	647.5101
	(60.3951)*	(60.6468)*
Mean yield	380.1443	384.0393
	(76.1169)*	(76.0770)*
Population	143.4149	144.5040
	(13.5695)*	(13.6336)*
Net returns	4.4082	4.4202
	(0.3004)*	(0.3011)*
Housing permits	14.2935	13.9893
	(4.9686)*	(5.0120)*
[Permits (t) − Permits$(t$-$1)]^2$	−1.2992	−1.2541
	(1.0080)	(1.0201)
R-square	0.2587	0.2594

[a]Number of observations is 11,207.
[b]Numbers in parentheses are standard errors.
*An asterisk indicates statistical significance at the 0.10 or smaller level.
Source: Authors' computations based on USDA/NASS/ARMS (USDA/ERS 2002) data.

deliberations clarify that price supports through LDPs, fixed decoupled payments, and other countercyclical measures will continue to be a fixture of future farm policies. FSRI preserved the loan-payment concept of farm payments but also included other measures known as countercyclical payments intended to provide additional support. Disaster-relief payments, at least those conveyed through the MLAPs that have characterized U.S. agricultural policy over the last few years, offer another important avenue for support within the farming sector. These payments, though of an ad hoc nature, most closely resemble a type of counter-cyclical support, and represent a response by legislators to unfavorable agricultural-market conditions. The payments, however, have been somewhat difficult to characterize, since eligibility for payments is not based on current production levels but rather it is based on an individual farmer's historically determined base acreage and yield, which are the same statistics used to determine the level of AMTAPs a farmer receives. An additional dollar of disaster-relief payments, delivered through MLAPs or other forms of disaster-relief assistance, would raise per-acre land values by U.S. $5.51 to U.S. $5.53. Thus it appears that current policy benefits are expected to continue and are reflected by their current capitalization into land values. The FSRI did not prove these expectations wrong. As mentioned above, a countercyclical-payment program, analogous to the ad hoc MLAP, was included as a formal program in the legislation of FSRI.

When considered as a group across all years, AMTAPs have a substantial effect on land values, raising the value of agricultural farmland by U.S. $4.06 per acre. Thus the extension of decoupled-program support through direct payments that occurred with FAIR seems to have been anticipated by land-market agents. The results of Model 2 (Table 6.2), which segregate the effects of AMTAP by year, suggest that the effect of AMTAP decreased in 1999 but rebounded substantially in 2000. Whether this AMTAP rebound is a reflection of the fact that policy discussions prior to the passage of FSRI caused farmers to suspect that such payments would be continued is unclear. It is clear, however, that land-market agents certainly did seem to anticipate a continuation of AMTAP-type benefits beyond FAIR.

The significance of other government payments [including conservation re-serve program payments, insurance-indemnity payments, and other forms of government support] is negative in both Model 1 and Model 2. These payments may be accompanied by land use restrictions (e.g. conservation programs) or they may signal other negative aspects of the value of agricultural land (e.g., higher insurance-indemnity payments may reflect more variable crop yields and thus may be associated with lower land values).

One of the major issues underlying an understanding of agricultural land values stems not from the profitability of the land resources in agriculture but rather from the potential value of the land for nonagricultural uses. Many land resources have exited from agricultural production along the rural-urban fringe in favor of more valuable uses to satisfy housing and commercial demands. We include two variables in our study that are intended to capture nonagricultural demands. The first variable is the total value of housing permits issued in the county each year. The second variable is the squared change in this value over the preceding year. (This latter variable is intended to capture the uncertainty associated with growth.)

A given plot of agricultural land may have an option value associated with the possibility of making the irreversible decision to convert the land to nonagricultural uses, including housing. As expected, the estimated coefficients for housing permits, population, and the urban dummy variables suggest that construction pressures have a significant influence on agricultural land values. In particular, each additional U.S. $10 million in new construction under county permits raises the per-acre value of agricultural land in that county by U.S. $13.98 to U.S. $14.29. Thus new construction pressures are likely to play an important role in shaping agricultural land values.

The discrete urban-indicator variables suggest that, as expected, land values escalate with urbanization. We found that land in a county that is in the Urban 2 category is worth a per-acre average of U.S. $351 more than the same land would be if it were located in a rural county. In the Urban 3 and Urban 4&5 categories, this value would increase to U.S. $579 and U.S. $647, respectively. Population growth also appears to bid up agricultural land values. We found that each additional 10 percent of population growth adds about U.S. $15 per acre to agricultural land values. We also found that population growth, the value of housing permits, and urban-indicator variables are correlated positively thus it may be difficult to identify the contribution of each to agricultural land values. In addition, we found that the variation in the number of housing starts represented by the squared difference of current and lagged values is not a significant determinant of the value of farmland.

Our results confirm that land that is more productive and profitable in agricultural activities will be more valuable than land that is less productive. One measure of productivity considers the profitability of agriculture (net of policy benefits) in the county and another measure of productivity considers the yield-performance of a farm relative to others in the county. In both cases, the coefficients are positive and highly significant, confirming expectations that more productive farmland will command a higher per-acre price. We again note that our measure of relative farm-yield performance may reflect both farm and operator characteristics, because it is not possible to separate factors influencing farm-yield performance tied to land quality from farm-yield performance tied to the abilities of the operator.

In all, our results indicate that each government farm-subsidy program has a different effect on the value of agricultural land. Thus the vague and overly general characterizations of the effects of program benefits on land values that have been common in the economic literature to date may not provide an accurate assessment of how present-day and future programs may affect land values.

Our results also confirm that, despite rhetoric to the contrary, decoupled payments with a fixed termination date (i.e., AMTAPs) continued to have substantial impacts on land values, even as the termination of the FAIR approached. Likewise, we found that land price and market-support programs substantially increased agricultural land values, but the greatest impact came from LDPs. There can be little doubt that government-program benefits are likely to persist and be a part of U.S. farm policy for some time to come. Naturally, land markets respond to expected future government-program payments, and those farms that received higher payments had more valuable land assets.

Estimates suggest that government policy adds as much as 69 percent to the value of land in some parts of the nation. Using our estimated parameters reported for Model 2 in Table 6.2, we estimated the contribution of the three main avenues of government support considered in our study—AMTAPs, LDPs, and disaster-relief payments (including MLAPs) for 2000. As a word of caution, it should be noted that our two models only measured marginal effects, and our analysis assumed market returns as constant.

The data in Figure 6.1 illustrate the percentages of land values in each state accounted for by AMTAPs in 1998. The effect of AMTAPs on land values was significant—generally less than 3 percent, although they were as high as 9 percent in some areas. The main influence we found were in the Corn Belt, the Northern Great Plains, and the Mississippi Delta regions.

Data in Figure 6.2 give the percentage effects on land values brought about by LDPs. The effects of such payments were most significant in the Corn Belt and Upper Great Plains regions. These payments were most significant for soybeans, thus it is not surprising that the greatest effects were realized in the soybean-producing areas. For our entire sample, LDPs accounted for about 3.9 percent of the increase in agricultural land values, although this figure in 2000 was as high as 21 percent in a few areas.

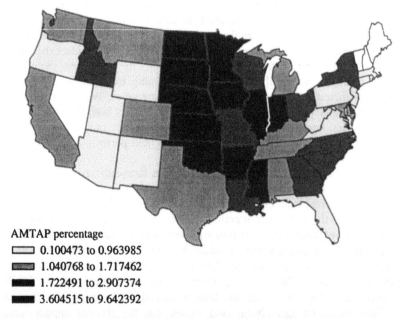

AMTAP percentage
▭ 0.100473 to 0.963985
▨ 1.040768 to 1.717462
▮ 1.722491 to 2.907374
▬ 3.604515 to 9.642392

Figure 6.1 Percentage of Land Values Accounted for by AMTAPs (1998)

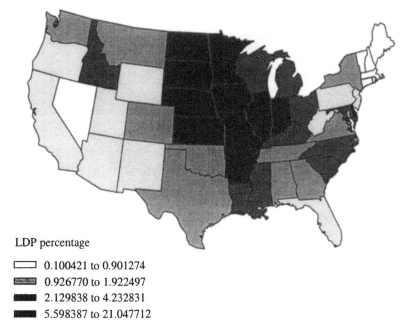

LDP percentage

▭ 0.100421 to 0.901274
▨ 0.926770 to 1.922497
▩ 2.129838 to 4.232831
▨ 5.598387 to 21.047712

Figure 6.2 Percentage of Land Values Accounted for by LDP Payments (2000)

Data in Figure 6.3 show the effect of disaster-relief payments on land values. Recall that the largest proportion of disaster-relief payments over the period of our study came in the form of MLAPs. For our sample as a whole, 3.4 percent of agricultural land values were accounted for by disaster-relief payments. Again, considerable variation existed across the United States, with the largest effects being realized in the Upper Great Plains and Mississippi Delta regions.

SUMMARY AND CONCLUSIONS

Our analysis investigated the effects of a variety of government programs on agricultural land values. In contrast to most existing research, we utilized USDA/NASS/ARMS farm-level data and accounted for the fact that farm programs convey benefits to producers through a variety of different policy mechanisms. Most researchers aggregate all government payments and reach general conclusions about the effects of government-program payments on land values.

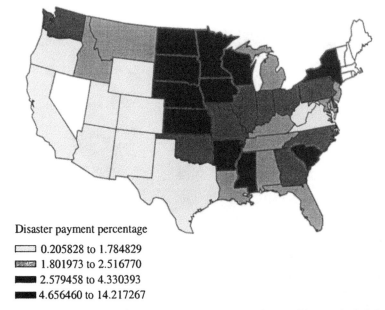

Disaster payment percentage
☐ 0.205828 to 1.784829
▨ 1.801973 to 2.516770
■ 2.579458 to 4.330393
■ 4.656460 to 14.217267

Figure 6.3 Percentage of Land Values Accounted for by Disaster-Relief Payments (2000)

This distinction was more important under FAIR, which was characterized by price supports, fixed decoupled payments, and ad hoc disaster-relief assistance. The results of our study confirm that different policy mechanisms have different effects on land values. Government market support programs conveyed through LDP price supports have had the greatest effect on land values. Ad hoc disaster-relief assistance, including the large MLAPs of the last few years, also had a large effect on land values. An important implication of this result is that producers may not view ad hoc payments as transitory policy phenomena, since such benefits seem to be capitalized into higher land values. The fixed, decoupled AMTAP benefits also appear to have a significant effect on land values. Under FSRI, this effect is greater than it would have been if these program benefits had ended with FAIR in 2002, suggesting that the continuation of such benefits into FSRI was anticipated.

In all, our results demonstrate that different policy mechanisms have different effects on land values. Without a doubt, the considerable support provided to U.S. farmers from 1998 to 2000 through government transfers increased agricultural land values and rents. However, the effects differed substantially across regions and years. Our empirical estimates found smaller effects on land values than is common in other empirical literature. This may be a reflection of our modeling approach, which gives careful consideration to nonagricultural demands for land. Our modeling approach also desegregated policy benefits by different types of programs. In addition, the 1998 to 2000 post-FAIR years contained a different set of agricultural policies than was common over much of the preceding seventy years.

REFERENCES

Barnard, C.H. (2000) "Urbanization Affects a Large Share of Farmland." *Rural Conditions and Trends*. 10(2): 57-63, Washington DC: USDA/ERS.

Barnard, C.H., G. Whittaker, D. Wesenberger and M. Ahearn. (1997) "Evidence of Capitalization of Direct Government Payments into U.S. Cropland Values." *American Journal of Agricultural Economics* 79(5): 1642–50.

Barnard, C.H., K. Wiebe, and V. Breneman. (2003) See Chapter 18, this volume.

Belsie, L. (2002) "Farm Subsidies Prop Up Midwest Land Values." *Christian Science Monitor*. Boston, National Edition (January 4).

Burt, O.R. (1986) "Econometric Modeling of the Capitalization Formula for Farmland Prices." *American Journal of Agricultural Economics* 68(1): 10–26.

Clark, J.S., M. Fulton, and J.T. Scott. (1993) "The Inconsistency of Land Values, Land Rents, and Capitalization Formulas." *American Journal of Agricultural Economics* 75 (1): 147–55.

Deaton, A. S. (1997) *The Analysis of Household Surveys: A Microeconometric Approach to Development Policy*. Baltimore: Johns Hopkins University Press and the World Bank.

DiCiccio, T.J. and B. Efron. (1996) "Bootstrap Confidence Intervals" *Statistical Science* 11(3): 189–212.

Featherstone, A.M. and T.G. Baker. (1987) "An Examination of Farm Sector Real Asset Dynamics 1910–85." *American Journal of Agricultural Economics* 69(3): 532–46.

Goodwin, B.K. and F. Ortalo-Magné. (1992) "The Capitalization of Wheat Subsidies into Agricultural Land Values." *Canadian Journal of Agricultural Economics* 40(1): 37–54.

Jones, H.L. (1968) "The Analysis of Variance of Data from a Stratified Sample." *Journal of the American Statistical Association* 63(321): 64–86.

Melichar, E. (1979) "Capital Gains versus Current Income in the Farming Sector." *American Journal of Agricultural Economics* 61(5): 1085–92.

Raup, P.M. (2003) See Chapter 2, this volume.

Reynolds, J.E. and W.E. Johnston. (2003) See Chapter 20, this volume.

Ryan, J., C. Barnard, R. Collender, and K Erickson. (2001) "Government Payments to Farmers Contribute to Rising Land Values." *Agricultural Outlook*. USDA/ERS (June-July): 22–26.

Shertz, L. and W. Johnston. (1997) "Managing Farm Resources in the Era of the 1996 Farm Bill," Staff Paper No. AGES 9711 (December). Washington, DC: USDA/ERS.

Shoemaker, R., M. Anderson, and J. Hrubovcak. (1990) "U.S. Farm Programs and Agricultural Resources." Agriculture Information Bulletin No. 1990 (September) Washington, DC: USDA/ERS.

USDA/ERS (U.S. Department of Agriculture). (2002) "Agricultural Resource Management Survey (ARMS): Detailed Documentation." Internet Website: www.ers.usda.gov/briefing/ARMS/detaileddocumentation.htm.

USDA/NASS (U.S. Department of Agriculture, National Agriculture Statistics Service). (2001) Agricultural Economics and Land Ownership Survey, (AELOS 1999). Internet Website: www.nass.usda.gov/census/census97/aelos/aelos.htm

Vantreese, V.L., M.R. Reed, and J.R. Skees. (1989) "Mandatory Production Controls and Asset Values: A Case Study of Burley-Tobacco Quotas." *American Journal of Agricultural Economics* 70(2): 319–25.

Weersink, A., S. Clark, C.G. Turvey, and R. Sarker. (1999) "The Effect of Agricultural Policy on Farmland Values." *Land Economics* 75(August): 425–39.

ENDNOTES

[*] The opinions expressed here are those of the authors and not necessarily those of the U.S. Department of Agriculture.

[i] Such payment caps have been a part of farm policy for some time, although the limits were typically not a constraint to most producers. Ultimately, the new 2002 Farm Bill maintained the three-entity provision that was often used to circumvent payment caps and introduced commodity certificates that effectively circumvented any attempts to limit payments.

[ii] MLAPs were generally tied to a farm's level of AMTAPs, and thus are sometimes known as double-AMTAPs, though the terms is misleading since the market loss adjustments were not necessarily double the AMTAPs levels. In our data, market loss assistance is aggregated with other forms of disaster relief (e.g., drought assistance payments).

[iii] The regions in this study are the ten USDA/ERS production regions: The Northeast includes Connecticut, Delaware, Massachusetts, Maryland, Maine, New Hampshire, New Jersey, New York, Pennsylvania, Rhode Island, and Vermont; the Lake States include Michigan, Minnesota, and Wisconsin; the Corn Belt includes Illinois, Indiana, Iowa, Missouri, and Ohio; the Northern Great Plains include Kansas, Nebraska, North Dakota, and South Dakota; Appalachia includes Kentucky, North Carolina, Tennessee, Virginia, and West Virginia; the Southeast includes Alabama, Florida, Georgia, and South Carolina; the Mississippi Delta includes Arkansas, Louisiana, and Mississippi; the Southern Great Plains include Oklahoma and Texas; the Mountain States include Arizona, Colorado, Idaho, Montana, Nevada, New Mexico, Utah, and Wyoming; and the Pacific States include California, Oregon, and Washington.

[iv] An excellent review of the conceptual framework underlying most land-value studies is presented in Weersink et al. (1999). Our approach follows their general outline.

[v] The discount factor b is derived from the mathematical expression

$$\sum_{i=1}^{\infty} \frac{1}{(1+r)^i} = \frac{1}{r} = b.$$

[vi] Estimation programs created by ERS use 15 subsamples.

[vii] We utilize the Surveyselect procedure of the Statistical Analysis System language in our estimation, though alternative approaches are also possible.

[viii] To be precise, if observation i represents ni farms out of the total of M farms in the population, the likelihood that observation i is drawn on any given draw is ni/M. It should be acknowledged that our approach may result in less efficient estimates than would be the case if sampling were used from individual strata. This could occur in cases in which inferences are being made about variables used in designing the stratification scheme in that such information is being ignored by not drawing from individual strata. To the extent that this is relevant to our analysis, the t-ratios reported in this chapter represent conservative estimates.

[ix] We utilize 1,500 replications in the applications that follow. The number of replications is necessarily limited by the computer-intensive nature of sampling according to the probability rule. If one were interested in estimating probability values rather than standard errors, a larger number of replications may be more desirable.

Chapter 7

Do Direct Payments Have Inter-Temporal Effects on U.S. Agriculture?

Terry Roe, Agapi Somwaru, and Xinshen Diao
University of Minnesota
U.S. Department of Agriculture, Economic Research Service
International Food Policy Research Institute

INTRODUCTION

The question of whether production flexibility contract payments (PFCPs) to farmers are likely to be minimally trade distorting is considered in an inter-temporal and economy-wide context. We show the circumstances, over time, under which a minimally trade-distorting result is likely to be obtained. If agricultural capital markets are complete, we find that payments have long run effects on land values and on land-rental rates, but they have no effect on production. If capital markets are not complete, we find production effects, but they are small in the short run (0.2 percent) and disappear in the long run. The only permanent effects of PFCPs are on land-rental rates and land values that increase by about 10 percent in the short run and taper off to slightly above 8 percent in the long run.*

THE ISSUE

The 1996 Federal Agriculture Improvement and Reform Act (FAIR) introduced new instruments of producer supports, including fixed payments, known as production flexibility contract payments, that are tied to historical base acreage and yields (Orden et. al. 1999). The general issue in this chapter is whether or not the

115

PFCPs to farmers have inter-temporal effects on resource allocation and production to the extent that they may not be in compliance with the minimally trade-distorting criteria for the green-box policy designation of the Uruguay Round Agreement on Agriculture (URAA). PFCPs have been viewed as decoupled from production, and have been designated green-box policies in contrast to amber-box policies, which exempts them from payment limits under the World Trade Organization (WTO). The key concern is not that PFCPs change the consumption patterns of recipients of program payments, which they almost surely do. Nor is the question whether or not payments change their investment patterns or labor-leisure choices. Instead, the question is whether or not PFCPs have effects on agricultural markets. The purpose of this chapter is to contribute to the PFCP debate by considering the inter-temporal effects of decoupled payments on market behavior.

We begin with a discussion of basic concepts to help clarify the nature of market linkages between the taxpayers and PFCP recipients. The second section of this chapter reports the results from calibrating to U.S. data a simple inter-temporal model of the Ramsey variety (Barro and Sala-I-Martin 1995: Chapters 2 and 3) with adaptations to account for sector-specific factors of production, multiple sectors, and segmented capital markets.

We suggest that if markets are complete, and taxed and recipient households have similar preferences, then PFCP effects on market outcomes, even over time, likely meet the minimally trade-distorting criteria of the URAA green-box designation. A possible exception is the indirect effect of PFCPs on land values that can potentially increase farmer's access to credit. In the real economy, farmers may hold expectations about the nature of future farm programs that can couple PFCP to production decisions. Further, markets are not complete because of reasons including the presence of fixed costs, of absent and incomplete risk markets, and of the fact that agricultural capital markets differ from capital markets for the corporate manufacturing and service sectors of the economy. For these and other reasons, this chapter contributes to the debate about the effects of PFCPs on the U.S. agricultural sector as opposed to answering the question about whether PFCPs are minimally trade distorting.

BASIC CONCEPTS

Over time, the recipients of policy-transfer payments are likely to consume more goods, including leisure, and to increase their savings. However, whether these individual decisions affect resource allocation and supply at the market level depends on the behavior of those who are taxed to provide the transfer.

The reason for the broader perspective is that the investment and consumption effects of those taxed can be exactly offset by the recipients of the PFCPs, post transfer, resource allocation and production at the market level are unaffected. In principle, this outcome is expected to prevail when capital markets work perfectly in allocating savings to investors in all sectors of the economy, when risk markets are complete in the sense of opportunities to insure against future contingencies, and when the representative rural household (or recipient of the policy transfer) has beliefs and consumption-savings preferences that are indistinguishable from other households. Under these circumstances, the wealth effect of a transfer on recipient behavior is offset by the negative wealth effect of those taxed to provide the transfer. If individuals vary in their beliefs and consumption-savings preferences, then direct payments can, in principle, have market-level effects because the decrease in savings of individuals taxed can depart from the increase in savings of recipients. (More specifically, if individual preferences are identical but non-homothetic, then marginal propensities to consume and save can differ among individuals of different income levels. In this case, the behavior of recipients can differ from that of those taxed with the result that transfer payments can affect market allocations over time.)

Of course, in real economies ideal market conditions do not prevail. Should the lack of ideal conditions render what is in principle a decoupled instrument an amber-box designation? A policy instrument might cause trade distortions due to market failures, which if corrected, could render them green-box designations. This argument has received some acceptance in trade negotiations. For example, public support of agricultural research and development (R&D) has a green-box designation, presumably because of the widely recognized fact that market forces alone lead to under-investment. Still, other market inefficiencies are endemic to even the most advanced economies, including information, risk, and capital markets.

As Stiglitz (1985: 21) argues, markets fail in the optimal provision of information, and theory is not "robust to slight alterations in informational assumptions." Empirical estimates of the value of information in risky markets by Antonovitz and Roe (1986) support subsequent work that agent's subjective forecasts of future events—the importance of information—dominates the small effects of risk preferences on production decisions. From the time of Sandmo (1971) for the case of the individual risk-averse agent, and Hirshleifer (1988) for the case of the market, it has been known that individual and market behavior under risk is affected by specific features of capital markets, such as the presence of liquidity constraints. Another form of market failure is that of fixed costs, which is the point raised by Chau and de Gorter (2000). Further, the presence of other amber-box policies helps to place the question of whether PFCPs are minimally trade distorting in the world of the second-best (Mas-Colell et al. 1995: 710). Then, policies that are viewed as trade distorting in an ideal economy can be welfare enhancing in a real economy. In our view, these conditioning factors should, in principle, be evaluated when analyzing whether or not an instrument is likely to be minimally trade distorting.

In perhaps the most serious econometric analysis to date, Goodwin et al. (2003) conclude from their analysis of the U.S. Department of Agriculture, Economic Research Service, Agricultural Resource Management Survey (USDA/NASS/ARMS) data that PFCPs seem to have modest effects on production for most farms. The mechanism through which this occurs, and how the conditioning factors mentioned above affect this result, is difficult to address with these data. Their analysis also suggests that the effects of PFCPs on the decision to purchase land are positive and statistically significant. This may suggest that, all else equal, farmers have a preference for investing in agriculture from the incremental increase in their savings account due to PFCPs relative to the rest of the economy.

In this chapter, we investigate two additional factors that may cause PFCPs to distort markets. One factor is the fact that agricultural-capital markets differ from nonfarm-capital markets. The other factor is that PFCPs are linked to land that was formerly planted to program crops.

Agricultural-capital markets differ from nonfarm-capital markets in that farmers cannot issue securities or bonds to finance farm activities as corporations can; instead they must rely more heavily on land and on other assets for collateral. Most corporations tend not to invest directly in the production of program crops, although contract production in broiler, egg, and hog production is common. Thus, the effect of the incremental tax increase on investments made by individuals outside of agriculture can differ from the investments made by recipients of PFCPs in agriculture. The net effect on capital markets can be greater if, all else equal, farmers face liquidity constraints or if they have a preference for investing in agriculture the proportion of PFCPs not allocated to their consumption uses, as the results of Goodwin et al. (2003) suggest. The difference in these markets does not imply that returns to capital in agriculture depart from returns in other sectors of the economy, at least in the longer run, since farm households also invest in stocks, bonds, and other economy-wide financial instruments (Collender and Morehart 2002). In the short run, the increase in agriculture's capital stock should have output, hence market effects. The question is whether these effects are large.

Another issue is that PFCPs are linked to land that was formerly planted to program crops. This linkage is important because land is an asset. In principle, individuals should allocate their savings among investment alternatives so as to equate returns between these alternatives with adjustments to account for risk and deferred capital gains. This adjusted rate of expected return should then be approximately equal to the expected rate of return to a dollar invested in a unit of land plus the gain (loss) due to changes in land prices. This condition is expected to prevail among assets when capital markets work perfectly in maximizing returns to savings (Barro and Sala-I-Martin 1995: 99). Since direct payments are targeted to land formerly planted to program crops, the cash-rental rate that a tenant is willing to pay for an acre of land is affected by the amount of his PFCPs, and consequently, so is the price of land. A change in the price of land affects its value as an asset, which affects wealth, and can consequently impact investment and consumption behavior. Since land is used as collateral, payments can (in principle) increase farmer's access to credit. Goodwin et al. (2003) find that PFCPs have small effects on land values, ranging from 2 percent to 6 percent in the Northern Great Plains and Corn Belt regions of the United States. Our analysis suggests even greater effects.

ANALYSIS

General Procedures

We first evaluate M innesota d ata on land values and government payments t o see if any linkage is suggested. Then, since the above discussion implies that an economy-wide approach is required to assess whether or not direct payments are likely to affect resource allocation and production, we report the results from an inter-temporal multi-sector model of the 1997 U.S. economy (details of the inter-temporal multi-sector model are presented in Appendix 7.A). Savings are endogenous, and assets are aggregated into three broad categories, capital inside of agriculture, capital outside of agriculture, and land. We fit two versions of the inter-temporal multi-sector model to our data. In the first version, we presume that capital markets for agriculture and for the rest of the economy are perfectly integrated so that any differences in their short-run rates of return to capital and land are instantly arbitraged to zero. In the second version, we relax this assumption so that the arbitrage condition only holds in the longer run. Otherwise the models are identical. Households are presumed to hold identical homothetic preferences over their consumption of goods and services. Most of the parameters of the model are based on the year 1997, while rates of growth in total factor productivity, growth in the U.S. labor force, and selected other parameters are taken from other research (Roe 2001, details of the calibration of the model are presented in Appendix 7.B). We find that our model reproduces some of the key outcomes observed for the actual economy for the years 1997 to 2001.

Land values appear to be linked to government payments

Using data from statistical reporting districts in Minnesota for the 1994 to 2000 period, we chart the average change in the value of land planted to crops (over 8 million acres), and the change in total government payments associated with these lands (Figure 7.1). Nominal land values have appreciated at the rate of about 6.6 percent per year. The years in which land appreciation was the highest were the two years following the enactment of FAIR.

The positive correlation shown by the data in Figure 7.1 is confirmed by a simple regression analysis. Analysis suggests that between 1994 and 2000, a 10 percent change in government payments tended to cause a 3.24 percent change in land values. S ince t he data in Figure 7.1 indicates that t his positive correlation persisted after the enactment of FAIR, we conclude that transfer payments affect land values. This appreciation affects wealth and can consequently affect investment and consumption behavior. In principle, since land is used as collateral, payments can increase access to credit for the farmer. Our simple model, however, does not capture this important link.

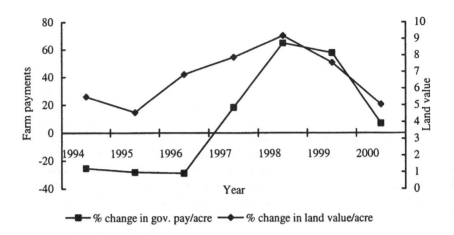

—■— % change in gov. pay/acre —◆— % change in land value/acre

Figure 7.1 Direct Payments Affect Land Values—Annual Growth Rates in Minnesota Land Values and Farm Payments

Effects of PFCPs on resource allocation are surprisingly small

We make the assumption that PFCPs, equal to U.S. $6.11 billion in 1997, are made to farmers in each period of time from 1997 in perpetuity. Thus, the results from this exercise should be interpreted as suggesting the directional effects of direct payments as opposed to placing undue emphasis on the magnitude of the direct payments. All of the reported results are compared to the base, which is the path of the economy in the absence of direct payments to farmers.

The case of integrated capital markets

This analysis presumes that investors allocate savings at each instant of time so as to arbitrage away any differences in rents to the three assets of capital in the rest of the economy, capital in agriculture, and land. Effectively, at each instant of time, the rate of return to agricultural capital is equated to the returns to capital in the rest of the U.S. economy. Since preferences are identical, the consumption and investment behavior of the recipients of PFCPs are exactly counterbalanced so that no net resource allocation effects are observed. This is the case in which payments are completely decoupled, even inter-temporally. Since payments are linked to land planted to program crops, however, land values are also affected, which supports the result reported in Figure 7.1 and the work of

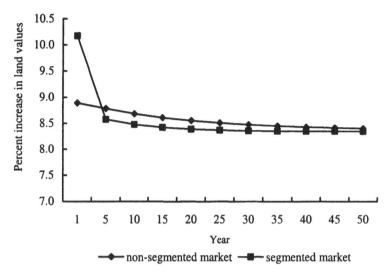

Figure 7.2 Effects of PFCPs on Land Values are Unchanged by Segmented Capital Markets (Percent Increase in Land Values)

Goodwin et al. (2003) (Figure7.2). We find that the U.S. $6.11 billion dollar payment, in the short run, causes land values to exceed their values in the base run by almost 9 percent, and then causes them to taper off to about 8.3 percent above their long-run base value.

These effects are due solely to PFCPs. Competition for land, and thus a right to transfer it, causes renters to pay higher rental rates to owners. If the land is sold, the buyer is willing to pay more if the PFCPs remain tied to the land. Since the base run also accounts for growth in agriculture's total factor productivity and capital deepening over the period, this reported escalation of land values is due only to government payments.

Of course, PFCPs and the escalation of land values change recipient-consumption patterns and level of assets (Figure 7.3). In the short run, our results suggest that asset values of recipient households increase by about 2 percent above their base values, due mostly to the elevation in land values. Most of the PFCPs are spent purchasing final goods. The spending proportion increases over time while the saved proportion falls. Total consumption expenditures are about 0.8 percent higher than are the expenditures in the absence of transfers. The rise in recipient household-asset holdings should also increase their access to credit. If liquidity constraints are binding, the increase in the value of land due to PFCPs should help to relax these constraints. This effect will tend to reduce the extent to which PFCPs are decoupled from production incentives.

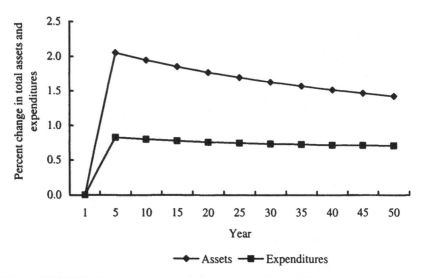

Figure 7.3 PFCPs Increase Assets and Consumption Expenditures

The case of segmented capital markets

The analysis above reproduces the directional effects that Goodwin et al. (2003) find on land values, but it does not reproduce the changes in production that they find. We repeat the analysis, but no longer do we allow agricultural capital markets to be perfectly arbitraged with capital markets in the rest of the economy at each instant of time, although they are arbitraged in the longer run. Within agriculture, and within the rest of the economy, the analysis allows all capital rents to be arbitraged.

Figure 7.2 shows that, in the short run, when the rural and rest of the economy capital markets are not perfectly arbitraged, the value of land is roughly one percent higher compared to when these two markets are arbitraged. The reason for this result is evident in Figures 7.4 and 7.5.

The data in Figure 7.4 show the percent change from its base of PFCPs on the capital-rental rate outside of agriculture, on the rental rate in agriculture, and, more generally, on the index of wages and other prices. The results of the data show that, within the first ten years of payments in equal amounts, the rental rate on agricultural capital declines by a modest 0.1 percent below the capital-rental rate observed in the base solution. This rate is 6.48 percent in year five. The effect on the capital-rental rate outside of agriculture, and on the price index of

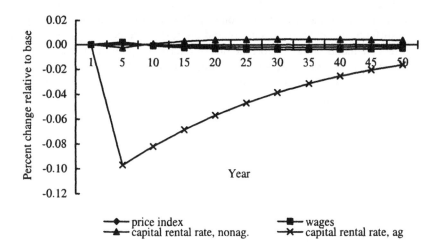

Figure 7.4 PFCPs Have Negligible Effects on Capital Rental Rates in Agriculture: The Segmented Case

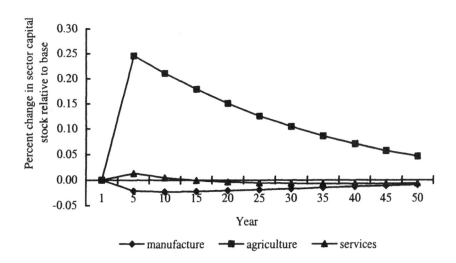

Figure 7.5 PFCPs have Small Effects on the Stock of Capital: The Segmented Case

goods, is almost imperceptible. Notice that even though the direct payments in equal amounts continue in perpetuity, agriculture's capital-rental rate slowly converges to that of the rest of the economy. In other words, in spite of the presumed differences between agriculture and the rest of the economy, in the long run, direct payments do not distort the rate of return to capital in agriculture.

The decline in agriculture's rate of return to its capital stock affects the price of land via the market-clearing condition for maximizing returns to savings. This condition amounts to equating, at each instant in time t, the rate of return to agriculture's capital stock $r(t)$ to the ratio of land rent, including PFCPs to the price of land, plus the rate of change in the price of land, or

$$r(t) = \frac{\pi(t)}{P(t)_{land}} + \frac{\dot{P}(t)_{land}}{P(t)_{land}},$$

where $\pi(t)$ is the land-rental rate including PFCPs, P_{land} is the price of land, and $\dot{P}_{land}$ is the change in the price of land. The solution to this differential equation yields the evolution in the price of land over time. As the interest rate falls, all else constant, the rental rate of land rises due to capital deepening, which in turn causes the value of land to increase. Since this analysis captures farmers' preferences for investing some of their savings in agriculture relative to the rest of the economy (or equivalently, if PFCPs help to relax the otherwise binding liquidity constraints), farmers will tend to increase agriculture's capital stock at a slightly higher rate. All else constant, the diminishing returns to the growth in agriculture's capital stock (Figure 7.5) causes the rate of return to land to decline and land prices to rise to a greater extent than in the case in which capital markets are presumed to be non-segmented (Figure 7.2).

We show why direct payments cause the rental rate of agricultural capital to decline (Figure 7.5). In early periods, farmers tend to allocate a relatively larger proportion of their payments to investment in agricultural capital than they do in latter periods. In the short run, the amount of capital invested in agriculture reaches a modest maximum of about 0.25 percent of the capital stock that would otherwise be accumulated relative to the base. As additional capital investments diminish returns to capital stock, farmers save less and spend a larger share of their PFCPs on final goods. In the long run, the amount of capital employed in agriculture is equal to the amount that would be employed in the absence of transfer payments (i.e., payments do not affect the long-run level of capital stock in the sector). Nevertheless, the half-life of the adjustment is about twenty years because the depreciation rate for buildings and structures is relatively small. The effect on capital stocks in the rest of the economy is almost imperceptible.

Relative to the base-case scenario, as farmers increase their level of capital stock, more labor hours are also allocated to production (Figure 7.6). The increased hours accrue from a decrease in leisure time and/or from an increase in hired labor. (Since leisure is found and treated here to be a normal good, the combination of wealth and price effects leaves the average level of leisure consumed by farmers to be virtually unchanged.) The slight increase in labor, relative to the base, comes from the labor market. Nevertheless, in absolute terms in

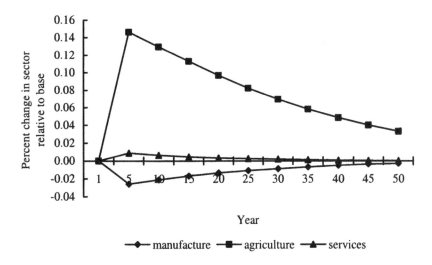

Figure 7.6 PFCPs Cause a Small Increase in Agricultural Employment: The Segmented Case

all of the analyses, there is an out migration of labor from agriculture.) Again, the magnitudes of these adjustments are relatively small. However, PFCPs encourage the use of capital relative to labor (Figure 7.7). That is, the capital to labor ratio rises relative to the base, because the presumed preference for investing in agriculture causes the rate of return to capital to fall slightly relative to the change in wages. The change in this ratio encourages an increase in the substitution of capital for labor relative to the base. In the long run, the ratio converges to the level expected in the absence of payments.

Finally, do the resource reallocation effects of PFCPs affect aggregate agricultural production? The data in Figure 7.8 reveal that U.S. agricultural production increases by a maximum of about 0.18 percent of its base value in the short run, and then in the longer run, it returns to levels that would prevail in the absence of payments. Even if payments were made into the indefinite future to farmers at approximately the levels they were in 1997, the payments would have no long-run effects on production. The effects that prevail into the long run are the elevated price of land (Figure 7.2), and the increased land-rental rates (Figure 7.8).

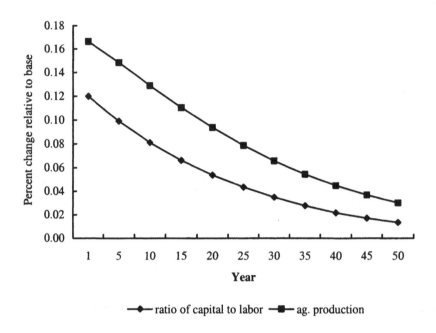

Figure 7.7 PFCPs Induce a Small Short-Run Increase in Agricultural Production and in the Use of Capital Relative to Labor: The Segmented Case

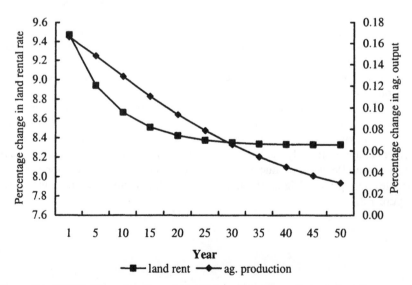

Figure 7.8 PFCPs Have Small and Declining Effects on Output but Large and Lasting Effects on Land Rental Rates: The Segmented Case

CONCLUSIONS

The general question addressed in this chapter is whether PFCPs to farmers are likely to cause market effects that are in excess of what might be termed minimally trade distorting. We consider this question in an economy-wide context, because the market effects of those taxed to provide the transfer might just offset the market effects of the recipients of the transfer. If this result sustains itself in the long run, the PFCP can be thought of as an efficient policy instrument that can be used to transfer resources from one segment of the population to another with no dead-weight losses, and with minimal trade distortions. The real economy is obviously complicated and encumbered with incomplete markets which make addressing the issue of whether PFCPs are truly minimally trade distorting a complex undertaking. Our contribution to addressing this issue lies in showing the circumstances over time under which a minimally trade-distorting result is, in principle, likely to result. We also attempt to account for the possibility that the capital markets in agriculture might differ from those serving firms in the rest of the economy, and then assess the extent to which this difference may cause payments to be more trade distorting.

We find empirical evidence, using data from Minnesota farms, to support the notion that direct payments to farmers affect land prices—a result obtained recently by others using a broader more comprehensive data set (USDA/ERS 2001a). Our economy-wide analysis finds that if agricultural capital markets are perfectly integrated with capital markets in the rest of the economy, and if the taxed and the recipients of the PFCPs hold identical and homothetic preferences over goods and services, then the key effects of payments over time are to increase the value of land by about 8 percent. Of course, this will also increase the wealth and expenditures by the program recipients on the consumption of final goods. None of these effects are trade distorting. The exception, which we do not investigate, is that land is an asset that can be collateralized, and as such, it potentially provides farmers access to more credit than would otherwise be the case.

All else constant, if we presume that farmers prefer to invest in agriculture, the increment of the PFCPs not spent on consumption goods, which they may because of any number of reasons including the presence of liquidity constraints, then we will find some evidence that payments cause resource allocation and output effects. But, these effects are small, and they persist only in the short run.

In this case, in the short to intermediate run, direct payments tend to cause capital deepening, increase the employment of labor, and increase agricultural output. However, these effects, even though they are extremely small, may cause aggregate agricultural production to increase by less than 0.2 percent in the short run. In the long run, payments cause no resource allocation and output effects. The only long-term effect of payments is to increase land values and land-rental rates.

We conclude that direct payments are relatively efficient policy instruments for transferring income from the rest of the economy to farmers. They are efficient in the sense that they have relatively small effects on agricultural resource allocation and production. As other analyses have shown, for example USDA/ERS (2001b), instruments affecting market access, export subsidies, and

farm support that directly influence farmers' incentives are far more distortion-
ary thus they are far less efficient instruments used to transfer income to farmers.

APPENDIX 7.A

The Analytical Model

The analytical model underlying the results presented in the various figures is
briefly sketched here. Neither comparative-static results nor proof of existence is
presented. The basic model is presented first, followed by a discussion of the
two adaptations to account for PFCPs, and to account for PFCPs in the presence
of segregated capital markets.

The Environment

The model we use for our analysis depicts an open economy in which agents
consume and produce at each instant of time a manufacturing good, an agricultural good, and a home good. The manufacturing good can also be allocated to
capital. The agricultural and manufacturing good can be traded internationally at
given prices p_a and p_m. Labor services are not traded internationally and domestic residents own the entire stock of domestic assets. The home good is only
traded in the domestic economy at endogenously determined price p_s. Agricultural, manufacturing, and home goods are indexed $j = m, a$, and s, respectively.
Households are of two types, indexed $i = u, r$. They may be thought of as denoting urban households that do not own land, and as other mostly rural households
that own land. The only feature-distinguishing type is their endowments of labor,
and the assets of capital and agricultural land. Their utility functions describe
identical preference relations. Households purchase goods C_{ij}, for consumption,
and earn income from providing labor services L_i in exchange for wages w that
earn interest income at rate r on capital assets A_i and receive rents π_i from
agriculture's sector-specific resource, land T. The manufacturing and home-
good sectors employ labor and capital services, while agriculture employs the
service of land. Two basic versions of the model are considered, one in which
the arbitrage condition between assets is presumed to hold, and another in which
the market for agricultural capital can clear at a rate of return that is different
from the capital employed in the manufacturing and service sectors of the economy.

Firms

The manufacturing and home-good sectors ($j = m, s$) employ constant returns-
to-scale technologies that, at the sector level, can be expressed as

$$Y(t)_j = F^j\left[A(t)L(t)_j, K(t)_j\right], j = m, s, \tag{7.A1}$$

where labor productivity grows at the exogenous rate x

$$A(t) = e^{xt}.$$

Omitting the (t) notation, except for emphasis, it is convenient to express the production functions of Equation (7.A1) in efficiency units per worker (or units per effective worker)

$$\hat{y}_j = \frac{Y_j}{A(t)L} = l_j f^j\left(\hat{k}\right) = \frac{L_j}{L} F^j(1, \frac{K_j}{A(t)L_j}). \tag{7.A2}$$

Output $\hat{y}_j$ is now expressed in units per effective worker in the economy where L is total labor supply, l_j is the share of total labor employed in sector j, and $\hat{k}_j$ denotes the amount of capital stock per effective worker employed in sector j. For purposes here, the technologies are assumed to satisfy the conditions

$$\frac{\partial f^j\left(\hat{k}_j\right)}{\partial \hat{k}_j} \to \infty \text{ as } \hat{k}_j \to 0 \text{ and } \frac{\partial f^j\left(\hat{k}_j\right)}{\partial \hat{k}_j} \to 0 \text{ as } \hat{k}_j \to \infty, j = m, s. \tag{7.A3}$$

The sector-level technology of agriculture is taken to be constant returns to scale (CRS) in all of its arguments, although land T, which can be rented among farmers, is assumed to be specific to the sector. The technology

$$Y(t)_a = F^a\left[A(t)L_a, K_a, A_a(t)T\right]$$

can be expressed in per-capita terms as

$$\hat{y}_a = l_a f^a\left(\hat{k}_a, \hat{T}_a\right), \tag{7.A4}$$

where $l_a = L_a/L$ is the share of total labor employed in agriculture, and $T_a = A_a(t)T/A(t)L_a$ is land in effective units per worker in agriculture. Thus in addition to exogenous growth in the productivity of labor at the same rate as other sectors $A(t)$, the productivity of land can also grow exogenously (Solow's residual) as determined by[i]

$$A_a(t) = e^{\eta t}.$$

As above, (7.A4) satisfies (7.A3) for $j = a$ and

$$\frac{\partial f^{j}\left(\hat{k}_{a},\hat{T}_{a}\right)}{\partial \hat{T}_{a}} \rightarrow \infty \text{ as } \hat{T}_{a} \rightarrow 0 \text{ and } \frac{\partial f^{j}\left(\hat{k}_{a},\hat{T}_{a}\right)}{\partial \hat{T}_{a}} \rightarrow 0 \text{ as } \hat{T}_{a} \rightarrow \infty.$$

Households

For reasons that are apparent later, it is useful to also express choice of households variables in efficiency units. The typical i^{th} household utility from consuming the sequence $(\hat{c}_{im}, \hat{c}_{ia}, \hat{c}_{is})_{t=\infty}^{t=\infty}$ is expressed as a weighted sum of all future flows of utility

$$\int_{t=0}^{t=\infty} \frac{u\left(\hat{c}_{im}, \hat{c}_{ia}, \hat{c}_{is}\right)^{1-\theta} - 1}{1-\theta} e^{(n-p)t} dt \qquad (7.A5)$$

where goods, $\hat{c} = C_{ij}/A(t)l_i$, are expressed in efficiency units per household member. The number of members is assumed to grow at the exogenously given positive rate n

$$l_i = e^{nt}, i + u, r,$$

and to discount future consumption at the rate $\rho > 0$. The elasticity of intertemporal substitution is given by $1/\theta$, where $1 \geq \theta > 0$. For the purpose of this analysis, we specify a CRS Cobb-Douglas form of $u\left(\hat{c}_{im}, \hat{c}_{ia}, \hat{c}_{is}\right)$, and normalize the number of household members in such a way as to equal the number of workers.

   The budget constraint in the cash flow of each household expresses savings $\dot{A}_i$ at an instant of time as the difference between income and expenditure E_i on goods. The budget constraint of the flow of the urban household is

$$\dot{A}_u = wl_u + rA_u - E_u, \qquad (7.A6)$$

where A_u is total household assets. Since capital assets are not traded internationally, $A_u = K_u$ and $\dot{A}_u = \dot{K}_u$. Expressing this constraint in terms of units per effective household member/worker, we obtain[ii]

$$\dot{\hat{k}}_u = \hat{w} + \hat{k}_u(r - x - n) - \hat{E}_u, \qquad (7.A7)$$

where $\hat{w} = w/A(t)$, $\hat{k}_u = K_u/A(t)l_u$ and $\hat{E}_u = E_u/A(t)l_u$. (This result follows from solving $\dot{\hat{k}}_u = \frac{d}{dt}\left(\dot{K}_u/Al_u\right)$ for $\dot{K}_u/Al_u$, recognizing that $\dot{A}/a = x$, and $\dot{l}_u/l_u = n$, and substituting the result into the budget constraint.)

The budget constraint of the flow of the rural household is

$$\dot{A}_r = wl_r + rA_r + \pi T - E_r,$$ (7.A8)

where π is the land-rental rate. In terms of units per effective household member/worker, we obtain

$$\dot{\hat{k}}_r = \hat{w} + \hat{k}_r \left(r - x - n \right) + \pi \hat{T} - \hat{E}_r,$$ (7.A9)

where $\hat{T} = T / A(t) l_r$.

Household expenditure at an instant of time is defined in the typical way as

$$\hat{E}_i = \mu \left(p_m, p_a, p_s \right) \hat{c}_i \equiv \underset{(c_{ij} \geq 0)}{Min} \left\{ \sum_j p_j \hat{c}_{ij} \middle| \hat{c}_i \leq u \left(\hat{c}_{im}, \hat{c}_{ia}, \hat{c}_{is} \right) \right\}.$$

Behavior of Households and Firms

Households choose positive values of the sequence $\left(\hat{c}_{im}, \hat{c}_{ia}, \hat{c}_{is} \right)_{t=0}^{t=\infty}$ to maximize Equation (7.A5) subject to the respective budget constraint in Equation (7.A7) or Equation (9), the stock of initial assets $\hat{k}_i(0), T$, and a limitation on borrowing. The first-order conditions obtained from the present-value Hamiltonian (Barro and Sala-I-Martin 1995: 63) yield the following Euler (Barro and Sala-I-Martin 1995: 65) equation

$$\frac{\dot{\hat{E}}_i}{\hat{E}_i} = \frac{1}{\theta} (r - \rho - x), \ i = u, r,$$ (7.A10)

that describes the path of expenditures over time for each household. If, at an instant of time, returns to capital services $(r - \rho - x) > 0$ are relatively high, the household forgoes expenditures $\hat{E}_i$ to accumulate assets for future consumption, the magnitude of which depends on the elasticity of inter-temporal substitution, $1/\theta$. In the long run, we expect $(r - \rho - x) = 0$. The transversality condition places a limit on borrowing and assures that the maxim and is bounded

$$\lim_{t \to \infty} \left[v(t)_i \hat{k}_i(t) \right] = 0$$

where the co-state variable $v(t)_i$ is the present value shadow price of income (Barro and Sala-I-Martin 1995: 63).

Competition in factor markets among firms in manufacturing and services implies that the cost function for each sector $j = m, s$ is given by

$$c^j\left(\hat{w}, r\right)\hat{y}_j \equiv \underset{l_j, \hat{k}_j}{\text{Min}} : \left[l_j\left(\hat{w} + r\hat{k}_j\right) \middle| \hat{y}_j \geq l_j f^j\left(\hat{k}_j\right) \right],$$

where, as above, $\hat{w} = w/A(t)$ is the effective wage rate.

Competition among firms in agriculture implies that gross returns are just sufficient to cover total factor cost, including returns to land. In this case, the sector's gross domestic product (GDP) function, in units of effective total labor, can be expressed as

$$\pi = \pi\left(p_a, \hat{w}, r\right)\hat{T} \equiv \underset{(i_a, \hat{k}_a)}{\text{Max}} 1_a \left[p_a f^a\left(\hat{k}_a, \hat{T}_a\right) - \hat{w} - r\hat{k}_a \right],$$

where $\hat{T} = A_a(t)T/A(t)L$. The land-rental rate $\pi\left(p_a, \hat{w}, r\right)$ is the rate per effective unit of land per capita required for the rental market among farmers to clear. The gradients of $\pi\left(p_2, \hat{\omega}, r\right)$ yield agricultural supply, and labor and capital demand per effective unit of land per capita.

Equilibrium

Definition

A competitive equilibrium for this economy is a sequence of positive values for prices $\left(p_s^*, \hat{w}^*, r^*, P_{land}^*\right)_{t=0}^{t=\infty}$, firm allocations $\left(\hat{y}_m^*, \hat{y}_a^*, y_s^*, \hat{k}_m^*, \hat{k}_a^*, \hat{k}_s^*, l_m^*, l_a^*, l_s^*\right)_{t=0}^{t=\infty}$ and household allocations $\left(\hat{k}_u^*, \hat{k}_r^*, \hat{c}_{um}^*, \hat{c}_{ua}^*, \hat{c}_{us}^*, \hat{c}_{rm}^*, \hat{c}_{ra}^*, \hat{c}_{rs}^*\right)_{t=0}^{t=\infty}$ given economy-wide aggregates $\left[p_m, p_a, \hat{k}(0), \hat{T}(0)\right]$ such that for at each instant of time t,

1. Given prices, all firms maximize profits subject to their technologies, and yield zero profits.
2. The discounted present value of household utility, subject to the mentioned constraints, is maximized.
3. Markets clear for

    (a) Labor $\displaystyle\sum_{j=m,a,s} l_j^* = 1,$

    (b) Capital $\displaystyle\sum_{j=m,a,s} l_j^* \hat{k}_j^* = \sum_{i=u,r} \frac{l_i}{L}\hat{k}_i^* = \hat{k}$, and

    (c) Home goods $\displaystyle\hat{y}_s^* = \sum_{i=u,r} \frac{l_i^*}{L}\hat{c}_{is}^*.$

4. The value of excess demand for manufacturing goods equals the value of excess demand for agricultural goods [Walra's law (Bagwati et al. (1998: 585)][iii]

$$P_m\left[\hat{y}_m^* - \sum_{i=u,r}\frac{l_i}{L}\left(\hat{c}_{im}^* + \dot{\hat{k}}_i^*\right)\right] + P_a\left[\hat{y}_a^* - \sum_{i=u,r}\frac{l_i}{L}\hat{c}_{ia}^*\right] = 0, \text{ and}$$

5. The no-arbitrage condition between the assets of capital and land to assuring the optimal allocation of savings (i.e., that the returns to the two types of investment are equalized)

$$r^* = \frac{\pi\left(p_a,\hat{w}^*,r^*\right)}{P_{land}^*} + \frac{\dot{P}_{land}^*}{P_{land}^*}. \tag{7.A11}$$

Equation (7.A11) is implicit in the statement of the rth budget constraint of the household.[iv] It states that returns to savings are maximized when the income in $t+dt$ from one unit of income invested in physical capital equals r, and that it must equal the same return in $t+dt$ to a unit of income invested in land. The economic return to one unit of income invested in land is $\pi\left(p_a,\hat{w},r\right)/P_{land}$ plus capital gains in the amount of per-unit $\dot{P}_{land}/P_{land}$ of land—the steady state.

Characterization

Given the endogenous sequence $\left(k,\hat{E}_u,\hat{E}_r\right)_{t=0}^{t=\infty}$, the five-variable sequence of positive values $\left(\hat{w},r,\hat{y}_m,\hat{y}_s,p_s\right)_{t=0}^{t=\infty}$ must satisfy the following five intra-temporal conditions at each instant of time.

1. Zero profits in manufacturing and services

$$c^j\left(\hat{w},r\right) = p_j, j = m, s. \tag{7.A12}$$

2. Clearing of the labor market

$$\sum_{j=m,s}\frac{\partial}{\partial\hat{w}}c^j\left(\hat{w},r\right)\hat{y}_j - \frac{\partial}{\partial\hat{w}}\pi\left(p_a,\hat{w},r\right)\hat{T} = 1. \tag{7.A13}$$

3. Clearing of the capital market

$$\sum_{j=m,s}\frac{\partial}{\partial r}c^j\left(\hat{w},r\right)\hat{y}_j - \frac{\partial}{\partial r}\pi\left(p_a,\hat{w},r\right)\hat{T} = \hat{k}. \tag{7.A14}$$

4. Clearing of the market for home goods

$$\sum_{i=u,r} \frac{\partial}{\partial p_s} \hat{E}_i \frac{l_i}{L} = \hat{y}_s. \tag{7.A15}$$

This system of five endogenous variables in five equations is similar to a static general-equilibrium model. In principle, the model can be solved to express each endogenous variable $\left(\hat{w}, r, \hat{y}_m, \hat{y}_s, p_s\right)$ as a function of the exogenous variables $\left(p_m, p_a, \hat{T}\right)$ and the remaining endogenous variables $\left(\hat{k}, \hat{E}_u, \hat{E}_r\right)$.

We now derive the system for the remaining endogenous variables that, together with Equations (7.A12) to (7.A14), will constitute a solution to the entire sequence of endogenous variables.

We use Equation (7.A12) to express $\hat{w}$ and r as a function of p_s. We omit exogenous variables to minimize clutter. Call this result

$$\hat{w} = W\left(p_s\right) \tag{7.A16}$$

and

$$r = R\left(p_s\right). \tag{7.A17}$$

We substitute Equations (7.A16) and (7.A17) for $\hat{w}$ and r into the factor market clearing the conditions of Equations (7.A13) and (7.A14) and solve these conditions for $\hat{y}_m$ and $\hat{y}_s$. Denote the solution for $\hat{y}_s$ as a function of the endogenous variables p_s and $\hat{k}$.

$$\hat{y}_s = Y^s\left(p_s, \hat{k}\right). \tag{7.A18}$$

Then, we substitute Equation (7.A18) for $\hat{y}_s$ in the home-good, market-clearing condition of Equation (7.A15). Totally differentiate the result with respect to time. The remaining terms include $\dot{\hat{E}}_i / \hat{E}_i$, $1 = u, r$, and the total change in capital stock $\dot{\hat{k}}$. We also substitute the Euler conditions of Equation (7.A10) for $\dot{\hat{E}}_i / \hat{E}_i$, $i = u, r$, and the household budget constraints of Equations (7.A7) and (7.A9) for $\dot{\hat{k}}$. We simplify to obtain the differential equation for the price of home-goods as a function of the endogenous variables p_s and $\dot{\hat{k}}$, the exogenous variables $\left(p_m, p_a, \hat{T}\right)$, and the parameters of the system, including θ, ρ and x. We denote the result as

$$\dot{p}_s = p^s\left(P_s, \hat{k}\right). \tag{7.A19}$$

The budget constraint of the economy is given by the sum of the budget constraints of the household in Equations (7.A6) and (7.A8), expressed in efficiency units per worker

$$\dot{k} = w + r\left(\frac{l_u}{L}\hat{k}_u + \frac{l_r}{L}\hat{k}_r - x - n\right) + \pi(p_a, \hat{w}, r)\hat{T} - \sum_{i=u,r}\frac{l_i}{L}\hat{E}_i. \quad (7.A20)$$

Notice that the home-good, market-clearing condition of Equation (7.A15) implies

$$\frac{\lambda_s}{p_s}\sum_{i=u,r}\frac{l_i}{L}\hat{E}_i = Y^s\left(p_s, \hat{k}\right), \quad (7.A21)$$

where λ_s is the share of expenditures on home goods. We solve Equation (7.A21) for $\hat{E}_i$ and substitute this result into Equation (7.A20). Finally, we also substitute Equations (7.A16) and (7.A17) for $\hat{w}$ and r, respectively, into Equation (7.A20). Express the resulting differential equation as

$$\dot{k} = k\left(p_s, \hat{k}\right). \quad (7.A22)$$

The two differential Equations (7.A19) and (7.A22) are non-autonomous and thus are difficult to solve directly. The procedure is to use the time-elimination method that converts the system, without loss of generality, into an easily solved autonomous system (Barro and Sala-I-Martin 1995: 488–91). The solution yields the sequence $\left[p_s^*, \hat{k}^*\right]_{t=0}^{t=\infty}$ given by

$$p_s = P^s(t) \quad (7.A23)$$

and

$$\hat{k} = K(t). \quad (7.A24)$$

Knowing Equations (7.A23) and (7.A24) allows the derivation of $\left[\hat{E}_u^*, \hat{E}_r^*\right]_{t=0}^{t=\infty}$. Together with the intra-temporal system, the remaining sequence of factor payments, firm and household allocations are determined. Finally, knowing these sequences allows us to use Equation (7.A11) and obtain that sequence of land prices $\left[P_{land}\right]_{t=0}^{t=\infty}$.

The Steady State

The steady-state solution can be found in two ways, both of which are useful in checking for analytical and computational errors. The first is to recognize that if a steady state exists, then the Euler conditions of Equations (7.A10) and (7.A11) imply

$$r_{ss} = \rho + x,$$

which in turn implies from Equations (7.A16) and (7.A17) that

$$\dot{p}_{s,ss} = R^{-1}(\rho + x) \text{ and } \hat{w}_{ss} = W(p_{s,ss}),$$

where the subscript ss denotes steady state. From these values, all of the remaining variables can be computed.

The second method is to find the roots $\left(p_{s,ss}, \hat{k}_{ss}\right)$ satisfying Equations (7.A19) and (7.A22) for the case where $\dot{p}_s = \dot{\hat{k}} = 0$, and then, computing the remaining variables of the system. The two methods should give exactly the same results.

If the steady state exists, the rate of growth in all levels of variables (expenditure, income, output supply, and factor demand) in the long run is given by, for example,

$$\left(\frac{\dot{\hat{k}}}{\hat{k}}\right)_{ss} = \frac{d}{dt}\left[\hat{k}_{ss}\right] = \frac{d}{dt}\ln\left[\frac{K}{A(t)L}\right] = \frac{\dot{K}}{K} - (x+n) = 0 \Rightarrow \frac{\dot{K}}{K} = x+n,$$

while $\dot{w}/w = x$ and $\dot{r}/r = \dot{p}_s/p_s = \dot{P}_{land}/P_{land} = 0$.

These long-run growth rates have important implications to resource adjustments in agriculture relative to the rest of the economy and to the extent to which farm policy may serve to hold excess resources in agriculture.

APPENDIX 7.B

Accounting for Direct Payments

The model in this chapter is calibrated to U.S. data for 1997 and solved under the assumption of no PFCPs. This generates a base sequence of the endogenous variables defined above. Then, PFCPs are added to the model. The urban- and rural-household budget constraints of Equations (7.A6) and (7.A8) are changed according to

$$\dot{A}_u = wl_u + rA_u - PFC - E_u \text{ and} \tag{7.B1}$$

$$\dot{A}_r = wl_r + rA_r + \left(\pi + \frac{PFC}{T}\right)T - E_r, \tag{7.B2}$$

respectively. This presumes a lump-sum transfer of PFC from non-landowning households to landowning households. It also presumes that the transfer is tied to the ownership of land; and that this payment is made at each instant of time in perpetuity.

The no-arbitrage condition among the assets of Equation (7.A11) now becomes

$$r = \frac{\pi(p_a, \hat{w}, r) + PFC/T}{P_{land}} + \frac{\dot{P}_{land}}{P_{land}}. \tag{7.B3}$$

Through the household-budget constraints, the transfer term PFC also enters the model's system of equations. While key household variables, such as the sequence $\left[\hat{k}_i, \hat{c}_{ij}\right]_{t=0}^{t=\infty}$ are changed by the payments, the negative effects on urban households are just offset throughout the sequence by positive effects on rural households, as reported in the figures. This results because household preferences are identical and homothetic (although their consumption and expenditure levels vary) and no market failures are present. Thus the sequence of key variables $\left[\hat{w}, r, \hat{k}_j, p_s, \hat{y}_m, \hat{y}_a, \hat{y}_s\right]_{t=0}^{t=\infty}$ remain the same as in the base solution. The only affected variable is the price of land obtained from Equation (7.B3).

The next experiment entails segregating the capital market of agriculture from that of the rest of the economy. Effectively, this means that the non-landowning urban households do not invest in agricultural capital $\hat{k}_a$ over the period of analysis. In addition to changes in household budget constraints, the market-clearing equation for capital in Equation (7.A14) is replaced by the following two equations

$$\sum_{j=m,s} \frac{\partial}{\partial r} c^j \left(\hat{w}, r_{ms} \right) \hat{y}_j = \hat{k}_{ms}, \text{ and} \qquad (7.B4)$$

$$-\frac{\partial}{\partial r} \pi \left(p_a, \hat{w}, r_a \right) \hat{T} = \hat{k}_a. \qquad (7.B5)$$

The first is the capital market-clearing equation for the manufacturing and home-good sectors while the latter is the market-clearing condition for agriculture alone. Notice that these two markets can clear at different interest rates, r_{ms} and r_a. The Euler conditions of Equation (7.A10) now become

$$\frac{\dot{\hat{E}}_u}{\hat{E}_u} = \frac{1}{\theta} \left(r_{ms} - \rho - x \right), \text{ and}$$

$$\frac{\dot{\hat{E}}_r}{\hat{E}_r} = \frac{1}{\theta} (r_r - \rho - x)$$

The no-arbitrage condition of Equation (7.B3) becomes

$$r_a = \frac{\pi \left(p_a, \hat{w}, r_a \right) + PFC/T}{P_{land}} + \frac{\dot{P}_{land}}{P_{land}}.$$

The system's two main differential Equations (7.A19) and (7.A22) are now increased by two additional equations, and the system is solved to generate results appearing in Figure 7.2 through Figure 7.8 in the steady state $r_{ms} = r_r = \rho + x$.

REFERENCES

Antonovitz, F. and T. Roe. (1986) "A Theoretical and Empirical Approach to the Value of Information in Risky Markets." *The Review of Economics and Statistics* LXVIII (1): 105–14.

Barro, R. and X. Sala-I-Martin. (1995) *Economic Growth.* McGraw-Hill.

Bhagwati, J.N., A.P. Panagariya, and T.N. Srinivasan (1998) *Lectures on International Trade, Second Edition.* Cambridge: MIT Press.

Chau, N. and H. de Gorter. (2000) "Disentangling the Production and Export Consequences of Direct Farm-Income Payments." Paper presented at the 2000 AAEA Meetings, Tampa, Florida.

Collender, R and M. Morehart. (2002) Decoupled Payments to Farmers, Capital Markets, and Supply Effects. U.S. Department of Agriculture, Economic Research Service, Draft (January). Washington DC: USDA/ERS.

Goodwin, B.K, A.K. Mishra and F.N. Ortalo-Magné. (2003) See Chapter 6, this volume.

Hirshleifer, D. (1988) "Risk, Futures Prices, and the Organization of Production in Commodity Markets." *Journal of Political Economy* 96(61): 1206–20.

Mas-Colell, A.M, M. Whinston, and J. Green. (1995) *Microeconomic Theory.* New York: Oxford University Press.

Orden, D., R. Paarlberg, and T. Roe. (1999) *Policy Reform in American Agriculture: Analysis and Prognosis.* Chicago: University of Chicago Press.

Roe, T. (2001) "A Three-Sector Growth Model With Three Assets and Sectoral Differences in TFP Growth." Economic Development Center Working Paper, Department of Economics and Department of Applied Economics, University of Minnesota (July).

Sandmo, A. (1971) "On the Theory of the Competitive Firm Under Price Uncertainty." *American Economic Review.* 61(1): 65-73

Stiglitz, J.E. (1985) "Information and Economic Analysis: A Perspective." *The Economic Journal* 95(Supplement): 21–41.

USDA/ERS (U.S. Department of Agriculture, Economic Research Service). (2001a) "Higher Cropland Value from Farm Program Payments: Who Gains?" *Agricultural Outlook* Washington, DC: USDA/ERS (November).

_____. (2001b) *Agricultural Policy Reform in the WTO: The Road Ahead.* Agricultural Economic Report No. 802. Washington, DC: USDA/ERS (May).

_____. (2002) "Agricultural Resource Management Survey (ARMS): Detailed documentation." Internet Website: www.ers.usda.gov/briefing/ARMS/detaileddocumentation.htm.

ENDNOTES

* This research was funded by USDA/ERS. The views expressed herein are those of the authors and not necessary those of the Economic Research Service, USDA. Appreciation is expressed to members of the project and others for their reviews and comments, and to the feedback provided by participants of seminars at the University of Minnesota, and USDA/ERS.

[i] Thus the framework allows growth in agriculture's total factor productivity (i.e. Solow's residual) to equal or exceed that of manufacturing and services, as has been found in other studies.

[ii] This result follows from solving $\dot{k}_u = \dfrac{d}{dt}(\dot{K}_u / Al_u)$ for K_u / Al_u, recognizing that $\dot{A}/A = x$, and $\dot{l}_u / l_u = n$, and substituting the result into the budget constraint.

[iii] This presumes a country has balanced trade, a condition which the data typically reject. The closure rule chosen is determined in the calibration of the model to data.

[iv] An equivalent statement of the rural household's budget constraint is
$$\dot{\hat{a}}_r = \hat{w} + \hat{a}_r(r - x - n) - \hat{E}_r,$$
where $\hat{a}_r = \hat{k}_r + P_{land}\hat{T}_a$. Then, use the no-arbitrage constraint (7.A11) to obtain the budget constraint (7.A9).

Chapter 8

Economies of Farm Size, Government Payments, and Land Costs

Luther Tweeten and Jeffrey Hopkins[*]
Professor Emeritus, The Ohio State University
U.S. Department of Agriculture, Economic Research Service

INTRODUCTION

Authors of a rich and varied literature have examined the impact of government programs on the structure of farming, especially on the size and number of farms. Tweeten (1993: 337–38) reviews numerous studies by economists who variously reached three conflicting conclusions: "in the absence of government programs, the U.S. today would have (1) fewer and larger farms, (2) more and smaller farms, and (3) the same number of farms." One objective of this chapter is to ascertain which of the three above-stated positions is supported by recent evidence on how commodity-program payments contribute to economies of farm size.

Analysis of farm policy and its impact on the structure of agriculture has evolved since the studies noted above. The 1996 and 2002 farm bills made direct government payments (rather than price supports) the principal instrument of economic support for farmers. We quantify economies of size by type of farm from recent data collected in the U.S. Department of Agiculture, Agricultural Resource Management Survey [USDA/NASS/ARMS (USDA/ERS 2002a)], and measure how such size economies of farm operators are influenced by passthrough of government payments to landowners through rents and land prices.

CONCEPTUAL FRAMEWORK

The economies of farm size, measured by resource (economic) cost for farms of different sales size, are important determinants of farm size. Presence of economies of larger size, as evidenced by lower combined operating and annualized overhead cost per unit of output on larger than on smaller farms, is an inducement for farms to grow in size. Given the nation's somewhat fixed farmland area, an increase in farm size in output and acres corresponds with falling farm numbers.

Prices for inputs and outputs are likely to be at least as favorable for large farms with greater input purchase and product sales volume than they are for small farms. Such market economies may be real or pecuniary. Pecuniary economies accrue to large farms if they are successful in exerting market power to obtain lower input prices and/or higher product prices from agribusinesses. Real economies accrue to large farms successful in securing benefits from agribusinesses passing on to farms the lower real costs per unit to supply input and to market output in volume. With prices likely to be at least as favorable on large as on small farms, economies of larger production size translate into higher rates of return and net incomes on larger farms.

With government payments as the principal instrument to enhance farm income through farm-commodity programs, one fact is undisputed: large farms receive bigger government payments than do small farms, other things equal. That fact has led some laypersons to conclude that larger payments to big farms encourage farms to become large. By that line of thinking, the wheat enterprise causes farms to get larger because wheat receipts are biggest on large wheat farms. (By the same line of thinking, one could conclude alternatively that wheat sales cause farms to get smaller because total wheat production costs are the greatest on large wheat farms.)

Examining government payments per unit of farm output on farms of different sizes circumvents such misconception. Other things equal, if payments are larger per unit of farm output on larger farms than they are on small farms, then payments encourage structural adjustment to fewer, larger farms. If payments are lower per unit of output on larger farms, then programs encourage movement to more and smaller farms, other things equal.

The market for land, the most fixed production resource, complicates the impact of government payments on economies of size. Labor and capital resources on well-managed commercial farms adjust between sectors so as to earn what they would earn if employed in other sectors, thus any excess profit accrues to land. With well-functioning markets, renters bid program benefits into rents, thereby "passing through" government payments to landowners. Similarly, the expected future value of government programs is capitalized into the value of land when real estate is sold, hence benefits of farm programs are retained by initial owners and lost to new owners. Even if government payment per unit of commodity is precisely the same for all farm classes, economies of size facing farm operators are influenced if government payments are passed from operators to owners through rents and land sales at different rates by farm size.

In short, government-commodity supports, including loan deficiency payments (LDPs), Agricultural Market Transition Act Payments (AMTAPs), and market loss assistance payments (MLAPs), are based on current production (e.g., LDP) or historic production (e.g., AMTAP, MLAP), hence they are expected to be proportional to output. Some types of environmental programs making up a smaller share of total direct payments are negatively correlated with an operation's production (land retirement through the Conservation Reserve Program) while others may have no relation with production levels (environmentally friendly production-management practices). More importantly, payment limitations (and measures to avoid them such as switching to non-program enterprises), and rapid passthrough of payments from operators to owners are expected to provide less payment per unit of output to operators of large farms. The following analysis examines the impacts of all direct payments on farm operators' economies of size.

ANALYSIS

In this chapter, we quantify economies of farm size as measured by all resource costs of production per unit of output per farm with and without government payments. Output is defined as the sum of the value of crops and livestock sold plus the value sold under contract for a farm, less inter-farm sales. Based on the proportion of land that is rented among farm sizes and the passthrough of payment benefits to landowners through rents and land sales, we quantify how the land market affects economies of size for farm operators.

Economies of Size

Before examining empirical estimates of total resource costs per unit of farm output, we briefly review key methodological issues of measuring economies of farm size.

First, our empirical results show returns to size rather than returns to scale. The latter measures the response of output when all resources are increased proportionately. Because farms do not change size by proportionally changing all inputs, the concept of returns to scale is academic. The concept of economies of size employed in this chapter reveals what happens to costs per unit of output when resources are allowed to vary in least-cost combinations.

Two methodologies traditionally have been employed to measure economies of size. One is "engineering" analysis where total cost per unit of production is found for a typical or representative type of farm. Linear programming or a related mathematical tool is employed to find the least-cost or profit-maximizing combination of inputs for farms of various sizes. Management is held constant across all sizes of farms. Because commodity prices are generally assumed to be constant over all farm sizes, the farm that produces the lowest total resource cost per unit of commodity output (sales) is assumed to be the most efficient size.

A second approach, the one employed in this chapter, measures all resource costs per dollar of all commodity output per farm based on cross-sectional data gathered from actual farms (Hallam 1993; Tweeten 1984 and 1989).[i] Such posi-

tivistic analysis based on what farms actually do rather than on synthesized behavior, assumes that farms choose least-cost or profit-maximizing combinations
of inputs and outputs. Hence management and enterprises are not held constant
when going from one size of farm to another when considering costs per unit of
output (sales) across farms of various sizes.

Besides having more comprehensive and timely data, this study differs from
previous positivistic studies in several ways. Previous empirical studies have
depicted economies of size as a graphic plot or freehand curve of average actual
cost of production at the mean or midpoint for each farm-size class (Tweeten
1989: 124). Such analysis is sensitive to the choice of farm size class intervals,
especially for very large and very small size classes, and is not sensitive to differences in the population of farms within each sales class. Our study avoids that
problem by fitting statistically a population-weighted mathematical function to
individual farm observations from the USDA/NASS/ARMS for 2001.
The methods used when collecting cost and return information in
USDA/NASS/ARMS are consistent with the recommendation of a task force set
up by the American Agricultural Economics Association on commodity costs
and returns. Thus, compared to previous studies that looked at mean cost per
dollar of sales for each farm size class within each type of farm, the data and
cost curves presented herein more accurately depict marginal costs of production
by type of farm.[ii]

Previous empirical studies and plots of USDA/NASS/ARMS data for 2001
indicate that a mathematical function expressing economies of size needs to be
decreasing throughout the range of size but at a decreasing rate. Cost per unit
does not appear to turn up for large farms within the range of sampled farms, but
unit costs decline very little if at all for very large farms.[iii] These considerations
suggest that economies of farm size can be depicted by a double logarithm
mathematical equation of the form

$$\ln(C) = \ln(a) + b\ln(S) + \ln(U),\qquad\qquad\qquad(1)$$

where C is total farm resource cost per dollar of sales of crops and livestock, S
is farm size in total sales of crops and livestock, U is error, $\ln(.)$ is natural
logarithm, and a and b are coefficients estimated by using ordinary least
squares regression analysis.

The double logarithm function in Equation (1), although fitting the data as
well or better than other mathematical forms based on actual data plots and
based on previous studies, does rigidly assume that the elasticity of unit cost
with respect to farm sales is constant over all farm sizes. However, the function
does not force elasticities to be equal among farm types nor does it force economies of farm size—unit costs can rise at a decreasing rate with larger farm size if
the data dictates.

Data are cross-sectional individual farm observations from the USDA/
NASS/ARMS for the United States in year 2001. We estimate the grain, cotton,
and oilseed farm equation from 2,371 observations; the other-crops farm equation from 1,364 observations; the livestock farm equation from 3,492 observations; and the all-farms equation from all of these observations. The sample data

are weighted so as to be representative of the 1,761,295 farms reporting positive output in 2001 (i.e., if any of the estimated 2,149,683 farms had zero output they were excluded from the analysis.)

Resource costs include annual operating costs incurred by the principal operator and all other partners in the operation (excluding the landlord), plus the annualized value of operator overhead costs. The latter include annual rent on leased land, interest and depreciation on overhead capital, and a 5 percent charge for operator management on farm crop and livestock sales less inter-farm sales. We charge the operator and family labor the hired-labor wage rate and equity capital is charged a 4 percent opportunity cost. The equity-capital charge is greater than the 10-year Treasury bond interest rate of 3.83 percent in late 2002, but includes a 1-percentage point charge for risk along with a 3-percentage point real interest rate.

Where total direct government payments to the farm operation were included, such payments were aggregated with crop and livestock receipts under the label "sales". Otherwise, only the operation's production inputs and outputs were included.

Economies of size dominate each of the 11 types of farms shown in Figure 8.1. Costs per dollar of output exceed U.S. $2 on small farms of each type. On

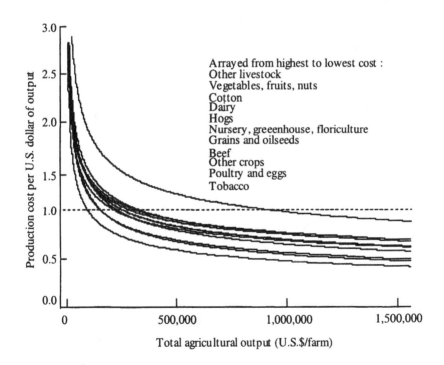

Source: Authors' compilation based on USDA/NASS/ARMS (USDA/ERS 2002a) data.

Figure 8.1 Total 2001 Resource Cost per Dollar of Output for 11 Types of U.S. Farms

very large farms, the marginal cost of producing another dollar of output ranges from approximately U.S. $1 on "other livestock" and on cotton farms to less than U.S. $0.50 on poultry and egg farms.

One would expect land rents and prices to be bid up or down as necessary so that U.S. $1 of resource cost would be required to produce U.S. $1 of crop and livestock receipts (including government payments) on all types of efficiently sized farms in Figure 8.1. The obvious and statistically significant disparity in cost per unit of output among efficiently sized farms in Figure 8.1 can be explained partly by our omission of transitory deviations of commodity prices and yields from expected equilibrium. For example, the cotton cost per unit is high in part because cotton price was very low in 2001, which was an average of 65 percent of its 1990 to 1992 price (USDA/ERS 2002b: 43). The poultry- and egg-unit cost of production curve in Figure 8.1 is unusually low in part because the denominator (the receipts) is inflated by 2001 poultry and egg prices, which were 116 percent of their 1990 to 1992 average price. The cost curves differ mainly by intercept a in Equation (1) rather than by slope or elasticity b, the latter reflecting economies of size.

No consistent relationship is apparent in Figure 8.1 between the level of the unit-cost curves and the receipt of government payments. Poultry and egg farms and hog farms display low costs and receive little income support from government. In contrast, cotton and dairy farms display high costs and receive income support from government.

Use of "normal" prices would reduce the vertical distances among the cost curves in Figure 8.1. But an important conclusion would remain: Economies of size characterize the many types of farms found in today's agriculture.

Results for Aggregated Crop Farms and for Livestock Farms With and Without Payments

To reduce the influence of transitory prices and yields and to simplify exposition, we present the remaining results for only three broad types of farms: (1) grain, cotton, and oilseed (i.e., program) farms, the major government-payment recipients; (2) other-crop farms; and (3) livestock farms. The latter, like other-crop farms, receive a relatively small share of government payments.

As background for estimating the impact of government payments on economies of size, data in Table 8.1 show the share of each of these broad types of farms in farm numbers (population), value of production, government payments, and rented land. Livestock farms (which may also produce crops although livestock enterprises predominate) account for approximately one-half of farm numbers, overall farm-commodity production, and rented farmland. They account for only one-fifth of government payments. The remaining farm numbers and production are split about evenly between program farms and other farms. Grain, cotton, and oilseed farms account for 40 percent of U.S. rented farmland while other-crop farms account for only 5 percent (Table 8.1).

Unit costs of production are shown for the three broad types of farms and for all U.S. farms in Figure 8.2. Little difference is noted between the curves for

Table 8.1 Farm Types as Share of all U.S. Farms (2001)

Variable	Grain and oilseed farms[a]	Other-crop farms	Livestock farms
	Percent of all farms		
Share of farms	20	21	58
Share of value of production	23	28	48
Net rented acres	40	5	55
Direct govt. payments	71	9	20

[a]Grains and oilseed farms include corn, sorghum, small grains, rice, soybeans, dry beans, dry peas, sunflowers, flaxseed, popcorn, grain silage and forage, grains and oilseeds for seed, and cotton.
Source: Authors' calculation from USDA/NASS/ARMS (2002a) data.

livestock farms with and without the inclusion of payments. These two curves would lie very close to the crop curves if the relatively high livestock prices in 2001 were normalized to a more typical year. For example, including payments, a U.S. $500,000 sales livestock farm is predicted to incur a resource cost of U.S. $0.77 to produce U.S. $1.00 of output in 2001, commensurate with the marginal cost of U.S. $0.77 for all farms, and slightly below the U.S. $.80 required to produce U.S. $1.00 of output on other farm types. The unusually high price for livestock in 2001, 6 percent above the 1990 to 1992 average, accounts largely for the low unit cost of production for livestock in that year. The conclusion is that, after correcting for transitory prices and for capitalization of government payments, the economies of size curves converge into essentially one curve in Figure 8.2.

The breakeven size at which farms cover all costs with payments included are somewhat similar among the broad farm types in Figure 8.2—U.S. $383,590 for the grain, cotton, and oilseed farms; U.S. $388,997 for other-crop farms; U.S. $384,530 for livestock farms; and U.S. $373,317 for the all-farms category. Expressed in corn-farm equivalence, the breakeven size for the all-farms category is 1,420 acres for a farm yielding 135 bushels per acre in 2001.

The breakeven grain, cotton, and oilseed farm would have been U.S. $659,684 (or 2,443 acres of corn equivalent) in 2001 without government payments. That high number traced partly to constant land rents and to prices that were bid up by the pool of current and prospective renters and owners in anticipation of program payments. This bidding process raised actual or opportunity costs of land for all farmers, not just for those farmers receiving payments. Also, both land costs and breakeven size would decrease in a no-payment equilibrium.

Each of the very large farms on each broad type of farm and on the all-farms category incurred costs on average of just under U.S. $1 to produce U.S. $1 of output in 2001. Thus the data provide evidence that markets work to equalize costs per dollar of sales among various farm types of a given size. Public support which lowers costs, raises returns to producers, and reduces the size of an economic unit are bid into asset values until costs equal receipts (including payments) on efficiently sized farms.

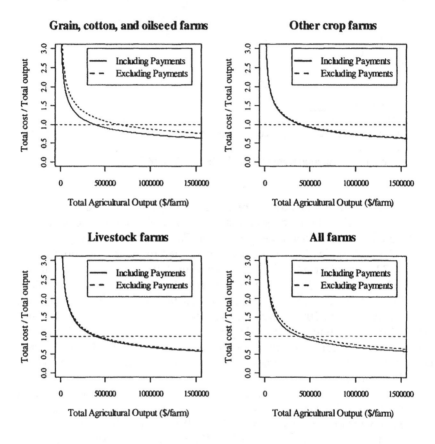

Figure 8.2 Total 2001 Resource Cost per Unit of U.S. Farm Output with and without Government Payments by Broad Type of Farm

Measuring the Impact of Government Payments on Economies of Size

If government payments in aggregate were directed more to smaller than to larger farms, they could flatten unit cost curves, reduce economies of size, and diminish incentives for farms to grow larger and to decline in number. That outcome is not the case according to the four panels of Table 8.2 showing the distribution of government payments, farm numbers, and farm sales among sales classes (class boundaries do not include government payments) within farm types in 2001. The table also shows unit costs of production with and without government payments along with the coefficients and statistical properties of equations to express resource cost per unit of output.

Table 8.2 indicates that payments are more concentrated on large program farms (Table 8.2, Panel A) than on large crop or livestock operations. Farms of that type with sales more than U.S. $1 million in 2001 accounted for only 1 percent of program farms and for 13 percent of sales but garnered one-half of all program payments to such farms. Other-crop farms and livestock farms received few payments in 2001 overall (Table 8.1), and those payments were not as unequally distributed as the payments were for grain, cotton, and oilseed farms (Table 8.2 Panels A, B and C).

To gain additional insight into the distribution of payments among farms and their impact on economies of size, the predicted cost of production by size with and without payments is shown in the last two columns of Table 8.2. The equation R^2 s, measuring the proportion of variation in cost per unit of output explained by farm size, ranges from 0.53 for program crops without payments (Table 8.2, Panel A) to 0.73 for livestock farms with payments (Table 8.2, Panel C). The remaining or "unexplained" proportion of variation in cost per unit of output among farms that is not accounted for by farm size, probably arises from unique managerial capabilities of operators, variation in weather and other natural phenomenon among farms, in addition to sampling and reporting error.

The effective payment per dollar of 2001 farm output, as measured by the predicted per-unit cost without payments less the per-unit cost with payments, ranged from U.S. $0.60 on small program farms to U.S. $0.13 on large program farms according to Table 8.2, Panel A. As expected, predicted effective payment rates were smaller for livestock and "other crop farms" in Table 8.2.

For each type of farm, however, payment per dollar of output was predicted to be larger for small farms than it was for large farms. Thus payments were predicted by the equations to flatten the unit cost curve and to diminish the economies of size.

Payments dampened economies of size in only a small way. The elasticity of unit cost with respect to farm size (i.e., the coefficient on the output variable in Table 8.2, Panel A), indicates that a 1 percent increase in farm size decreases unit costs by about 0.32 percent with or without payments.

Measuring Impacts on Size Economies with Actual Payments

Based on the R^2s and on plots of data, the double logarithm Equation (1) acceptably depicts the systematic relationship between resource cost per unit and sales of crops and livestock without payments (Table 8.2). *Payments per dollar of output are not monotonic with respect to farm size, however, and therefore the double logarithm Equation (1) is inappropriate to express unit cost with payments.* At the least, government payments and unit production costs are not expected to follow the same pattern with respect to farm size. Table 8.3 recognizes such considerations when demonstrating how passthrough of program payments to land costs influences economies of size facing farm operators.

In Table 8.3, Panel A, actual average program payments per dollar of crop and livestock receipts in 2001 are highest for farms with U.S. $10,000 to U.S. $49,999 of sales for program farms and tend to be uniformly quite high among

Table 8.2 Total Resource Predicted Cost per Dollar of Output, and Shares of Farm Numbers, Sales, Payments, and Rented Land by Type of U.S. Farm 2001

Sales category	Share of:				Cost per dollar of output	
	Farms	Sale	Payment	Land rental	[a]With payment	[b]Without payment
U.S. dollars			Percent			U.S. dollars
			A. Grain, cotton, and oilseed farms			
Less than 10,000	16	1	0	0	4.13	4.73
10,000–49,999	30	6	1	6	2.30	2.68
50,000–99,999	18	9	3	10	1.70	2.00
100,000–249,999	20	27	7	30	1.29	1.53
250,000–499,999	10	25	14	27	1.01	1.20
500,000–999,999	4	20	26	20	0.80	0.96
1,000,000 or more	1	13	48	6	0.64	0.77

[a]Cost per dollar of output on crop and oilseed farms (with payments) = $66.69*output^{-0.327}$, $SE(b) = 0.025$, $R^2 = 0.54$
[b]Cost per dollar of output on crop and oilseed farms (without payments) = $71.27*output^{-0.318}$, $SE(b)=0.0243$, $R^2 = 0.53$

			B. Other crop farms			
Less than 10,000	57	2	0	0	4.42	4.46
10,000–49,999	18	3	2	6	2.40	2.43
50,000–99,999	8	4	7	10	1.75	1.79
100,000–249,999	9	10	9	22	1.31	1.34
250,000–499,999	4	10	18	21	1.01	1.04
500,000–999,999	2	10	31	14	0.80	0.82
1,000,000 or more	2	61	33	26	0.63	0.65

[a]Cost per dollar of output on other-crop farms (with payments) = $80.99*output^{-0.341}$, $SE(b) = 0.02$, $R^2 = 0.66$
[b]Cost per dollar of output on other-crop farms (without payments) = $82.1*output^{-0.343}$, $SE(b)=0.02$, $R^2 = 0.66$

			C. Livestock farms			
Less than 10,000	54	2	0	6	5.36	5.43
10,000–49,999	24	6	2	21	2.68	2.74
50,000–99,999	7	5	5	17	1.88	1.93
100,000–249,999	8	16	9	23	1.36	1.40
250,000–499,999	4	14	20	17	1.01	1.05
500,000–999,999	2	14	34	10	0.77	0.80
1,000,000 or more	1	44	31	6	0.59	0.62

[a]Cost per dollar of output on livestock farms (with payments) = $144.13*output^{-0.39}$, $SE(b) = 0.007$, $R^2 = 0.74$
[b]Cost per dollar of output on livestock farms (without payments) = $140.142*output^{-0.382}$, $SE(b)=0.0073$, $R^2= 0.73$

Source: Authors' calculations from USDA/NASS/ARMS (USDA/ERS 2002a) data.

(Continued)

Table 8.2 Total Resource Predicted Cost per Dollar of Output, and Shares of Farm Numbers, Sales, Payments, and Rented Land by Type of U.S. Farm, 2001 (*Continued*)

Sales category	Share of:				Cost of output	
	Farms	Sale	Payment	Land rental	[a]With payment	[b]Without payment
U.S. dollars			Percent		U.S. dollars	
	D. All farms					
Less than 10,000	47	2	0	3	4.96	5.11
10,000–49,999	24	5	1	14	2.55	2.69
50,000–99,999	9	6	5	14	1.82	1.93
100,000–249,999	11	16	9	25	1.33	1.43
250,000–499,999	5	15	21	22	1.00	1.09
500,000–999,999	2	14	34	14	0.77	0.85
1,000,000 or more	2	42	30	8	0.60	0.66

[a]Cost per dollar of output on all farms (with payments) = $117.53*output^{-0.372}$, $SE(b) = 0.008$, $R^2 = 0.71$
[b]Cost per dollar of output on all farms (without payments) = $108.63*output^{-0.359}$, $SE(b) = 0.0076$, $R^2 = 0.692$

Source: Authors' calculations from USDA/NASS/ARMS (USDA/ERS 2002a) data.

farm size classes. For other-crop farms in Panel B of Table 8.3, payment per dollar of receipts is also higher for small farms and declines to less than U.S. $0.05 per dollar of output for farms bigger than U.S. $500,000 in sales. As expected, payments are small for livestock farms (Table 8.2, Panel C).

Other things equal, farmers can shift their resources and enterprise mix to types and sizes of farms that most nearly meet their needs. For that reason, the best single measure of impacts of government payments on economies of size and incentives for farm consolidation is payment per dollar of output among all farms as shown in Table 8.3, Panel D. Payment per dollar of output averages highest at U.S. $0.18 on farms with sales of U.S. $50,000 to U.S. $99,999 and lowest at U.S. $0.05 on farms with sales of U.S. $1 million or more. Payments are less for larger farms not because of payment limitations (Table 8.2, Panel A indicates that large program farms received about the same payment per dollar of output as small farms) but because large farms tend to emphasize livestock, fruits, vegetables, and other enterprises that do not receive government payments. The implication is that, other things equal, payments create small incentives for more and smaller farms.

The second issue examined in Table 8.3 is how passthrough of payment benefits into land costs affects economies of farm size over time. Farm operators pass payment benefits to landowners in higher rents and land prices. Payments retained by operators depend on the size of payments, proportions of land rented and owned, and the size and timing of passthrough of payments from renters to landowners. The payments retained by farm operators for three time horizons are calculated and used to adjust cost per unit of output to show the impact of

Table 8.3 Total 2001 Resource Cost per Dollar of Output with Government Payment Passthrough, by Farm Type and Total for United States

Sales category	Rent/ total acres	Payment/ output	Cost of output					
			Short term		One year		Five year	
	%	U.S.$	U.S.$	% de-cline	U.S.$	% de-cline	U.S.$	% decline
U.S. dollars			A. Grain, cotton, and oilseed farms					
Less than 10,000	25	0.22	4.52	--	4.55	--	4.58	--
10,000–49,999	47	0.36	2.32	49	2.42	47	2.50	45
50,000–99,999	49	0.35	1.65	29	1.75	28	1.84	27
100,000–249,999	63	0.33	1.20	27	1.32	24	1.42	23
250,000–499,999	70	0.30	0.90	25	1.02	23	1.12	21
500,000–999,999	74	0.29	0.67	25	0.80	22	0.90	20
1,000,000 or more	75	0.25	0.52	22	0.63	21	0.72	20
			B. Other crop farms					
Less than 10,000	10	0.05	4.41	--	4.41	--	4.42	--
10,000–49,999	22	0.12	2.32	47	2.33	47	2.35	47
50,000–99,999	33	0.15	1.63	30	1.66	29	1.69	28
100,000–249,999	28	0.06	1.28	22	1.29	22	1.30	23
250,000–499,999	47	0.09	0.95	26	0.97	25	0.99	24
500,000–999,999	49	0.04	0.78	18	0.79	19	0.80	19
1,000,000 or more	48	0.02	0.63	19	0.64	19	0.64	20
			C. Livestock farms					
Less than 10,000	19	0.03	5.40	--	5.40	--	5.41	--
10,000–49,999	29	0.08	2.66	51	2.68	51	2.69	50
50,000–99,999	40	0.08	1.86	30	1.87	30	1.89	30
100,000–249,999	38	0.06	1.34	28	1.35	28	1.36	28
250,000–499,999	39	0.06	0.99	26	1.00	26	1.01	26
500,000–999,999	39	0.05	0.76	24	0.77	23	0.78	23
1,000,000 or more	32	0.02	0.60	21	0.60	22	0.60	22
			D. All farms					
Less than 10,000	17	0.05	5.06	--	5.07	--	5.07	--
10,000–49,999	32	0.16	2.53	50	2.56	50	2.59	49
50,000–99,999	42	0.20	1.73	32	1.78	30	1.83	29
100,000–249,999	46	0.16	1.27	27	1.31	27	1.35	26
250,000–499,999	54	0.17	0.92	27	0.97	26	1.02	25
500,000–999,999	54	0.14	0.71	23	0.75	22	0.79	22
1,000,000 or more	43	0.05	0.61	14	0.62	18	0.63	20

Source: Authors' calculations from USDA/NASS/ARMS (2002a) data and Table 8.2.

payment passthrough on economies of size. Owner-operators retain all payments over all lengths of run, but renters and new owner-operators do not.

In the short-run scenario, the full amount of the actual average payment per dollar of farm output from Table 8.3 is subtracted from the cost per dollar of farm output excluding payments (Table 8.2, last column) to form the first (short run) measure of economies of size with payments in Table 8.3. With operators retaining the entire payment regardless of whether they own or rent their land, the result is that costs are U.S. $4 lower on the largest farms than on the smallest farms for program farms (Table 8.3, panel A) and U.S. $4.45 lower for all farms (Table 8.3, panel D). The respective reductions would have been U.S. $3.96 and U.S. $4.45 without payments, hence payments had little impact on economies of size in the short run.

The second measure of economies of size in Table 8.3 applies to one year—a period long enough for some payment benefits to be passed from renters to landowners. Goodwin et al. (2003) find that approximately 60 percent of LDPs and AMTAPs are transferred from renters to landowners through higher rents in one year. Passthrough of other payments such as for disasters and market loss assistance was less. The analysts did not have data for the 2002 farm bill. However, market loss and disaster assistance ad hoc programs were effectively institutionalized by formula into the 2002 bill as countercyclical payments. Thus farmers are likely to view them to be nearly as permanent as LDPs and AM-TAPs, increasing the passthrough of countercyclical payments to landowners. Consequently, we assume in Table 8.3 that 60 percent in one year and 100 percent in five years and longer of government payments from all program sources are passed from renters to landowners.

Goodwin et al. (2003) have no data available to measure the difference in passthrough rate for renters on small versus large farms. However, 2001 USDA/NASS/ARMS data do show a much larger incidence of land renting by operators of large farms than by small farm operators, creating a differential passthrough rate by size of farm that affects economies of size for farm operators. Table 8.3 shows that large farm operators rent one-half to three-fourths of their land whereas small farm operators rent only one-tenth to one-fourth of their land, when rental share is defined as net rent (in which land rented in is subtracted from land rented out) divided by land operated.

For the one-year length-of-run scenario, the payment passthrough to the landowner is 60 percent times the payment rate times the rented share of land. The impact is to raise cost per dollar of output to operators. However, the cost advantage of the lowest cost large farm over the high cost small farm is reduced from the previous "short-term" case of U.S. $3.99 to the "one-year" case of U.S. $3.91 per dollar of output (Table 8.3, Panel A). The "percent-decline" columns in Table 8.3 show how the diseconomies created by payments are distributed among farms of different size.

The 5-year scenario in Table 8.3 recognizes loss of program payment benefits through rents and land sales. That scenario assumes that payment benefits are fully passed from renters to landowners. In addition, the 1 percent of land that is sold each year by farm operators to off-farm landowners [based on the 1998 Agricultural Economics and Land Ownership Survey, (AELOS) (USDA/NASS 2001)] is assumed to accumulate to 5 percent in five years. The rental share plus 5 percent is multiplied times the payment rate and added to the short-run cost per dollar of output to form the five-year column. The largest farm-cost advantage over the smallest farm is cut to U.S. $3.86 for program farms in Table 8.3, panel A.

A fourth long-run scenario of all program-payment benefits passed from an entirely new set of landowners to current landowners could be added to Table 8.3. It would be redundant, however, because it would be the same as the without-payment column on the far right in Table 8.2. The long-run cost advantage shown in Table 8.3, Panel A of large over small farms, at U.S. $3.96, is more like the scenario in which passthrough does not occur. Thus program-payment passthrough reduces economies of size in the short and intermediate run but not in the long run. Caution is appropriate when reaching conclusions, however, because any phasing down or out of government payments could lead to major adjustments that could change the entire structure of resource costs and could change the mix of crop and livestock enterprises.

CONCLUSIONS

Several conclusions follow from the analysis in this chapter regarding economies of size, government payments, and passthrough of payment benefits from tenants to landowners.

1. Economies of size are large and permeate American agriculture. Full economic costs of production and hence claims on resources per unit of food and fiber production on small farms are several times as great as those on larger farms. Thus enormous economic incentives exist to expand farm size by farm consolidation, increased capital intensity, or other means. Those incentives along with rapid technological improvements have massively restructured agriculture toward larger and fewer farms.

2. Costs decline at a decreasing rate with greater farm size, but per-unit cost of production continues to fall with the expansion of large farms. Unit costs undoubtedly begin to rise after farms reach some very large size, but data are currently inadequate to determine where diseconomies of size begin to predominate.

3. Plots of data used in this analysis, not shown in this chapter, illustrate two points. One is that the mathematical equations used in this analysis express quite well the central tendencies of individual farm observations regarding unit costs of production. However, per-unit cost of output on large farms is somewhat underestimated by the double-logarithm equations employed in this study. Actual unit costs are less than but nearer U.S. $1 to produce U.S. $1 of output than are predicted by the equations. The second point is that the primary data show considerable variation by size and type of farm. While such dispersion of primary data suggests that a well-managed, medium-sized farm can produce more efficiently than a poorly managed large-sized farm, we are inhibited in making such inferences because of inability to judge whether dispersion of data arises from nature (weather, pestilence, etc.), management, marketing, or reporting error.

4. Because of falling per-unit costs as farm size increases, efforts to intervene to slow or stop farm enlargement can be costly. Based on data reported in the far right column of Panel D of Table 8.2, the cost of producing the actual 2001 farm food and fiber output of U.S. $203 billion would be U.S. $173 billion with farms between U.S. $500,000 to U.S. $1,000,000 in sales. Cost of producing the actual 2001 output would be an estimated 2.43 times greater if produced solely by farms between U.S. $10,000 to U.S. $50,000 in sales but would be 34 percent lower than 2001 costs if produced solely by farms bigger than U.S. $1,000,000 in sales.

5. *Ceteris paribus,* government payments modestly diminish economies of size because payments are less per dollar of output on large than on small- and intermediate-sized farms. Payments are lower per unit of sales on large farms not so much because of payment limits but rather because such farms tend to produce commodities not receiving payments—livestock, fruits, and vegetables. It is notable that after adjustment for government payments and unique prices, economies of size are similar on program and nonprogram commodity types of farms. Of course, other things are not equal. Security and capital provided operators by commodity payments may encourage leveraging of equity, enabling operators to consolidate into fewer and larger farms.

6. Passthrough of commodity-program payments from renters to landowners is rapid, about 60 percent in one year. (Passthrough of government payments from new farmland owners to former farmland owners is much less rapid, about 1 percent per year.) Because operators of large farms rent a much higher proportion of their land than do operators of small farms, operators of large farms lose benefits of commodity programs most rapidly. Thus the lower payments per unit of output on larger farms and the rapid loss of payments to landowners tend to somewhat flatten unit cost curves facing operators of large farms, discouraging size expansion.

7. In the long term, all program benefits are lost to operators on small and large farms alike as land is sold at prices including the expected value of all future program benefits. One result is to create uncertainty—those who purchase land at inflated prices are subject to severe financial shock at such time as government payments are unexpectedly curtailed. Also, land prices inflated by government-payment expectations constitute a financial barrier to entry of capital-short, would-be farm owner-operators. The alternative is to rent land, but because of passthrough costs such decisions must be based on market rather than on policy conditions.

8. As the ultimate beneficiaries of farm commodity-program benefits, farm landowners appear to be of keen interest to policymakers. Data indicate that farm-operator households have somewhat higher incomes and much higher wealth than the average nonfarm household (USDA/ERS 2002b: 63). Perhaps landowner households, given their obvious asset position and income from agriculture, have even higher income and wealth than do farm-operator households, but USDA/NASS/ARMS data do not permit an answer.

9. The coexistence of large numbers of small farms with U.S. $3 or more of costs to produce U.S. $1 of receipts alongside a few larger farms with U.S. $1 or less of costs to produce U.S. $1 of receipts implies egregious extant economic disequilibrium. However, any conclusion from results of this study that farmers do not adjust to economic incentives would be premature. Farmers have responded massively to disequilibrium in the past, and the great farm-to-urban exodus appears over—farm numbers and total labor hours have changed little in recent years. Rural amenities and tax advantages apparently constitute a psychic or social return on small farms that, when added to economic returns, make those farms competitive with larger farms. Operators of small farms may use off-farm income to pay for the privilege of farming at a loss, but government payments have little or no bearing on such decisions.

10. This study did not consider environmental externalities that, if internalized, could change the unit social cost curve profiles. The topic needs much more study, but to date evidence is lacking that inclusion of all costs of protecting the environment (internalization of negative environmental externalities) would diminish economies of size (Boehlje et al. 1999; Martin and Zering 1997).

REFERENCES

Boehlje, M., S. Hofing, and C. Schroeder. (1999) *Farming in the 21ˢᵗ Century*. Staff Paper 99-9. West Lafayette: Department of Agricultural Economics, Purdue.

Goodwin, B., A. Mishra, and F. Ortalo-Magné. (2003) See Chapter 6, this volume.

Hallam, A. (1993)"Empirical Studies of Size, Structure, and Efficiency in Agriculture." In Arne Hallam (ed.), Chapter 8, *Size, Structure, and the Changing Face of American Agriculture*. Boulder: Westview Press.

Martin, L. and K. Zering. (1997) *Relationships between Industrialized Agriculture and Environmental Consequences*. Staff Paper 97-6. East Lansing: Department of Agricultural Economics, Michigan State University.

Tweeten, L. (1993) "Government Commodity Program Impacts on Farm Numbers." In Arne Hallam (ed.), Chapter 13, *Size, Structure and the Changing Face of American Agriculture*. Boulder: Westview Press.

_____. (1989) *Farm Policy Analysis*. Boulder: Westview Press.

_____. (1984) "Causes and Consequences of Structural Change in the Farming Industry." NPA Report No. 207. Washington, DC: National Planning Association.

USDA/ERS (U.S. Department of Agriculture, Economic Research Service). (2002a) "Agricultural Resource Management Survey (ARMS): Detailed documentation." Internet Website: www.ers.usda.gov/briefing/ARMS/detaileddocumentation.htm.

_____. (2002b) *Agricultural Outlook*. AGO-296. Washington, DC: USDA (November).

USDA/NASS (U.S. Department of Agriculture, National Agriculture Statistics Service). (2001) Agricultural Economics and Land Ownership Survey, (AELOS 1999). Internet Website: www.nass.usda.gov/census97/aelos/aelos.htm.

ENDNOTES

[*] Comments of Mary Ahearn, Mitch Morehart, Carl Zulauf, and Nigel Key are much appreciated. Views herein are solely those of the authors and not necessarily those of reviewers, Ohio State University, or the U.S. Department of Agriculture.

[i] Environmental externalities are not included.

[ii] Industry average costs of production are not shown in Figure 8.1 but lie above the marginal cost curves because marginal costs are falling throughout the range of the available data. Average cost would exceed one dollar per dollar of output throughout the data for each farm type, implying that the farming industry on the whole and for each farm type is losing money. That is, opportunity cost is not covered. Losses are concentrated on small farms.

[iii] The few sample observations for various large farms constrain either statistical or informal conclusions; hence inferences for very large farms should be guarded.

Section III:
Capital Markets and Farmland Values

Chapter 9

Capital Markets, Land Values, and Boom-Bust Cycles

Allen M. Featherstone and Charles B. Moss
Kansas State University
University of Florida

INTRODUCTION

Using a stochastic-trend representation of farmland values, we examine the tendency of farmland values to increase or decrease as a response to changes within the financial arena of the U.S. agricultural sector. The results of most economic models of farmland prices imply that the value of farmland is determined by discounted anticipated future rents or by net returns to farmland (Alston 1986; Burt 1986; Featherstone and Baker 1987). However, the results of Schmitz (1995) and of Falk and Lee (1998) indicate that while the values of agricultural assets are determined by market fundamentals (i.e., returns to farmland and interest rates), in the short run significant deviations from the discounted value formulation occur.

That farmland values increase more than would be appropriate in response to an increase in returns has been well documented (Featherstone and Baker 1987; Irwin and Colling 1990; Falk 1991; Clark et al. 1993; Schmitz 1995). These periods of overreaction are referred to as boom or bust cycles. A boom period of time is one in which farmland increases in value above its fundamental value (i.e., the present value of expected future returns to farmland) while the bust portion of the cycle is a period of time in which farmland values decrease below their fundamental value. Sustained (intermediate or long-term) periods of either over-valuation or under-valuation are inconsistent with efficient markets for farmland. Given that farmland values have accounted for roughly 75 percent of the assets of U.S. agriculture, boom-bust cycles have a profound effect on the well being of the sector's position of wealth and access to credit. Hence, when the U.S. agricultural sector experiences a boom-bust cycle, some of the more productive producers are driven out of the industry. Thus efficient producers may tend to borrow more heavily against farmland to acquire productive assets such as equipment resulting in a relatively higher business risk (Bierlen and Featherstone 1998). In addition to the effect of boom-bust cycles on the U.S. agricultural sector, Schmitz (1995) links these cycles to a host of financial difficulties for farmers and for rural communities.

In light of the fact that the identification of boom-bust cycles within the land market is well documented, the causes for these cycles remain issues of considerable debate in the economic literature. Recognized causes for boom-bust cycles include quasi-rationality or bubbles[i] (Featherstone and Baker 1987), time-varying risk premiums (Hanson and Myers 1995), overreaction (Burt 1986; Irwin and Colling 1990), fads (Falk and Lee 1998), risk aversion and transaction costs (Just and Miranowski 1993; Chavas and Thomas 1999; Lence and Miller 1999; Lence 2001).

CAPITAL-ASSET PRICING AND BOOM-BUST CYCLES

The theory underlying boom-bust cycles has been developed mostly in the literature on finance (Shiller 1981; LeRoy and Porter 1981). In addition, Tirole (1985) identifies three factors leading to price bubbles—durability of the asset, scarcity of the asset, and common belief in future returns. Durability implies that the asset is used for several production periods (crops in the case of farmland). Decisions involving assets that are used up in the short-run are easily changed (e.g., the level of nitrogen changed in each time period). Decisions about more durable assets cannot be varied as readily (e.g., the size of a barn cannot be varied from year to year). Scarcity in this case can most easily be described by the old farmer adage "they aren't making any more farmland." Thus unlike other agricultural assets (e.g., equipment), the total level of farmland in the United States is fixed in the long run. Finally, common beliefs imply that farmers as a whole have roughly the same expectations about future returns. In the case of farmland in the United States, farmers form their expectations about future returns based on markets (both futures and cash markets) for the same commodities. Thus everyone may observe the futures markets for corn increasing and conclude that the return to farmland will increase in the short run. Further, intermediate and long-run expectations for most farmers will probably be based (directly or indirectly) on baseline projections by the U.S. Department of Agriculture. Thus, farmland values are affected by these three conditions, but so are other classes of assets. Arbitrage could mitigate the effects of boom-bust cycles of farmland valuation, however, factors such as the absence of short selling and transaction costs make arbitrage quite risky (Chavas 2003; Lence 2003; Miller 2003).

Market Overreaction, Fads Arbitrage, and Bubbles

The issue of whether or not the farmland market in the United States responds to market fundamentals is not only of interest in the land market, but it is also of interest in the stock-pricing literature. Many equity investments such as stocks and bonds possess the three factors that could lead to asset bubbles (i.e., durability, scarcity, and common beliefs). However, such bubbles could imply inefficient allocation of capital across industries or firms within an industry. Shiller (1981) and LeRoy and Porter (1981) propose a variance-bounds test to examine whether stock-price movements are too variable to be justified by dividend movements. Irwin and Colling (1990) use this methodology to analyze whether the volatility in farmland prices was consistent with the variability in returns to farmland. They find that the hypothesis of excess volatility could not be rejected (that the variability in the return to farmland was potentially larger than that implied by the variability of farmland prices). The variance-bounds methodology has been criticized due to the potential nonstationarity[ii] of the data series (Kleidon 1986) and of the data's small-sample bias (Flavin 1983).

The next generation of research in the financial literature is the development of the test of the present-value model, which accounts for nonstationary data (Campbell and Shiller 1987). Campbell and Shiller reject the present-value model for stock data. Falk (1991) uses this methodology to reject the present-value in farmland markets [i.e., Falk (1991) does not find a stationary relationship between farmland values and returns to farmland]. Lee (1998) argues that the rejection of the variance-bounds tests and the present-value model imply that either stock prices react to nonmarket fundamentals or to time-varying-discount factors. Hanson and Myers (1995), when testing for a time-varying discount rate (in the present value formula for farmland), find that some but not all variation in farmland values can be explained by a time-varying-discount rate, thus leaving open the possibility of movement in farmland prices due to nonfundamentals.

A related issue to the effect of nonstationarity in asset prices is Fama's (1970) observation of negative autocorrelation in stock-market returns that may be a violation of market efficiency (i.e., a constant negative autocorrelation would imply consistent abnormal profits from public information). Fama and French (1988) and Poterba and Summers (1988) argue that separating movement of stock returns into temporary and permanent components can explain this autocorrelation. Lee (1995) further examines the issue of temporary and permanent components of stock returns and finds that both temporary and permanent shocks in stock dividends significantly affect the price of stocks. In addition, Lee (1995) finds that the initial response to both temporary and permanent shocks in stock returns is strong.

Lee (1998) proposes a method to examine the relative importance of various components of prices due to permanent and temporary changes in income, discount factors, and nonfundamental factors. Using a vector-autoregressive model (VAR) (modeling the values of a current group of variables as a function of past values for the same set of variables), Lee decomposes the response of stock prices to permanent and temporary changes in earnings and dividends, changes in discount factors, and nonfundamental factors. He finds that about 50 percent of the variation in stock prices is not related to earnings or dividends. Thus while the long-term trend in stock movements is due to permanent changes in fundamentals (i.e., changes in stock dividends or discount rates), the short-term volatility of stock prices is due to time-varying-discount rates and to nonfundamental factors. Lee (1998) identifies an overreaction and mean-reversion process in the stock market. Because the deviation of price from its fundamental value declines over time, he argues that the deviation is more consistent with fads rather than with bubbles.

Falk and Lee (1998) apply the methodology proposed by Lee (1998) to examine farmland prices. They find that fads and overreactions play an important role in explaining short-run pricing behavior, while long-run price movements are explainable by permanent fundamental shocks.

Transaction Costs

Another related anomaly in the finance literature has led to the examination of transaction costs. Over the last 100 years, the average real rate of return has been more than 6 percent higher than that on Treasury bills (Kocherlakota 1996). Using a representative-agent paradigm from the macroeconomic literature, Mehra and Prescott (1985) demonstrate that a risk premium of 6 percent or higher would be plausible only if the investor were risk averse and if he had a relative risk-aversion coefficient near 20 (Kocherlakota 1996).

Several models have been proposed that modify this preference structure to account for the equity risk premium in excess of six percent. Epstein and Zin (1991) use a generalized-expected-utility function that does not require the coefficient of relative risk aversion and the elasticity of substitution to be reciprocals of each other. A model of habit formation was used by Constantinides (1990) and a model of "keeping up with the Joneses" was used by Abel (1990). Kocherlakota (1996) argues that while sensitivity analysis to the preference structure is useful for explaining some issues, it still does not explain the equity-premium puzzle without assuming extreme risk aversion to marginal variation in consumption.

Kocherlakota (1996) indicates that incomplete markets and trading costs have also been proposed as possible solutions to the equity-premium puzzle. The trading-cost literature has considered borrowing and short-sales constraints, transaction costs, and market segmentation as a possible explanation. Kocherlakota (1996: 66) argues that while these may be possibilities for explaining the equity-premium puzzle, the premium on equity (stocks) is not compensation for risk as is suggested by financial theory; instead, it is transaction costs. In addition, the source of sizable transaction costs is not clear. Kocherlakota (1996) concludes that "Just as people in real life are able to find ways around government regulations of the economy, rational individuals are able to find ways around barriers to trade."

Transaction costs have been used to examine the behavior of farmland markets. Just and Miranowski (1993) develop a structural model of farmland prices based on inflation and changes in the opportunity cost of capital. They argue that changes in farmland values in the late 1970s and the early 1980s were due to inflation-rate changes and to changes in the real returns on alternative uses of capital.

Chavas and Thomas (1999) use the Epstein and Zin (1991) model incorporating transaction costs to examine the dynamic nature of land prices. They find that risk neutrality and the static capital-asset pricing model (CAPM) are rejected indicating that both risk aversion and transaction costs are important determinates to the price of land. Chavas and Thomas further argue that transaction costs for the sale and purchase of farmland range from 7.5 percent to 13 percent.

Lence (2001) cautions about interpreting the Just and Miranowski (1993) and Chavas and Thomas (1999) results because there is an incorrect derivation of the first-order-optimization criterion and of the use of nonstationary data in the Just and Miranowski estimation process.

Kocherlakota (1996: 67) concludes the equity-premium-puzzle discussion with the following quote,

> With this in mind, we cannot hope to find a resolution to the equity-premium puzzle by continuing in our current mode of patching the standard models of asset exchange with transaction costs here and risk aversion there. Instead, we must seek to identify what fundamental features of goods and assets markets lead to large risk-adjusted price differences between stocks and bonds.

In light of the above literature, abandoning the stock-pricing literature and specifying models of transaction costs and risk aversion will remain unlikely to explain boom-bust behavior that occurs in the land markets.

Model of Transitory and Permanent Components

Falk and Lee (1998) estimate a time-series model based on a vector-moving-average (VMA) representation that tests for the effect of permanent fundamental, temporary fundamental, and nonfundamental innovations on farmland values. Permanent-fundamental changes are changes in the long-run returns to farming (e.g., changes in technology or agricultural policy). Temporary-fundamental changes are those changes that are related to only short-run profitability (e.g., fluctuations due to weather or pests). Nonfundamental innovations are changes in agriculture or in the general economy that may affect farmland values but may fail to affect the long-run profitability of agriculture.

Falk and Lee's (1998) formulation of farmland values is based on the Campbell and Shiller (1987) dividend-price-ratio estimation of stock prices, which is a dynamic version of the Gordon formulation for stock prices. In the Gordon formulation, the dividend-to-price ratio of stock prices equals the market-determined-discount rate of the returns for assets of similar risk less a constant for future growth in dividends. In this formulation, expected future-dividend growth increases the price of the stock compared to other stocks of the same relative risk. Based on the Gordon model, Campbell and Shiller (1987) propose estimation of a time-series representation of the natural logarithm of the dividend-price ratio. This dividend-price ratio should be cointegrated with the discount rate of similar assets, given some adjustments for future changes in the dividends. Any short-lived deviation that has a finite-lived moving-average representation in the dividend-price-ratio/discount-rate relationship of similar assets is consistent with the long-run equilibrium. The VMA relationship is then estimated using a VAR formulation of the natural logarithm of the dividend price ratio, and using the difference between the logarithmic change in dividends and the observed interest rate. Falk and Lee (1998) restrict the relationship between these two time series to test for a long-run equilibrium or for cointegrating relationship.

As discussed above, Falk and Lee (1998) use this formulation to test the present-value model (PVM) of farmland. They find that the present-value relationship (i.e., the dividend-price ratio that is cointegrated with the difference of the change in dividends and interest rate) holds in the long run, but that significant short-run noise exists in the fundamental relationship. Decomposing the variance of the forecast error in the VAR representation, Falk and Lee conclude that most of the short-run noise is attributable to temporary nonfundamental innovations. Thus, they conclude that fads are more important when explaining farmland price movements than are bubbles.

The empirical model we apply uses fundamental temporary versus permanent (or borrowing from macroeconomic literature transitory versus permanent) formulation for modeling income. However, instead of relying on the dividend-price relationship used in the stock market literature, we estimate directly a transitory versus a permanent representation for the return to farmland. The decomposition of returns to farmland into permanent and transitory components allows us to analyze the effect of permanent change on income for farmland values through time. And it is consistent with the PVM. Mathematically, the stochastic-trend model is similar to the VMA process as derived from the VAR. Differences between the stochastic trend and the VAR approach arise in the imposition of the cointegration constraints on the time-series process.

As discussed above, most theoretical models of farmland prices hypothesize that the value of an asset (e.g., farmland) is determined by the returns to that asset (e.g., cash rents or returns to farmland), which arise from the use of the asset in the future. However, this definition is immediately fraught with practical questions. Most notably, future returns are not directly observable. This difficulty is not onerous to theoretical formulations since the expected cash flow is substituted effortlessly for the future cash flow. In the real world, however, the task of ascertaining future cash flows is dramatically more difficult for farmers, who bid on farmland, and for empirical researchers, who work with actual data. In fact, one could suggest that changes in farmland should actually be used to infer changes in expected future returns, since changes in farmland values are observable directly.

We express changes in farmland values as

$$\Delta V_t = -E_t CF_t + \frac{r_t}{1+r_t} V_t + \gamma_t, \tag{1}$$

in which ΔV_t is the observed change in farmland values ($\Delta V_t = V_t - V_{t-1}$), $E_t CF_t$ is the expected return on farmland in period t based on information available at the start of period t, r_t is the discount rate for the farm sector in period t, and γ_t is a term that represents changes in expectations about future returns (Schmitz 1995). Most previous research (including Schmitz 1995) substituted observed cash rents or imputed cash returns to farmland for expected cash flows. This substitution implies

$$\Delta V_t = -\left(CF_t + v_t\right) + \frac{r_t}{1+r_t} V_t + \gamma_t, \tag{2}$$

in which v_t is defined as the difference between observed cash flows and expected cash flows. This substitution then yields a composite error term

$$\Delta V_t = -CF_t + \frac{r_t}{1+r_t} V_t + \left(\gamma_t - v_t\right). \tag{3}$$

Without additional information it is impossible to separate the short-run errors, which are due to the substitution of observed cash flows from expected cash flows, and from the changes in long-run expectations.

In his empirical results, Schmitz (1995) finds that changes in asset values over time are consistent with Equation (3), but that the composite error term ($\gamma_t - v_t$) exhibits autocorrelation in the short run. Schmitz (1995) links this autocorrelation with boom-bust cycles in farmland prices. However, this conclusion leaves the cause of these boom-bust cycles undetermined. Specifically, it is impossible to determine if the persistence in short-run disequilibria is attributable to the inability to recognize changes in the distribution of future returns γ_t, or if the persistence in short-run disequilibria is simply a short-run phenomenon due to the parameterization of expected returns v_t.

Regardless of the exact nature of the residual the general formulation in Equation (3) gives a framework for a detailed discussion of boom-bust cycles. Rearranging Equation (3), we value farmland in period t as

$$V_t = \left[-\left(1+r_t\right) E_t CF_t + r_t V_t\right] + V_{t-1} + \gamma_t^*. \tag{4}$$

This expression implies that as long as farmland yields an equilibrium rate of return of

$$-\left(1+r_t\right) E_t CF_t + r_t V_t = 0, \tag{5}$$

farmland values will remain the same. If farmers expect cash flows to increase in the future ($\gamma_t^* > 0$) then farmland values will also increase.

Next, returning to the notion of the observable, an observed increase in farmland prices could be interpreted as an indication of increased future earnings. Such an event would be consistent with a boom cycle (or a positive-rational bubble). Based on Equation (1), the changes in expected-future returns are expressed as

$$\gamma_t = E_t CF_t - \left[V_{t-1} - \frac{1}{1+r_t} V_t \right]. \tag{6}$$

Under this scenario, increased farmland values in excess of the equilibrium rate of return expressed in Equation (5) would imply increased expected future returns. Within this context, a positive shock due to unanticipated inflation (as experienced in the last half of the 1970s) or due to additional increases in the returns to farming (e.g., the expansion of the export markets) could set the U.S. agricultural sector on an economic roller coaster.

This story, while reasonable, does not fit the traditional notion of efficient asset markets treasured by financial economists. Specifically, under the efficient-market hypothesis, investors would not be able to systematically make a profit based on publicly available information. Under this hypothesis, investors could force the asset market back into equilibrium through arbitrage. In the case of a boom cycle, investors could put downward pressure on farmland values by selling short on farmland (i.e., selling farmland that they do not actually own, which is a common practice in equity markets). Short sales are common in the stock market because the short sale can be easily offset by the purchase of stock at any point in time. However, such short sales are difficult in farmland markets because farmland is not always available to offset the short sale. Also, farmland markets tend to be thinner and nonhomogeneous in quality. And since the agricultural sector is typically composed of sole proprietorships, the equity transaction is not readily available. Thus the lack of liquid assets and equity markets in agriculture limits the market disciplines that rectify deviations in other, more liquid, commodity markets. Hence, the illiquidity of farmland and of agricultural-equity markets contributes to boom-bust cycles for farmland (Tirole, 1985).

The liquidity story is similar for bust cycles. A short-run contraction in farmland markets will be validated by the inability of agriculture to draw on equity markets and by the unwillingness of outside investors to invest in farmland markets.

EMPIRICAL EVIDENCE OF BOOM-BUST CYCLES FOR U.S. FARMLAND

Given the Schmitz (1995) formulation, we develop a model of farmland prices that allows for both temporary and permanent fluctuations in returns to farmland. A detailed description of the model is presented in Appendix 9.A of this chapter, but the model is based on a multivariate stochastic-trend model estimated using a Kalman filter (Harvey 1989; Moss and Shonkwiler 1993). Using the Kalman filter, we estimate the state-space model for Florida, Kansas, Illinois, and Indiana based on USDA data for 1950 through 1995 (USDA/ERS 2002a,

2002b). The actual and predicted per-acre values for farmland prices and returns to farmland are presented in Table 9.1.

In general, the estimated farmland values using the stochastic-trend specification follow the actual values fairly closely. Land values in Florida (Figure 9.1) follow a rather steady upward trend that stalls in the early 1980s,while Kansas (Figure 9.2), Illinois (Figure 9.3), and Indiana (Figure 9.4) each follow the pattern of increasing farmland prices throughout the 1960s and 1970s and then decline significantly in farmland values during the period of financial stress in the early- and mid-1980s. This decline in farmland values then gives way to a period of increasing farmland values at a rate that is slightly less rapid than the increase observed in the 1960s and 1970s. Differences between the Florida land-value cycle and those observed in the other states follow from two factors. First, farmland values in Florida are significantly affected by urbanization. Second, agriculture in Florida is less reliant on program crops, which were particularly hard hit in the other states during the financial stress of the early- and mid-1980s.

Apart from following the general trend in farmland values, the deviation between observed and predicted farmland values in Figures 9.1 through 9.4 is consistent with the rational-bubble hypothesis. Specifically, the predicted values of farmland exceeded actual farmland values in each state in the early 1980s. This result could be interpreted as continued expectations from the late 1970s. However, when the stochastic trend corrects, the expected value of farmland falls below the observed market value after 1985. Thus, the empirical results of our empirical model support the Schmitz (1995) boom-bust conclusions. The question then becomes: Why do farmland values exhibit boom-bust cycles?

In order to investigate the cause of the boom-bust cycles, we begin to decompose the variance (or errors from the stochastic trend model) into permanent and temporary (or transitory) innovations using the Kalman filter model. As presented in Appendix 9.A, the Kalman-filter specification of the stochastic-trend model decomposes the variance into long-run and short-run volatility. The long-run volatility relates to permanent changes in an endogenous variable, while the short-run volatility indicates the presence of transitory changes. Roughly speaking, these changes are also related to stationary and nonstationary time-series processes. Thus, changes in the return to farmland could be permanent in which case they would affect the price of farmland, or temporary in which case they would not affect the price of farmland because the price of farmland is based on future returns.

In Table 9.2, we present the decomposition of the variance for each series into long-run and short-run components. These decompositions are particularly significant for traditional farmland models. Specifically, in Florida, Illinois, Indiana, and Kansas, the long-run volatility dominates the total volatility for both farmland values and interest rates. However, the volatility in returns to farmland appears to be mostly short-run in nature (the share of total variance which is short run is larger than for either farmland values or real interest rates). These results can be interpreted in several ways. Using the permanent/transitory dichotomy, most of the fluctuation in returns to farmland appears to be transitory. This result is consistent with the nature of U.S. agriculture, because returns to farmland are notoriously risky. Specifically returns to farmland are functions of crop price, of risky crop yield, and cash expenses.

Table 9.1 Actual and Predicted per-acre Values of Farmland and Returns to Farmland in Florida, Kansas, Illinois, and Indiana

	Florida				Kansas			
Year	Actual land value	Predicted land value	Actual income	Predicted income	Actual land value	Predicted land value	Actual income	Predicted income
				U.S. dollars				
1965	279.14	292.54	24.78	29.30	125.37	118.70	10.61	8.01
1966	275.98	293.06	24.41	25.18	134.10	133.51	11.60	10.60
1967	290.75	283.14	26.31	23.89	145.68	144.33	9.92	12.45
1968	299.93	297.12	29.10	26.22	151.27	157.20	9.96	10.24
1969	333.11	308.68	34.78	30.26	148.89	162.00	12.05	9.80
1970	353.06	349.85	30.24	38.12	152.04	155.18	15.00	12.41
1971	377.36	376.53	40.23	32.11	163.30	155.36	17.33	16.79
1972	434.75	403.76	51.03	43.33	186.85	168.35	24.87	20.23
1973	569.61	473.82	67.41	59.54	237.56	199.40	40.56	30.70
1974	642.19	657.78	63.80	83.47	277.94	269.55	28.74	55.60
1975	715.26	761.31	80.28	75.62	321.34	327.28	23.00	35.24
1976	807.72	832.04	76.02	92.05	373.97	378.76	16.30	22.60
1977	920.26	922.22	77.49	84.09	392.74	437.95	16.39	13.71
1978	1,078.69	1,044.55	97.35	81.20	471.05	447.10	16.24	13.58
1979	1,296.49	1,233.87	105.41	105.07	551.96	532.68	27.05	14.51
1980	1,473.31	1,509.35	115.13	117.61	582.24	636.13	7.54	29.05
1981	1,432.95	1,719.79	118.67	127.82	590.91	658.61	18.79	6.51
1982	1,491.67	1,592.82	139.59	128.85	565.70	642.95	28.49	16.52
1983	1,553.78	1,580.66	162.19	152.60	558.54	585.47	21.76	34.56
1984	1,506.97	1,621.03	160.13	182.60	454.75	557.92	28.34	25.99
1985	1,446.55	1,538.68	170.23	176.42	384.02	428.64	34.48	31.82
1986	1,508.05	1,436.61	182.08	181.66	343.32	339.91	34.82	40.10
1987	1,680.50	1,499.61	203.83	193.29	378.10	298.73	40.01	39.71
1988	1,716.31	1,732.31	254.63	219.21	387.60	348.94	38.98	44.57
1989	1,899.62	1,801.60	271.51	286.66	411.01	382.57	31.62	42.08
1990	1,922.45	2,016.40	216.90	309.87	412.34	420.52	48.30	31.09
1991	1,921.24	2,034.71	276.04	224.69	421.22	425.71	35.77	50.24
1992	1,887.28	1,989.96	281.69	282.18	424.33	431.84	49.87	37.05
1993	2,033.20	1,909.80	258.02	298.59	456.67	432.77	46.37	52.28
1994	2,065.70	2,069.23	242.56	263.93	479.73	470.26	49.81	49.79
1995	2,150.12	2,123.84	193.48	236.68	497.34	500.98	32.24	52.33

Source: Authors' computations. *(Continued)*

Table 9.1 Actual and Predicted per-acre Values of Farmland and Returns to Farmland in Florida, Kansas, Illinois, and Indiana (*Continued*)

	Illinois				Indiana			
Year	Actual land value	Predicted land value	Actual income	Predicted income	Actual land value	Predicted land value	Actual income	Predicted income
				U.S. dollars				
1965	386.36	359.88	36.62	26.23	320.09	297.67	34.98	19.09
1966	412.83	414.47	38.98	38.23	345.18	347.03	32.30	36.19
1967	431.72	447.87	37.36	43.38	363.86	378.24	29.74	36.20
1968	450.56	463.75	29.66	40.23	363.88	394.51	28.53	31.13
1969	446.90	477.22	38.06	28.87	353.55	384.15	39.32	28.23
1970	449.77	463.75	31.10	37.52	366.73	360.75	29.05	41.52
1971	477.49	457.64	37.90	30.77	380.54	369.85	42.15	29.83
1972	521.53	488.73	49.75	38.18	435.04	388.15	41.61	44.21
1973	665.53	546.56	90.76	55.48	524.85	458.33	94.84	45.74
1974	786.38	740.61	78.22	119.64	642.64	583.94	59.68	123.19
1975	992.40	918.44	107.11	101.46	797.48	748.15	80.81	73.97
1976	1,370.39	1,194.69	75.53	130.55	1,073.98	960.24	80.18	89.97
1977	1,535.59	1,727.86	75.31	81.36	1,235.20	1,343.91	60.55	88.20
1978	1,766.59	1,928.30	73.80	72.18	1,456.36	1,543.55	70.62	59.51
1979	1,951.46	2,126.20	102.80	70.13	1,719.34	1,768.45	74.54	67.66
1980	2,093.01	2,266.04	44.49	107.65	1,875.84	2,052.11	58.90	75.39
1981	1,936.09	2,342.25	98.10	40.19	1,668.49	2,177.72	61.14	57.04
1982	1,757.94	2,044.55	81.00	94.10	1,490.20	1,787.03	67.65	57.61
1983	1,757.95	1,728.38	19.33	87.85	1,509.62	1,457.16	17.92	67.14
1984	1,309.83	1,685.20	96.59	14.11	1,220.12	1,437.98	96.39	13.27
1985	1,162.55	1,190.06	119.46	86.02	1,048.84	1,128.97	88.83	91.42
1986	1,079.41	1,003.19	103.08	162.44	944.18	924.15	75.27	119.15
1987	1,179.81	943.55	98.34	133.08	1,019.80	825.03	93.41	89.85
1988	1,277.62	1,102.62	82.44	108.73	1,076.16	941.95	70.66	103.70
1989	1,305.68	1,287.50	132.99	80.72	1,090.30	1,065.37	109.64	71.93
1990	1,361.35	1,357.26	111.73	140.01	1,117.16	1,113.02	100.40	115.53
1991	1,426.53	1,418.31	79.36	121.68	1,166.15	1,145.33	56.30	110.20
1992	1,426.01	1,488.42	122.90	74.81	1,225.20	1,200.39	100.16	51.17
1993	1,566.65	1,476.05	109.09	121.18	1,303.30	1,274.13	106.11	95.15
1994	1,710.01	1,631.99	128.98	114.57	1,403.84	1,367.43	93.53	116.70
1995	1,899.01	1,820.60	68.14	136.78	1,545.66	1,486.81	65.65	99.32

Source: Authors' computations.

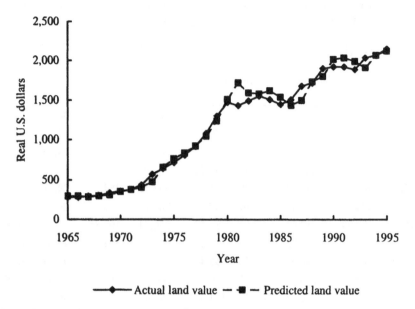

Figure 9.1 Actual and Predicted Farmland Values in Florida from 1965 to 1995

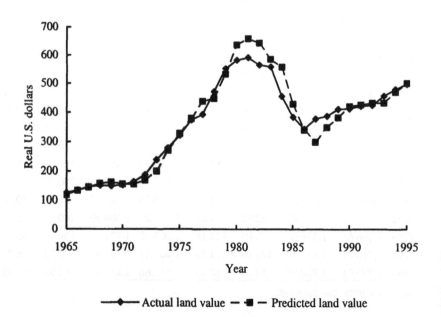

Figure 9.2 Actual and Predicted Farmland Values in Kansas from 1965 to 1995

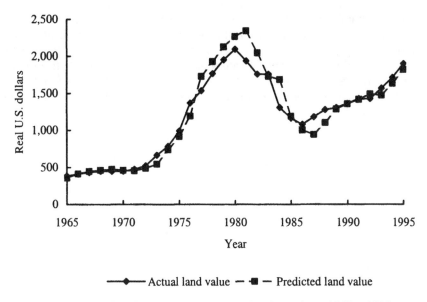

Figure 9.3 Actual and Predicted Farmland Values in Illinois from 1965 to 1995

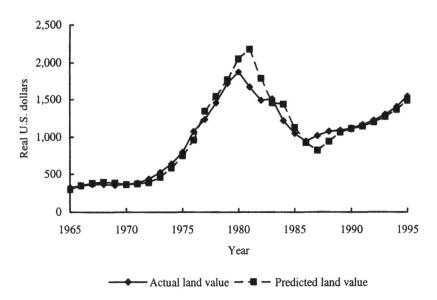

Figure 9.4 Actual and Predicted Farmland Values in Indiana from 1965 to 1995

Overall, crop price depends on factors, including domestic and foreign demand for agricultural outputs, which are sensitive to aggregate yields (both foreign and domestic), agricultural policies, and economic growth. Crop yields at the farm level are dependent on factors, including weather and pests. Compared to a farm's annual returns, farmland values themselves are less volatile from year to year. Thus, the results indicate that fluctuations in the return to farmland are highly transitory (or short-run) while changes in farmland values tend to be more permanent.

Table 9.2 Decomposition of System Variance for Farmland Prices, Returns to Farmland and Interest Rate Series

Variable	Total variance	Contemporaneous variance	Percent contemporaneous variance
		Florida	
Farmland prices	0.0040	0.0000	0.0000
Returns to farmland	0.0199	0.0070	0.3537
Interest rate	0.0105	0.0000	0.0000
		Kansas	
Farmland prices	0.0049	0.0000	0.0000
Returns to farmland	0.1349	0.0788	0.5844
Interest rate	0.0105	0.0000	0.0000
		Illinois	
Farmland prices	0.0058	0.0000	0.0000
Returns to farmland	0.1424	0.0567	0.3982
Interest rate	0.0105	0.0000	0.0000
		Indiana	
Farmland prices	0.0054	0.0000	0.0000
Returns to farmland	0.1201	0.0716	0.5956
Interest rate	0.0105	0.0000	0.0000

Source: Authors' computations derived from USDA/ERS (2002a and 2002b) data.

A second interpretation of the volatility results for returns to farmland and farmland prices involves the measurement-error decomposition presented in Equation (3). Based on this decomposition, there are two sources of residual information in the stochastic-trend model. One source arises from the substitution of observed cash returns for expected returns ν_t, while the second arises from changes in long-run expectations γ_t. In view of these two sources of residual information, the short-run variance measures the difference between observed and expected returns while the long-run variation measures changes in long-run expectations.

Finally, the difference in relative variance (long-run versus short-run) has a technical interpretation. Relying on the cointegration literature, long-run volatility (by its very nature) measures changes in long-memory time-series processes, which are common nonstationary trends. Short-run variations then represent noise around the cointegrating relationships (if, indeed, noise exists).

CONCLUDING THOUGHTS

Changes in farmland values have significant implications for the economic well being of the U.S. agricultural sector. Given the significance of farmland values to the agricultural-asset portfolio, variations in farmland values can significantly affect agricultural wealth and, hence, a farmer's access to credit. Significant declines in farmland values may result in productive producers leaving the sector due to the lack of credit market access. Changes in farmland values that are driven by alterations in market fundamentals (i.e., changes in returns to farmland or changes in discount rates) are consistent with economic efficiency, but modifications in farmland values (apart from divergence from asset fundamentals) have adverse implications for the economic efficiency of the U.S. agricultural sector. This study focused primarily on the possibility of boom-bust cycles in farmland prices, which are defined as sustained periods of over-valuations or under-valuations of farmland as a response to temporary shocks to the profitability of agriculture.

To examine the possibility of boom-bust cycles in farmland values, this study estimated a stochastic-trend model for four states (Florida, Kansas, Illinois, and Indiana) using the Kalman filter. This analysis indicated that land values are not affected by transitory factors. However, income has a very large transitory component. The estimated stochastic-trend specification we used reproduced the perceived boom-bust periods observed in each state during the 1970s and 1980s, which poses two questions: What triggers a boom-bust cycle? Can a cycle be determined ex ante? The answer to the second question is much more difficult in that anticipating boom-bust cycles in the land market of U.S. agriculture is just as difficult as anticipating a boom-bust cycle in its technology sector. However, awareness regarding the fact that asset markets are susceptible to boom-bust cycles will make boom-bust cycles less likely to occur.

APPENDIX 9.A

In order to estimate the movements in farmland values over time incorporating boom-bust cycles, we formulate a state space model for the states of Florida, Illinois, Indiana, and Kansas based on the theoretical results presented in Equation (4). Specifically, the state-space model depicts a set of observable variables (i.e., farmland price, returns to farmland, and interest rates) as functions of unobserved or latent variables that evolve over time. In our formulation, these latent variables represent the true or long-run values of the prices of farmland, returns to farmland, and interest rates.

The measurement equation depicts how the observed variables are related to latent variables. In our study

$$y_t = Z\alpha_t + \varepsilon_t$$

$$y_t = \begin{bmatrix} V_t \\ CF_t \\ r_r \end{bmatrix}, Z = \begin{bmatrix} 1 & \phi_1 & \phi_2 & 0 & 0 & 0 \\ 0 & 1 & \phi_3 & 0 & 0 & 0 \\ 0 & 0 & 1 & 0 & 0 & 0 \end{bmatrix}, \varepsilon_t = \begin{bmatrix} \varepsilon_{1t} \\ \varepsilon_{2t} \\ \varepsilon_{3t} \end{bmatrix}, \quad (9.A1)$$

$$\alpha_t = \begin{bmatrix} \mu_{1t} \\ \mu_{2t} \\ \mu_{3t} \\ \beta_{1t} \\ \beta_{2t} \\ \beta_{3t} \end{bmatrix}$$

in which V_t is the value of farmland, CF_t is the return to farmland, r_t is the agricultural interest rate, μ_{1t} is the long-run price of farmland, μ_{2t} is the long-run or expected return to farmland, μ_{3t} is the long-run or expected agricultural interest rate, ε_{it} are measurement errors and ϕ_is are estimated parameters. This measurement equation is identified by the unities in the diagonal of the parameter matrix. Thus, the latent variable for farmland values is defined by the first equation. Within this formulation, the observed price of farmland is a function of the expected returns to farmland and the expected interest rates while the observed returns to farmland are functions of the expected returns to farmland and the expected interest rate.

To complete the formulation, we add a system of transition equations that dictate how the unobserved variables change from one period to the next. The transition equation used in this study follows the doubly integrated random-walk formulation used by Moss and Shonkwiler (1993)

$$\alpha_t = T\alpha_{t-1} + \upsilon_t$$

$$T = \begin{bmatrix} 1 & 0 & 0 & 1 & 0 & 0 \\ 0 & 1 & 0 & 0 & 1 & 0 \\ 0 & 0 & 1 & 0 & 0 & 1 \\ 0 & 0 & 0 & 1 & 0 & 0 \\ 0 & 0 & 0 & 0 & 1 & 0 \\ 0 & 0 & 0 & 0 & 0 & 1 \end{bmatrix}, \upsilon_t = \begin{bmatrix} \upsilon_{1t} \\ \upsilon_{2t} \\ \upsilon_{3t} \\ \upsilon_{4t} \\ \upsilon_{5t} \\ \upsilon_{6t} \end{bmatrix}. \quad (9.A2)$$

Taken together, the measurement and transition equations allow for a variety of alternative time-series specifications. An important consideration in the analysis of farmland values is that this formulation allows for nonstationarity. Specifically, the transition equation presented in Equation 9.A2 allows the latent variables to be random walks. Thus, the observed variables may be functions of nonstationary latent variables. Any linear relationship between these nonstationary variables is then a representation of a cointegrating vector. Thus, this formulation includes the possibility of cointegrating relationships in land (e.g., Falk 1991).

In order to estimate the state-space formulation specified in Equations 9.A1 and 9.A2 as a Kalman filter (Harvey 1989), we start with a recursive representation of the variance matrix (a Ricati formulation)

$$
\begin{aligned}
F_t &= Z P_{t|t-1} Z' + \Sigma_\varepsilon \\
P_{t|t-1} &= T P_t T' + \Sigma_\upsilon
\end{aligned}
\tag{9.A3}
$$

in which F_t is the variance matrix for the observed variable y_t based on the measurement equation, a projected variance matrix for the latent variables $P_{t|t-1}$, and the covariance matrix for the error in the measurement equation Σ_ε. For identification purposes, we assume both variance matrices to be diagonal. The projected-variance matrix for the latent variables is then modeled as the variance of the transition equations based on last period latent variables. The predicted values of the latent variables of this period are simply the expectations of the transition equation conditioned on the latent variables of the last period

$$
\alpha_{t|t-1} = T\alpha_t .
\tag{9.A4}
$$

Given these predictions, the estimated value of the residual in the current period is

$$
\varepsilon_t = y_t - Z\alpha_{t|t-1},
\tag{9.A5}
$$

the updated value of the latent variables is

$$
\alpha_t = \alpha_{t|t-1} + P_{t|t-1} Z' F_t^{-1} Z P_{t|t-1},
\tag{9.A6}
$$

and the unconditional value of the variance matrix for the latent variables is

$$
P_t = P_{t|t-1} - P_{t|t-1} Z' F_t^{-1} Z P_{t|t-1}.
\tag{9.A7}
$$

The normal likelihood function for the state-space model within the Kalman filter formulation can then be specified as

$$
L = -\frac{1}{2} \sum_{t=1}^{T} \left[\ln|F_t| + \varepsilon_t' F_t^{-1} \varepsilon_t \right].
\tag{9.A8}
$$

Maximum likelihood estimation involves specifying an initial covariance matrix for latent variables P_0, and then selecting the set of parameters (ϕ_1, ϕ_2, ϕ_3, $\sigma_{\varepsilon\varepsilon,1}$, $\sigma_{\varepsilon\varepsilon,2}$, $\sigma_{\varepsilon\varepsilon,3}$, $\sigma_{\upsilon\upsilon,1}$, $\sigma_{\upsilon\upsilon,2}$, $\sigma_{\upsilon\upsilon,3}$, $\sigma_{\upsilon\upsilon,4}$, $\sigma_{\upsilon\upsilon,5}$, $\sigma_{\upsilon\upsilon,6}$, μ_{10}, μ_{20}, μ_{30}, β_{10}, β_{20}, β_{30}) that maximize the likelihood function in Equation 9.A8.

Featherstone and Moss

REFERENCES

Abel, A.B. (1990) "Asset Prices Under Habit Formation and Catching Up With the Joneses." *American Economic Review* 80(1): 38–42.

Alston, J.M. (1986) "An Analysis of Growth of U.S. Farmland Prices, 1963–82." *American Journal of Agricultural Economics* 68(1): 1–9.

Bierlen, R. and A. M. Featherstone. (1998) "Fundamental *q*, Cash Flow, and Investment: Evidence from Farm-Panel Data." *Review of Economics and Statistics* 80(3): 427–36.

Burt, O. (1986) "Econometric Modeling of the Capitalization Formula for Farmland Prices." *American Journal of Agricultural Economics* 68(1): 10–26.

Campbell, J.Y. and R.J. Shiller. (1987) "Cointegration and Tests of Present Value Models." *Journal of Political Economy* 95(5): 1062–88.

Chavas, J.P. and A. Thomas. (1999) "A Dynamic Analysis of Land Prices." *American Journal of Agricultural Economics* 81(4): 772–84.

Chavas, J.P. (2003) See Chapter 13, this volume.

Clark, J.S., M. Fulton, and J.T. Scott, Jr. (1993) "The Inconsistency of Land Values, Land Rents, and Capitalization Formulas." *American Journal of Agricultural Economics* 75(1): 147–55.

Cochrane, J.H. (1991) "Volatility Tests and Efficient Markets." *Journal of Monetary Economics* 27(3): 463–85.

Constantinides, G.M. (1990) "Habit Formation: A Resolution of the Equity Premium Puzzle." *Journal of Political Economy* 98(3): 519–43.

Epstein, L.G. and S.E. Zin. (1991) "Substitution, Risk Aversion, and the Temporal Behavior of Consumption Growth and Asset Returns II: An Empirical Analysis." *Journal of Political Economy* 99(2): 263–86.

Falk, B. (1991) "Formally Testing the Present Value Model of Farmland Prices." *American Journal of Agricultural Economics* 73(1): 1–10.

Falk, B. and B.S. Lee. (1998) "Fads versus Fundamentals in Farmland Prices." *American Journal of Agricultural Economics* 80(4): 696–707.

Fama, E.F. (1970) "Efficient Capital Markets: A Review and Theory of Empirical Work." *Journal of Finance* 25(2): 383–417.

Fama, E.F. and K.R. French. (1988) "Permanent and Temporary Components of Stock Prices." *Journal of Political Economy* 96(2): 246–67.

Featherstone, A.M. and T.G. Baker. (1987) "An Examination of Farm Sector Real Estate Dynamics: 1910-85." *American Journal of Agricultural Economics* 69(3): 532–46.

Flavin, M.A. (1983) "Excess Volatility in the Financial Markets: A Reassessment of the Empirical Evidence." *Journal of Political Economy* 91(6): 929–56.

Hanson, S.D. and R.J. Myers. (1995) "Testing for a Time-Varying Risk Premium in the Returns to U.S. Farmland." *Journal of Empirical Finance* 2(3): 265–76.

Harvey, A.C. (1989) *Forecasting Structural Time Series Models and the Kalman Filter* New York: Cambridge University Press.

Irwin, S.H. and R.L. Colling. (1990) "Are Farm Asset Values Too Volatile?" *Agricultural Finance Review* 50(1): 58–65.

Just, R.E. and J.A. Miranowski. (1993) "Understanding Farmland Price Changes." *American Journal of Agricultural Economics* 75(1): 156–68.

Kleidon, A.W. (1986) "Variance Bounds Tests and Stock Price Valuation Models." *Journal of Political Economy* 94(5): 953–1001.

Kocherlakota, N.R. (1996) "The Equity Premium: It's Still a Puzzle." *Journal of Economic Literature* 34(1): 42–71.

Lee, B.S. (1995) "The Response of Stock Prices to Permanent and Temporary Shocks to Dividends." *Journal of Financial and Quantitative Analysis* 30(1): 1–22.

_____ (1998) "Permanent, Temporary, and Non-Fundamental Components of Stock Prices." *Journal of Financial and Quantitative Analysis* 33(1): 1–32.

Lence, S.H. (2003) see Chapter 15, this volume.

_____. (2001) "Farmland Prices in the Presence of Transaction Costs: A Cautionary Note." *American Journal of Agricultural Economics* 83(4): 985–92.

Lence, S.H. and D.J. Miller. (1999) "Transactions Costs and the Present-Value Model of Farmland: Iowa, 1900–94." *American Journal of Agricultural Economics* 81(2): 257–72.

LeRoy, S.F. and R.D. Porter. (1981) "The Present Value Relation: Tests Based on Implied Variance Bounds." *Econometrica* 49(3): 555–74.

Mehra, R. and E.C. Prescott. (1985) "The Equity Premium: A Puzzle." *Journal of Monetary Economics* 15(2): 145–61.

Miller, D.J. (2003) See Chapter 14, this volume.

Moss, C.B. and J.S. Shonkwiler. (1993) "Estimating Yield Distributions with a Stochastic Trend and Nonnormal Errors." *American Journal of Agricultural Economics* 75(4): 1056–62.

Poterba, J.M. and L.H. Summers. (1988) "Mean Reversion in Stock Prices: Evidence and Implications." *Journal of Financial Economics* 22(1): 27–59.

Schmitz, A. (1995) "Boom-Bust Cycles and Ricardian Rent." *American Journal of Agricultural Economics* 77(5): 1110–25.

Shiller, R.J. (1981) "Do Stock Prices Move Too Much to be Justified by Subsequent Changes in Dividends?" *American Economic Review* 71(3): 421–36.

Stiglitz, J.E. (1990) "Symposium on Bubbles" *Journal of Economic Perspectives* 4(2): 13–18.

Tirole, J. (1985) "Asset Bubbles and Overlapping Generations." *Econometrica* 48(6): 1499–1528.

USDA/ERS (U.S. Department of Agriculture, Economic Research Service). (2002a) Farm Income Data. Internet Website: www.ers.usda.gov/Data/FarmIncome/.

_____. (2002b) Farm Balance Sheet Data. Internet Website: www.ers.usda.gov/Data/FarmBalanceSheet/.

ENDNOTES

[i] Cochrane (1991) identifies key differences between fads and bubbles along with defining each concept. Rational bubbles or bubbles are defined as the case where the price of an asset today is high only because investors believe that the selling price will be high tomorrow (Stiglitz 1990). Fads are often referred to as waves of irrational optimism and pessimism (Cochrane 1991). Fad-price deviations will be slowly reversed, whereas bubble-price deviations are expected to last forever. In both cases, price deviates from the asset pricing fundamentals due to noise, feedback trading, irrational expectations, or some other inefficiency.

ii Nonstationarity is typically defined based on the autoregressive process. For example, the first-order autoregressive process for the time series y_t is expressed as

$$y_t = \alpha_0 + \alpha_1 y_{t-1} + \varepsilon_t$$

where y_{t-1} is the lagged value of the time series (or the value in previous period), α_0 and α_1 are estimated parameters, and ε_t is an error term. The time series is stationary if $|\alpha_1| < 1$. If $|\alpha_1| = 1$ the process is said to possess unit roots, or be nonstationary.

Chapter 10

Hysteresis and the Value of Farmland: A Real-Options Approach to Farmland Valuation

Calum Turvey
Rutgers University

INTRODUCTION

Despite the prominence of the present-value criteria in determining land values, researchers have been unable to show a time-series correspondence between cash flow from agriculture and the market value of farmland (Featherstone and Baker 1987; Falk 1991; Clark et al. 1993; Hanson and Myers 1995; Falk and Lee 1998; Weersink et al. 1999). The recent debate between Roche (2001) and Falk et al. (2001) over whether departures of land prices from their fundamental values is caused by bubbles or fads illustrates the variety of explanations applied to, and the economic significance of, the farmland-price problem. The notion that land prices are not correlated with fundamental economic information is problematic when the present-value model (PVM) is used in economic or an applied context. The conventional approach to pricing farmland is to estimate current and future cash flows and discount those cash flows using an appropriate risk-adjusted discount rate to its present value (e.g., Baker et al. 1991). If the net present value (NPV) rule is not sufficiently robust to explain the most rudimentary of agricultural investments, then there is a need to explore a more general theory of asset valuation.

One of the key elements largely ignored in the land-capitalization formula is the impact of uncertainty on land prices. How the sellers and buyers of farmland deal with uncertainty plays an important role in the market pricing of farm assets. For example, Turvey et al. (1992) use a portfolio-choice model to show how the rental rate of farmland will increase as the expected marginal-value product of farmland increases with uncertainty. Featherstone and Baker (1988) have shown that land values increase with rental rates. If rental rates increase with uncertainty, and farmland values increase with rental rates, then it follows that farmland prices increase with uncertainty. However, neither the portfolio-selection model nor the econometric land-price model adequately explains the farmland-price puzzle caused by a persistent wedge between the PVM, which is believed to be economically rational, and actual farmland prices.

New models of capital investment based on the theory of real options have been able to explain many peculiar anomalies in investment theory. A real option has a broad definition as summarized in Trigeorgis (1993) and Amran and Kulatilaka (1999). In general, a real option represents the value to a firm of having the flexibility to accept, reject, or postpone new investment opportunities. As new information arrives, uncertainty about future cash flows gradually resolves and management may have significant flexibility to alter its strategy in order to capitalize on improved upside potential, while limiting or mitigating the downside relative to the manager's initial expectations (Trigeorgis 1993). This flexibility is represented by put or call options embedded in the investment opportunity with an underlying asset being the gross project value of expected operating cash flows (Trigeorgis 1993). The theory of real options can show how uncertainty explains the acceptance of projects that have a negative net present value (NPV) or the rejection of projects with a positive NPV(Dixit and Pindyck 1994).

A special case of the real-options framework includes a behavioral characteristic called hysteresis. While hysteresis has several economic interpretations (Katzner, 1999) it mostly relates to economic conditions in which the past is related to the present in a number of ways. Persistent or ergodic price structures, for example, can relate past prices to the present and into the future. The belief that a reduction in the price of a traded security will be followed eventually by a rise in its value is an example of a hysteresis phenomenon that would cause a seller to resist selling immediately. Bubbles and fads that result in extra fundamental valuations come from an extreme form of hysteresis in which rational beliefs about future outcomes become a self-fulfilling prophecy. The importance of hysteresis in a real-options framework is that it creates a zone of inactivity. For example, models by Dixit (1989), Krugman (1989), and Martzoukos (2001) explore models in which the presence of sunk costs, brought about by irreversible investments in the presence of large exchange-rate fluctuations, can affect firms' decisions about exit and entry into foreign markets that are not reversed when the exchange rate returns to its previous level. Ansic and Pugh (1999) note that a firm's exit decision from a market is determined not only by its current trading position, but also by its expected value of remaining in the market. While in the market, exiting means that expected future profits cannot fall below zero, but the firm also forgoes any opportunity to increase future firm value. In an industry influenced by exchange-rate uncertainty, this means that the higher the volatility in exchange rates the higher is the option to remain in the industry.

This chapter presents an economic framework in which owners of farmland, having already made a sunk-cost investment for the perpetual rights to cash flow from the land and from other property rights, own an option on future capital gains relative to current fundamental values. Like the Dixit (1989), Krugman (1989), Martzoukos (2001), and Ansic and Pugh (1999) trade models, the failure of farmers to sell land at the current present value of cash flows is determined by the opportunity costs associated with future capital gains. Hysteresis thus creates a wedge between the present value of current cash flows and the possibility of increased farmland values in the future. This put-option framework provides the owner with the right, but not the obligation, to sell the farmland at some future date when operating cash flows reach or exceed a specific value. If these options exist, the PVM usually applied to valuing farmland will be incorrectly specified, since it ignores this option value. The true value of farmland under a real-options framework is the fundamental (present) value of the land based on currently observed operating cash flows, including expected growth, plus an option on future capital gains above expectations. The farmland-price puzzle can plausibly be explained by an agricultural economy in which the buyers of farmland have to purchase all or part of the seller's option in order to induce a sale.

This chapter explores the proposition that the market value of land includes real options on future growth and capital gains. If uncertainty is a significant component of land value, the traditional present-value capitalization formula is misspecified. Using a Dixit and Pindyck (1994) framework, the real-options valuation of farmland is examined using agricultural data for Ontario, Canada. The present-value bid price of both land and its real-option value is calculated and then compared to observed land values. The results provide some support for a new theory of real-asset pricing that can explain observed deviation of farmland-market values from the PVM, which includes the possibility of bubbles.

BACKGROUND

The application of real-options theory to farmland investments has previously been discussed. For example, Cappozza and Helsley (1990) and Capozza and Sick (1994) cast the urban-rural land-price relationship within a real-options framework. In their models, farmland prices increase with the real option to convert farmland to urban uses. Risk is determined by urban rents, and the real-option value diminishes as the distance to urban centers increase. Titman (1985) and Quigg (1993) provide examples for pricing options to develop undeveloped land by setting development costs as the strike price, and rents (or sale) as the state variable. Bailey (1991) examines how shares of rubber and palm-oil companies in Malaysia are valued when production can be temporarily suspended if marginal costs exceed marginal revenues. Under shutdown provisions the firm's present value is zero and no positive cash flow is generated, yet the option to resume production at some future date—when uncertainty drives prices above marginal costs—results in a positive-share value. In a similar context, Dixit (1992) notes that in the early 1980s there must have been many farm families with asset returns below their marginal cost of labor and other costs. These families did not immediately sell the land and exit farming; instead they kept the farm alive on the chance that future cash flow would increase. This is the basis of the hysteresis argument used in this chapter. Under the conventional present-value rule, negative cash flows will result in an asset that has no value. In agriculture, however, we do not observe zero-valued land assets. Even land taken out of production because of low productivity will be put into production if prices increase to some trigger level. One can view marginal costs as the strike price on an option to produce agricultural commodities: When prices fall below marginal costs, production is abandoned. But there is always the possibility that price will increase at some future date, so the option to produce has value. With this option in place, land has value in excess of its present value, which is why we do not observe landowners accepting zero-valued bids for farmland even when that land generates no cash flow. Likewise, when prices are above marginal costs and productive land has a positive present value, we still do not observe land being sold at its present-value-bid price, even with growth expectations included.

The variety of explanations available for the discrepancies takes the PVM as given and solves for possible external influences. Fads and bubbles result from econometric models that include conventional capitalization as part of the econometrics and then attribute sustained deviations from the model as fads or bubbles (Featherstone and Baker 1987; Falk et al. 2001; Roche 2001). Models that attribute government programs examine the incremental cash flow from stabilization policies and the possibility of a reduced risk premium in the discount rate as exogenous and endogenous solutions to the capitalization model (Featherstone and Baker 1988; Weersink et al. 1999). Other models that examine time-varying discount rates consider the problem as one of the PVM being correct and solving the discount rate as the internal rate of return on the market value given cash-flow expectations (Hanson and Myers 1995; Weersink et al. 1999).

None of the many studies on farmland values has explicitly considered uncertainty as a source of the discrepancy between observed land values and the land-capitalization model. Yet when uncertainty is included in the present-value framework in the form of real options, it is found that the conventional PVM fails to represent the full value of the asset or investment. Bailey's (1991) result is a case in point. The failure of idle land to sell for zero dollars is another. In a similar context, the observation that some Ontario dairy farmers have paid excessive premiums for farmland on the expectation that future environmental restrictions may tie herd size to their land base, suggests that the premium paid by these farmers can be interpreted as the value of a real option to produce milk at some future date. While not considered explicitly in the various studies on farmland values, a finding of fads or bubbles is entirely consistent with Dixit's (1992) contention that bubbles arise from an extreme form of real options arising from hysteresis. In fact, Roche (2001) states that a bubble arises when the anticipation of increasing prices includes more market participants in pursuit of short-term capital gains. If a fad or bubble is viewed as a form of hysteresis, then its fervor can only arise if the landowners postpone sales so that demand exceeds supply. Furthermore, since the conventional land-capitalization model includes growth expectations, then surely fads and bubbles can arise only from the speculation that actual growth will exceed expected growth. Such speculation can only arise from uncertainty in future outcomes, and from optimistic buyers and sellers who recognize these outcomes. Therefore, any model that concludes in favor of fads or bubbles must also admit that hysteresis, uncertainty, and the possibility of extremely good outcomes is not a trivial component of farmland pricing.

THE FARMLAND INVESTMENT PROBLEM

The bid price, or fundamental value of farmland, is often based on the present-value structure

$$V(\pi) = \frac{\pi^*}{r - \alpha}, \tag{1}$$

where π^* is a rational expectation of cash flows generated from the land, α is the anticipated growth rate in cash flow, and r is an appropriate discount rate. This model has been tested quite extensively in the agricultural economics literature under the null hypothesis that there is no significant difference between the land-capitalization model and the observed market price for farmland V^* (i.e., $Ho: V(\pi) = V^*$). Rejecting this null hypothesis implies that some other, unknown economic is driving market values of farmland. The correct but unknown model can be specified as

$$V_t^* = V(\pi_t) + F_t, \tag{2}$$

where F_t represents a persistent economic that drives a wedge between the fundamental value of farmland $V(\pi_t)$ and the market value V_t^* at some particular point in time t. The exact source of this wedge is unclear from the existing literature. Speculative bubbles and fads (Featherstone and Baker 1987, Falk and Lee 1998), time varying discount rates (Hanson and Myers 1995, Weersink et al. 1999), or government programs (Featherstone and Baker 1988, Weersink et al. 1999) have all been proposed, but none have provided clear-cut explanations. [i]

Real options may provide a plausible explanation for the farmland-price wedge. In essence, this chapter proposes that the difference between market values and fundamental values is a problem of the optimal timing of a market transaction, given uncertainty about future values of the asset (or the underlying state variable from which asset values are derived). The key proposition of this chapter is that owners of land have a positively valued option to postpone the sale of land in the hopes of higher future capital gains. In order to induce a market transaction, the buyer must purchase all or part of this option from the seller. The option value will be above and beyond the land's fundamental value, and therefore the proposition offers a plausible explanation for the land-price puzzle. The key question asked in this chapter is whether uncertainty over future outcomes (cash flow) can cause the owners of capital to delay the sale of capital in the hope of increasing capital gains and real wealth.

To put the timing problem into perspective, reconsider what is being assumed in the classical farmland bid-price model defined by Equation (1). In order for Equation (1) to be valid, it must be assumed that all variables are known with reasonable certainty. As such, the owner of the land has the right to receive the present-value benefit of all future cash flows from the land. With this present value known, there is no benefit to postponing the decision, since the present value of selling now or one year from now is the same in present-value terms.[ii] Likewise, if the buyer can invest the amount $V(\pi)$ at the rate r, then he will be indifferent toward buying now or later.

When future cash flows are uncertain, the PVM no longer holds, at least in terms of the timing of the sale. For example, if cash flows evolve randomly over time according to the Brownian motion

$$d\pi = \pi(\alpha dt + \sigma dZ),\tag{3}$$

where π is current cash flow, α is the expected natural per-unit growth rate in cash flows of time, σ is the volatility of cash flow as measured by the standard deviation of its percentage change, and Z is a standard Wiener process, then by Ito's lemma the value of farmland will evolve stochastically according to

$$dV(\pi) = \frac{\pi}{r-\alpha}(\alpha dt + \alpha dZ) = V(\pi)(\alpha dt = \sigma dZ).\tag{4}$$

For convenience, we denote the present value of cash flow as $V(\pi)$ and the change in this present value as $dV(\pi)$, given a change in cash-flow expectations (and other variables).[iii] Likewise, I and dI represent the value of farmland and the change in the value of farmland given an expectation and change in expectation of π. We can write the land-price dynamic as

$$dI = I(\alpha_I dt + \alpha_I dZ_I).$$

If $dI = 0$, then the value of the land is fixed even though $dV(\pi) \neq 0$. When I is a present value $\alpha_I = \alpha$, $\sigma_I = \sigma$, $dZ_I = dZ$, and $dI = dV$. The change in the value of the assets is then perfectly correlated with the change in the present value of the cash flow. The identity $dI = dV$ is the standard assumption of the PVM and the condition being evaluated in this chapter. We also leave open, for the purpose of discussion, the possibility that $dV(\pi)$ and dI follow correlated Brownian motions, but they do not follow the exact present values. In this case, $dI \neq dV(\pi) \neq 0$.

THE EMERGENCE OF REAL OPTIONS ON FARMLAND

The operating hypothesis of this chapter is that the owners of capital own an option to postpone the sale of the asset under conditions of risk. This option is similar to an American put option, which provides the owner with the right, without obligation, to sell the land at some (unspecified) future date.[iv] The option value arises from a number of economic conditions (Dixit and Pindyck 1994, Abel et al. 1996, McDonald and Seigel 1986, Trigeorgis 1993, Amran and Kulatilaka 1999). First, the landowner has already made a sunk-cost investment in the land, entitling him to all future capital gains. Second, the decision to sell land is irreversible in that it cannot be reversed without cost. Once the land is sold, all rights to unanticipated future capital gains are forgone. Third, the decision to defer the sale is reversible, which means that any (capital gain) losses that might arise from misjudging future cash flows and probabilities can be mitigated in a significant way economically. Fourth, the future sale of land will occur only if incremental capital gains exceed zero by a certain amount.

In the short run, the option to postpone does not guarantee a capital gain, and decreases in land values may be observed. In fact, the risk of capital losses is not totally eliminated until the land is sold. Under the third condition, however, there exists with known probability a level of cash flow x^* that optimizes the capital gain. Under the real-options hypothesis, landowners will delay the sale of land until x^* occurs—even if (in the short run) π, hence $V(\pi)$, declines. Dixit's (1992) hysteresis argument explains this behavior.[v] The value of the option to postpone the sale of land until π^* occurs is defined by $F(\pi^*)$. The problem stipulates only at what level of cash flow (given current levels) it is best to sell; it does not determine when π^* or the sale will occur. In this context, the real option is similar to a perpetual option as described by Merton (1973). Later we derive and calculate formulas that determine π^* and $F(\pi^*)$.

Let $I = V(\pi)$ define the current land value evaluated at current cash flow π, and $I(\pi^*)$ the present value of land evaluated at π^*. The capital gain when π^* occurs is $\left[I(\pi^*) - I \right] > 0$. The decision faced by the landowner is to sell the land immediately, an irreversible decision netting zero additional capital gains, or postpone the sale until π^* occurs and sell at that time. The value of the option is then determined by the boundary condition $F(\pi^*) = \text{Max}\left[0, I(\pi^*) - 1 \right]$.

THE OPTION TO BUY FARMLAND

To this point, we have ignored the buyer's position. In much of the literature on real options, the position is taken that the buyer-investor can postpone a capital purchase when future outcomes are uncertain. This is equivalent to an American call option on the real asset (the right to purchase). This literature requires the same conditions of irreversibility, or at least costly reversibility (Abel et al. 1996). For example, the investment in a plant is irreversible in that it cannot be used for any other purpose. If cash flows on the project decrease, then the investment will have lost value. Hence, by delaying the investment, future cash-flow uncertainties and ambiguities can be somewhat resolved (McDonald and Siegel 1986. If the market becomes too competitive, the postponement decision can be reversed and the investment can be made immediately (Dixit and Pindyck 1994). The risk to the buyer is that by postponing the investment, interim cash flow will have been lost.

The boundary condition for the real option to buy an asset is isomorphic to an option to sell (McDonald and Seigel 1986, Trigeorgis 1993), that is,

$$F\left(\pi^*\right) = \text{Max}\left[0, V\left(\pi^*\right) - I\right].$$

For the call option, the investment is made when the net present value (NPV) equals $V\left(\pi^*\right) - I$, or it is not made at all. For the put option, the sale is made when land prices exceed current prices by $V\left(\pi^*\right) - 1$, or when it is not sold.

The problem with the call option is that it assumes that the buyer acts as a monopolist and that the investment can be postponed indefinitely while I is held constant (Dixit and Pindyck 1994). That is, in reference to Equation (4), $dI = 0$ while $dV > 0$. If I is a present value, as is the case of capital assets like farmland, then I will not remain constant at all.[vi] In fact, when π^* occurs

$$I\left(\pi^*\right) = V\left(\pi^*\right) \text{ and } dI\left(\pi^*\right) = dV\left(\pi^*\right).$$

The call-option value is based on

$$F\left(\pi^*\right) = \text{Max}\left[0, V\left(\pi^*\right) - I\left(\pi^*\right)\right] = \text{Max}\left[0, V\left(\pi^*\right) - V\left(\pi^*\right)\right] = 0.$$

Hence, when the fundamental value of an asset is equal to the present value of future cash flows, given the prevailing economic conditions, there is no (or little) option value to waiting.[vii] In contrast, the value of the seller's put option is measured relative to

$$F\left(\pi^*\right) = \text{Max}\left[0, V\left(\pi^*\right) - I\right],$$

where I represents the current value of farmland (or a fixed point in time). The economic result suggests that in farmland markets the owners of land hold a valuable option to postpone the sale while buyers hold no such option to postpone the purchase.[viii]

REAL OPTIONS AND THE MARKET VALUE OF FARMLAND

The existence of the put option suggests that the value of the land to the owner is not simply $V(\pi)$, but $V(\pi)$ plus the option to future (unanticipated) capital gains, $V^* = V\left[(\pi) + F(\pi^*)\right]$, while the value to the buyer is $V^* = V(\pi)$. If both buyer and the seller have symmetric information and can agree upon $V(\pi)$, then negotiating a land transaction involves only $F(\pi^*)$. What is not known is how negotiations will evolve and what proportion of the option the buyer must purchase in order to induce a sale.

The farmland transaction can be viewed in terms of a Stackelberg leader-follower game with symmetric information (Lambrecht 2001). In this game, the seller (being the leader) will only agree to a sale if he receives a bid premium. The purchaser is the follower, and the first pass bid will equal V_t, which is the current fundamental (present) value of the land. The leader (as a monopolist) sets $V^*_t = V_t + F(\pi)$, where $F(\pi)$ is the current value of

$$F(\pi^*) = \text{Max}[0, E(V_{t+n} - V_t)] > 0,$$

as the initial offer.[ix] The bid-ask spread after the initial pass through is equal to the option value. Since neither V_t^* or V_t are acceptable equilibrium solutions, a second-best solution must be sought. The second-best solution emerges from a sequential solution of the reaction function of the counter party. The seller will internalize the requirement of the buyer for an immediate sale at some price $V_t < v_t < V_t^*$, while the buyer internalizes the requirement of the seller for some price $V_t > v_t \geq V_t^*$ [i.e., the second-best equilibrium will result for some $\lambda(0 < \lambda < 1)$, such that the seller receives a bid premium of $\lambda F(V)$]. A value of $\lambda = .5$ represents a Nash equilibrium, while a value of $\lambda > .5$ represents a greater degree of market power by the seller (the land-demand curve may be highly inelastic) and $\lambda < .5$ represents a greater degree of market power by the buyer (the land-demand curve may be highly elastic). With the bid premium $\lambda F(V)$, the transacted value of land will be

$$v_t = V_t + \lambda F(V_t).$$

The reduced-bid premium $\lambda F(V_t)$ represents the certainty equivalent value of receiving v_t immediately and with certainty, rather than V_t^* with uncertainty. From the perspective of the buyer, the seller has not fully extracted all of the option value. The buyer pays v_t for an asset with an expected worth of V_t^*. The difference $(1-\lambda)(FV)$ represents the expected value of the capital gain above v_t that accrues to the buyer.

Why the option that leads to such a negotiation or equilibrium exists for farmland is of course debatable. If commodity prices are low relative to history, farmers might see an increase as being imminent. When commodity prices are rising, farmers could examine the economic conditions prior to the increase and could conclude that the demand forces are more than transient. Optimistically, they could conclude that prices will continue to rise. In terms of stock-market behavior, Shiller (2000) argues that factors such as the Internet, baby-boom demand, and herd mentality all contributed to the rise of the DOW index in the late 1990s. In agriculture, new cost-reducing technologies, free-trade agreements, biotechnologies and pharmaceuticals, unfound nutritional values, and future conversion of lands outside of agriculture are all possibilities that can give rise to option values. In the 1970s and 1980s the common belief that land provided a hedge against inflation gave rise to an option on the real purchasing power of the dollar, and this option value became built into the price of land. One cannot discount the role that government programs might play either. For example, when prices fall, stabilization or even disaster relief could be available to limit the downside risk so that there will always be a floor to the price of land, leaving the upside to roam with the markets.[x]

MARKET PRICES, BID PRICES, AND REAL-OPTIONS VALUES

This section develops the real-options pricing model along the lines of Dixit and Pindyck (1994). (It is assumed throughout that $\lambda = 1$.) In their Chapter 6, Dixit and Pindyck (1994) provide a solution based on market arbitrage, contingent-claims analysis, and dynamic programming. The dynamic-programming approach is developed in this chapter. The solution to the real-options valuation involves two steps. The first step is to determine a formula that describes the value of land as a function of the stochastic variables. The value of land is denoted by $V(\pi)$ where π, defined as cash flow, is the stochastic variable that determines value. Once the basic land-price formula is determined, the second step determines the real-option value. It is assumed that cash flows evolve over time according to Equation (3); by Ito's lemma, the value of farmland evolves according to Equation (4).

The Dixit-Pindyck (1994) solution is based on the notion that the real-option value fluctuates with time and risk. Given that $V(\pi)$ describes the value of land given π, α, and r, the option value $F(\pi, t)$ is given by the Bellman equation

$$F(\pi,t) = E\left[F(\pi,t) + dF(\pi,t) \right] e^{-rdt}. \qquad (5)$$

Note that Equation (5) includes the time-designated variable t in the option value $F(\pi,t)$, whereas the original Dixit-Pindyck model does not. I will derive the general partial differential equation for option prices with t included. (This will be used later to price real options on finite-lived investments using a Black-Scholes (1973) option-pricing model. In Equation (5) the current options price is given by the expected value of the option price plus the change in its value over time. This is then discounted to the present. Applying Ito's lemma to Equation (5) using $d\pi$, and using the fact that $1 - rdt$ is equivalent to e^{-rdt},

$$F[F(\pi)] = F(\pi) = \left[\tfrac{1}{2}\sigma^2\pi^2 F'' = \alpha\pi F + F_t'\right](1-r)Fdt, \text{ or,}$$

$$\tfrac{1}{2}\sigma^2\pi^2 F'' + \alpha\pi F'' + F_t' - rF = 0. \tag{6}$$

Setting $F_t' = 0$ in Equation (6) provides the stochastic differential equation used to solve for the real-option price $F(\pi)$. [Note that $F(\pi) \neq F(\pi,t)$]. [xi] To obtain this solution we add three boundary conditions. These are

$$F(0) = 0, \tag{7}$$

$$F(\pi^*) = V(\pi^*) - V(\pi), \text{ and} \tag{8}$$

$$F'(\pi^*) = V'(\pi^*). \tag{9}$$

The condition in Equation (7) says that if cash flows are zero, the option will be zero. The condition in Equation (8) is a value-matching condition, which says that at some level of cash flow π^*, the value of the option must equal the present value of the investment. Also in Equation (8), $V(\pi) = I = \dfrac{\pi}{r-\alpha}$, which is the current present value of expected cash flows. Since V is the current reference point, the right-hand side of Equation (8) gives the capital gain when the π^* trigger is reached. The condition in Equation (9) is the smooth-pasting condition, which says that the optimal time to sell occurs for some π^* such that the incremental gain in options value exactly equals the incremental gain in NPV. The smooth-pasting condition ensures that, at some point, the option value will become tangent to the options-payoff curve (Figure 10.1).

The equations in Appendix 10.A show that the solution to Equation (6) is

$$F(\pi^*) = A\pi^{*\beta}, \tag{10}$$

with β given by Equation (10.A3) and A given by Equation (10.A5). The number given by π^*, and therefore the values of $F(\pi)$ and $F(\pi^*)$, depends on initial conditions. These initial conditions are represented or captured by A. This means that the value of $\pi_{t-1}^* \neq \pi_t^* \neq \pi_{t+1}^*$ if $\pi_{t-1} \neq \pi_t \neq \pi_{t+1}$. This implies that the real-option value is a continually changing value over time since the initial bid price changes with π. If π_t is high but then decreases, subsequent calculations of π^* and $F\left(\pi^*\right)$ will be lower, and if π increases in the next period, the values for π^* and $F\left(\pi^*\right)$ will also increase. It follows that the current value of the option to wait $F\left(\pi\right)$, will also be changing over time. Keep in mind that $F\left(\pi\right) \neq F\left(\pi^*\right)$. $F\left(\pi\right)$ is the value of the option to wait when current cash-flow expectations are π, (i.e., the current value of an option is to postpone the sale until π rises to π^*). In contrast, $F\left(\pi^*\right)$ is the intrinsic value of the option at π^*.

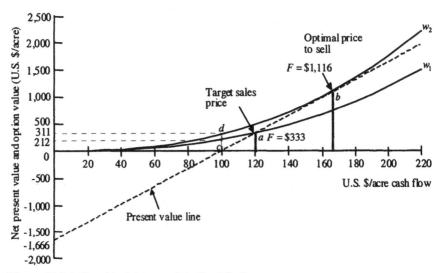

Figure 10.1 A Graphical Approach to Real Options

The calculus of option values (Dixit and Pindyck 1994) reveals that the option value will increase as risk increases, as growth rates increase, or as costs of capital decrease. As risk increases (due to commodity-price volatility for example), the optimal trigger π^* will be higher implying that the sale will be postponed even longer. Likewise, an increase in the expected growth rate in cash flow from farming or a decrease in the discount rate will increase the value of the option.

An Illustrative Example of Real-Option Values

How uncertainty creates the option value $F\left(\pi\right)$ is discussed in relation to Figure 10.1, in which there are two curved lines, w_1 and w_2, and a hatched line. The hatched line represents the NPV of the farmland investment for a given rate of discount and cash flows as indicated on the X-axis. Suppose that the per-acre current cash flow from one acre of land is $100 and the market is fair in the sense that the maximum bid price will simply be the perpetual present value of the cash flows with zero growth. For per-acre cash flows of $100, the NPV will equal zero (i.e., $NPV = \pi / r - V = 0$). From this position, cash flows in the future can increase or decrease. In Figure 10.1, the per-acre range of possible outcomes is $0 to $220. Using a discount rate of 6 percent and zero growth $\alpha = 0$, the bid price for land is $1,666 = $100/$0.06$. If cash flow falls to zero, the land is worthless and the value of the investment will be lost. This is why the present-value line intersects the Y-axis at –$1,666. On the positive side, the value of land could increase to $3,666 if per-acre cash flow increased to $220. This would imply a per-acre gain over the initial condition of $2,000.

Suppose that when per-acre cash flow is $100, a landowner is offered the maximum bid price of $1,666. The farmer must decide whether or not the offer price is fair enough to induce a sale. Assume that variance in the annual percentage change in cash flow $\alpha = .1637$. The farmer realizes that by selling immediately he eliminates the downside risk associated with a decrease in cash flow but recognizes that uncertainty can also cause an increase in the price of land, and by postponing the sale, there is a real possibility of capital gains.

The farmer's decision comprises a real option to postpone the sale of the land until some future date or trigger. This real option gives the farmer the right, but not the obligation, to sell the land at some price in the future. The curves w_1 and w_2, represent the value of this option under two separate and mutually exclusive possibilities. The first option sets a goal or target exogenously. The land will be sold if cash flows equal or exceed $120 per acre. At $120 per acre the value of land will increase to $2,000 per acre and the capital gain (the intrinsic value of the option) will be equal to $333 per acre.

But what is this option worth when the current conditions are $100 per acre? In Figure 10.1 at point c, which is read off the option curve w_1 and is calculated by setting π^* in Equation (10.A5) and (10.A6) or Equation (10) equal to $120, the per-acre option value is $212. The difference between the $333 and $212 is due to uncertainty. However, a conventional interpretation applies—if current expected cash flow is $100 per acre, the expected per-acre value of postponing the sale until expected cash flow increases to $120 is $212. This suggests that the farmland is not worth its $1,666 fundamental value, rather the farmland is worth the fundamental value plus the value of the option to wait. In other words, the value of the farmland is $1,878 (i.e., $1,666 + $212).

Setting a target at $120 is arbitrary. The curve w_1 intersects the present-value line at a, but the optimality condition from smooth pasting forces the slope of the option-value curve to equal the slope of the present-value curve. (In Figure 10.1, twist the option-value curve upwardly until it is tangent to the present-value line.) Using Equation (10.A4), the calculated per-acre value of π^* is $167. At this point of tangency, the per-acre value of farmland is $2,782 with an option value of $1,116. We verify the value of this put option by substituting $\pi^* = \$167$ into Equation (10.A5) and then solving Equation (10.A6) or Equation (10). In this instance, the farmer has the right, but not the obligation, to postpone selling the land until expected per-acre cash flows increase from $100 to $167.

If the optimal strategy for maximizing capital gains is to postpone the sale decision until per-acre cash flow equals $167, what is the value of the option when per-acre cash flow is $100? To calculate the current value of the option, solve Equation (10.A5) using π^*, which gives the value for A. Next, using this value for A, substitute $100 for π^* in Equation (10.A6) or in Equation (10). This real-option value is $311 and corresponds to d in Figure 10.1 when per-acre cash flow is $100.

The interpretation of this option value is as follows. The landowner has the right to postpone the sale of the land into the future. Given current conditions, it is optimal to postpone the sale until per-acre cash flow equals $167. But $167 occurs with uncertainty, and may not occur at all. Since per-acre cash flow of $167 will occur farther into the future than will any lesser value, the seller will want to know the equivalent price of the option in current dollars. The per-acre option price of $311 is the current value of receiving a gain of $1,116 in the future. That is, given per-acre current cash flows of $100, risk of 16.37 percent, zero growth, and a discount rate of 6 percent, the owner would be willing to give up a future gain of $1,116 for $311 current dollars. The per-acre value, $F(100) = \$311$, is therefore the opportunity cost of forgoing future capital gains by selling immediately.

Because the target at b is optimal, the real per-acre value of farmland when expected cash flows are $100 is equal to its fundamental value plus its real-option value (Figure 10.1). This equals $1,977 ($1,666 + $311), which means that the seller will be indifferent to receiving $1,977 today for the land, or will postpone the sale under uncertain conditions until expected per-acre cash flows approach $167.[xii]

Now consider the buyer's position. With per-acre cash flow of $100 the farmer's maximum bid price is $1,666, which is also equal to the present-value cash flows. It will not be lower because the seller has the same information as the buyer. The buyer's NPV is zero. Because of the option value, the seller does not have to accept the bid, but can wait to see if cash flows increase. If cash flow increases to $120 per acre, the present value and bid price will increase to $2,000. The seller has achieved a $334 per-acre capital gain, but the NPV to the buyer is still zero. If the seller holds off until cash flows increase to the optimum cash flow of $167 per acre, the bid price facing the buyer will be $2,783, the seller will have gained $1,117, and the NPV facing the buyer is still zero. The key point here is that the buyer is not advantaged by postponing the purchase, since he will always face (at most) a zero-NPV investment. The seller, on the other hand, has every incentive to postpone the sale. If cash flow falls below $100 per acre in the short run, with hysteresis the seller will delay the sale while waiting for cash flow and land value to increase.

The graphical approach indicates several key observations regarding the value of farmland. The most obvious is that the true value of farmland is not just the present value of cash flow; it must also include the option value of future capital gains. In Figure 10.1 the expected per-acre value of these gains is $311. Under this theory, a bid of $1,666 using conventional bid-pricing rules will understate the land's true value. That is, while the buyer may bid $1,666 per acre, the seller can justify asking $1,977. The real-option value has created a wedge between the bidding and asking prices. In order for a transaction to take place, the buyer will have to purchase at least part of the option to future capital gains from the seller. Consequently, the actual value at which farmland will transact will be in excess of the conventional bid price.

REAL OPTIONS AND THE FARMLAND INVESTMENT

In the previous section, it was argued that the seller of land owns an option to postpone his or her sale until cash flows increase. The option to wait has value, since by waiting future capital gains may be higher. When the strike value π^* is reached the seller will sell the land for $V\left(\pi^*\right)$, its exact present value, and the buyer will purchase the land for $V\left(\pi^*\right)$. Consequently the buyer has no option to postpone, since no matter what the level of cash flow is he will always pay its fundamental bid price value for a zero-NPV investment. It was then argued that, since the buyer had no option to wait while the seller did, to induce a sale the buyer would have to purchase part of the option from the seller. Buying the option, or part thereof, implies that observed land prices will always be higher, or at least no lower, than the fundamental value of future cash flows. In this section, this proposition is evaluated using historical land, cash flow, and interest rates for the province of Ontario.

The data for 1971 to 1998 are presented in Table 10.1. Columns (1), (2), (3), and (4), are the nominal per-acre value of land and cash income from farming. The prime rate is reported in nominal terms. Since the prime rate represents the best rate to borrowers, 3 percent was added to account for risk and to ensure a positive expected return to equity.[xiii]

Table 10.1, Column (5) is the deterministic present value $V(\pi)$. The first-bid price is calculated for 1975 and the expected cash income $E(\pi)$ equals a (.4, .3, .2, .1) weighted average of the previous four years [Table 10.1, Column (1)] with the most recent year having the highest weight. The discount rate is the rate in Table 10.1, Column (4) plus 3 percent. The annual growth rate α equaled the average value of $\ln\left(\pi_t / \pi_{t-1}\right)$ during the 1971 to 1998 period (4.6 percent). Likewise, volatility $\sigma = .164$, was computed as the standard deviation of $\ln\left(\pi_t / \pi_{t-1}\right)$ from 1971 to 1998.

Table 10.1, Column (6), calculates the difference between the market value of farmland [Table 10.1, Column (2)] and the fundamental value [Table 10.1, Column (5)]. The market value falls below the fundamental value of the land in only two of the years. On average, the per-acre market price exceeds the fundamental price by $618 with a maximum of $1,329. Table 10.1, Column (7) presents the value of the real option at current cash flows— $F(\pi) = A\pi^B$ —from Equation (10) and A is calculated from Equation (10.A5) using π^*. Table 10.1, Column (8), is the economic value of land, which is equal to the fundamental value plus the option value $V(\pi) + F(\pi)$ from Table 10.1, Columns (5) and (7).

Table 10.1: Real-Options Valuation of Ontario Farmland (Means and Standard Deviations from 1975 to 1998)

Year	Land price/ acre	Cash flow/acr $E(\pi)$	Prime rate (I)	PV bid price $V(\pi)=I$	Market - $V(\pi)$	Option value $F(\pi)$	$V(\pi)+F(\pi)$	Optimum put option strike value π^*	$\pi[E]*\pi]$	$V(\pi^*)$	Option value $F(\pi^*)$	$(V(\pi)+ F(\pi)) /V(\pi)$
(1)	(2)	(3)	(4)	(5)	(6)	(7)	(8)	(9)	(10)	(11)	(12)	(13)
1971	325	25.36	6.48									
1972	352	31.83	6.00									
1973	440	45.62	7.65									
1974	583	56.82	10.75									
1975	749	59.63	9.42	580.26	168.74	164.10	744.36	99.71	2.20	1276.76	696.49	1.28
1976	893	60.08	10.04	630.77	262.23	168.38	799.15	111.76	2.10	1325.01	694.24	1.27
1977	995	56.31	8.50	839.21	155.79	260.87	1100.09	138.31	2.39	2006.45	1167.24	1.31
1978	1138	64.77	9.69	719.69	418.31	198.38	918.08	125.31	2.15	1550.71	831.02	1.28
1979	1301	73.09	12.90	538.40	762.60	115.99	654.38	110.11	1.81	975.35	436.95	1.22
1980	1528	65.72	14.25	521.51	1006.49	103.54	625.05	113.87	1.73	900.64	379.13	1.20
1981	1695	77.27	19.29	377.71	1317.29	59.19	436.90	102.61	1.54	580.20	202.50	1.16
1982	1659	71.05	15.81	504.86	1154.14	92.25	597.11	118.47	1.65	833.94	329.08	1.18
1983	1542	85.65	11.17	753.73	788.27	183.36	937.09	141.34	1.96	1478.50	724.77	1.24
1984	1509	97.19	12.06	742.21	766.79	169.07	911.28	145.62	1.88	1392.75	650.54	1.23
1985	1402	73.62	10.58	963.73	438.27	245.62	1209.34	175.46	2.03	1954.71	990.98	1.25
1986	1288	107.56	10.52	929.34	358.66	238.08	1167.42	168.67	2.04	1892.20	962.86	1.26
1987	1288	103.61	9.52	1176.55	111.45	329.43	1505.98	203.20	2.18	2567.55	1391.01	1.28

(*Continued*)

Table 10.1: Real-Options Valuation of Ontario Farmland (Means and Standard Deviations from 1975 to 1998) (continued)

Year	Land price/ acre	Cash flow/acr $E(\pi)$	Prime rate (I)	PV bid price $V(\pi)=I$	Market - $V(\pi)$	Option value $F(\pi)$	$V(\pi)+F(\pi)$	Optimum put option strike value π^*	$\pi[E*\pi]$	$V(\pi^*)$	Option value $F(\pi^*)$	$(V(\pi)+F(\pi))/V(\pi)$
(1)	(2)	(3)	(4)	(5)	(6)	(7)	(8)	(9)	(10)	(11)	(12)	(13)
1988	1489	105.61	10.83	1063.81	425.19	265.66	1329.48	196.15	2.00	2125.95	1062.13	1.25
1989	1908	96.85	13.33	871.54	1036.46	182.66	1054.20	182.10	1.78	1552.90	681.36	1.21
1990	2147	90.70	14.06	818.11	1328.89	164.18	982.29	177.05	1.74	1421.43	603.32	1.20
1991	2303	81.63	9.94	1162.18	1140.82	313.13	1475.31	204.86	2.12	2459.09	1296.91	1.27
1992	2184	104.91	7.48	1529.05	654.95	537.73	2066.78	242.92	2.71	4136.65	2607.60	1.35
1993	2144	83.18	5.94	2176.91	-32.91	974.74	3151.65	346.81	3.68	8008.20	5831.29	1.45
1994	2134	76.54	6.88	1710.98	423.02	655.05	2366.03	268.67	2.98	5099.92	3388.94	1.38
1995	2188	84.41	8.65	1203.47	984.53	368.19	1571.65	199.59	2.36	2835.45	1631.98	1.31
1996	2384	92.38	6.06	1881.88	502.12	823.44	2705.33	298.17	3.56	6691.94	4810.05	1.44
1997	2471	92.17	4.96	2563.07	-92.07	1420.90	3983.97	462.05	5.38	13786.45	11223.38	1.55
1998	2538	87.97	6.60	1783.29	754.71	711.83	2495.12	279.73	3.14	5597.58	3814.29	1.40
Mean	1,703.21	82.99	10.35	1085.09	618.11	364.41	1449.50	192.19	2.38	3018.76	1933.67	1.29
Std. dev.	524.29	15.46	3.41	575.16	420.26	332.07	901.64	88.23	0.86	3010.79	2474.20	0.10
Minimum	325	25.36	4.96	377.71	-92.07	59.19	436.90	99.71	1.54	580.20	202.50	1.16
Maximum	2,538.00	107.56	19.29	2563.07	1328.89	1420.90	3983.97	462.05	5.38	13786.45	11223.38	1.55

Table 10.1, Column (9), presents the optimal cash-income level π^* that would trigger a sale. This is the smooth-pasting condition. It is calculated from Equation (10.A4) where the variable $V(\pi)$ is set equal to the prevailing fundamental land value as computed in Table 10.1, Column (5). In Table 10.1, Column (10), the ratio of $\pi^* / E(\pi)$ provides a multiple of the current expected cash income to induce the selling of land (optimally).

Table 10.1, Column (11), calculates $V\pi^*$, which is the value of land evaluated at π^*, and Table 10.1, Column (12) gives the real-option values $F(\pi^*)$. The option values are calculated from Equation (10) but it is easy to see that Table 10.1, Column (12), is simply the difference between Table 10.1, Column (11), and Table 10.1, Column (5). The results indicate that the per-acre trigger-cash level averages $192, while actual cash averages $83. Expected cash flow will have to increase by an average of 2.38 times, with a minimum of 1.54 and a maximum of 5.38, in order to induce an optimal sale. This result indicates that (relative to fundamental value) there has been a persistent real option on farmland in Ontario. The mean per-acre real-option value π^* is $1,933 with a range from $202 to $11,223.

Table 10.1, Column (13), calculates the ratio $\left[V(\pi) + F(\pi)\right]/V(\pi)]$ that measures the percentage increase in the fundamental land value due to real options. The results indicate that, on average, the option value will increase fundamental land prices by about 29 percent. But there is variability in the ratio, which ranged from a low of about 16 percent to a high of 55 percent. According to the Stackelberg theory, these values should represent a maximum since the transaction price is hypothesized to be lower than $V(\pi)=F(\pi)$ for $\lambda \neq 0$.

To interpret these results further, the data for 1998 in Table 10.1 tell the following story. In 1998, the market price of land was $2,538. Based on expected per-acre cash flows of $89.12 ($= .1 \cdot \$76.34 + .2 \cdot \$84.41 + 3 \cdot \$92.38 + .4 \cdot \$92.17$), a discount rate of 9.6 percent ($= 6.6$ percent $+ 3$ percent) and an anticipated growth of 4.6 percent, the fundamental value of land was calculated to be $1,783. There was a per-acre $754 wedge between the land's market and fundamental values. Based on current cash-flow expectations of $89.12, the optimal strategy for the landowner was to wait until cash flow increased to $279.73 before selling the land. The value of this option was $3,814.29. However, instead of waiting for cash flow to rise, the present value of the option given current cash flow expectations was $711.83. That is, given that the optimum-sell trigger at some future but unknown date would be $279.73, the value of the option at current cash-flow expectations was $576.36. In terms of conventional pricing of a perpetual American option, the option value at smooth pasting ($3,814) could be viewed as the intrinsic value of the option at expiration, while $711.83 would be the current value of the option. This would imply that the maximum value of farmland would be the current fundamental value of $1,783.29 plus the current value (or a negotiated portion thereof) of the real option on capital gains—$2,495.12 ($1,783.29 + $711.83). In 1998, the market value of land exceeded the economic value by $43 ($2,538 - $2,495).[xiv]

These values are of interest because they indicate the range of possible error in valuation when uncertainty is excluded from the capital-budgeting problem. The data in Figure 10.2 illustrates the relationship between $V(\pi), V(\pi) + F(\pi)$ and actual land values between 1975 and 1998. The graph shows that actual land values were more often than not higher than the economic value $\left[V(\pi) + F(\pi)\right]$. Since $V(\pi) + F(\pi)$ represents a theoretical maximum, any excess in market value above this sum could represent a speculative bubble. The graph indicates the possibility of bubbles between 1978 and 1986 and between 1988 and 1991. During these periods, the market price of farmland was well in excess of the real-optioned value, which suggests the existence of speculative bubbles.

REAL OPTIONS AND LAND-PRICE BUBBLES

The question now is to make some determination as to the source of the bubbles. To see how the bubble fits in a real-options framework, return to Figure 10.1, in which a speculative bubble will arise from option prices on the w_1 curve to the right of b. Mathematically, the interpretation is that a higher value is placed on the option to postpone than is warranted by the value of the investment at any π right of π^*. The higher the value of π, the higher the option value, with each incremental increase in expectation of π leading to successively larger bubbles. The economic interpretation is that the seller has expectations of extra-fundamental capital gains. The seller sees economic value where none exists and the evidence suggests that buyers agree. The belief that in times of inflation land will retain value regardless of cash-flow generation, the belief that government support will always be forthcoming no matter what happens, and demand from nonfarm sources driven by nonfarm valuations (e.g., commercial or residential construction) are all examples of things that could cause a bubble. The results indicate two extended bubbles (Figure 10.2). Even if it is assumed that market to value ratios of 1.5 or less are due to short-term shocks (as posited by Falk and Lee 1998), two distinct bubbles are still identifiable—one from 1979 to 1984 and one from 1989 to 1991.

Of course there are other explanations for the results. The values for a and σ as long-run variables were assumed constant in the Table 10.1 calculations. The assumed volatility for σ was 16.4 percent. An increase in volatility would increase $\pi^*, V(\pi^*), F(\pi^*)$ and $V(\pi) + F(\pi)$, but an examination of the data using rolling five-year volatility measures for period 1978 to 1985 was actually lower than 16.4 percent, with an average of about 12 percent, so volatility cannot explain the 1979 to 1984 bubble. The five-year rolling average volatility for 1989 to 1991 was about 22 percent, which does indicate a higher volatility than was used in Table 10.1. Therefore, it is possible that the 1989 to 1991 bubble did not really exist because the volatility was underestimated. However, the years

1987 and 1988 both had short-run volatility measures of 23 percent but there
were no bubbles.

The second possibility is the assumption of a 4.6 percent long-run expected-
growth rate. Using short run growth estimates (five-year rolling averages) we
find that the 1979 to 1984 rates are 11.8 percent, 7.9 percent, 2.4 percent, 4.3
percent, 2.8 percent, and 7.0 percent, respectively. Empirically, π^* decreases

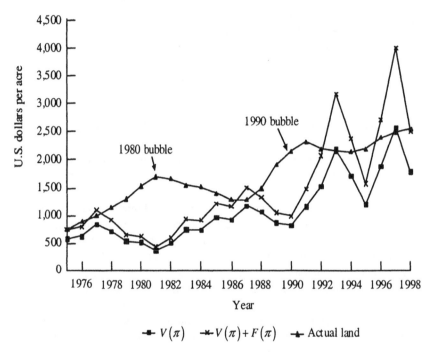

Figure 10.2 Real Options and the Value of Farmland in Ontario

with increased growth, so it is possible that underestimating growth led to the
bubble conclusion for 1979, 1980, and 1984. This cannot explain the bubble
from 1981 to 1983; so underestimating growth is not a good explanation for the
finding of the 1979 to 84 bubble, in which the growth rates were 2.0 percent,
−1.1 percent, and 1.7 percent. Since these growth rates were lower than the long-
run average, their effect increased π^* by shifting the smooth-pasting condition
to the right, but the curvature of the option-valuation function changed through
growth's effects on A and β, so the option value decreases overall. Therefore,
underestimates of growth cannot explain the 1989 to 1991 bubble. The indica-
tions are that in all likelihood there have been two speculative bubbles in Ontario
farmland prices since 1975.

DISCUSSION AND CONCLUSIONS

This chapter asks a very fundamental question: Can real options explain the discrepancies between observed land values and the land-capitalization model? The emerging theories and literature on hysteresis and real options provide a rich and reasonable approach to explaining many of the discrepancies that agricultural economists discovered when positive theories, such as the pricing of farmland, do not hold under rigorous empirical testing. Real options emerge when investors have the option to act or to postpone making an investment decision when the investment has uncertain outcomes or is derived from an underlying stochastic process. This real option comes about because the seller always has an option to sell immediately or to postpone the sale in the hopes that cash flow and hence land values and capital gains will ultimately increase. Options in agricultural land markets emerges when long-run positive growth trends and uncertainty signal an expectation of capital gains at some point in the future. Because of the growth trend, farmers may not be willing to sell even if cash flows decrease below the trend in the short run. For market-determined assets, such as land, this means that the observed land price does not always match the price predicted from the simple discounting of future cash flows.

The results tend to support the hypothesis. Using Ontario data from 1975 to 1998, it was shown for about 9 years that the market value of land was very close to its fundamental value plus the real-option value. In 15 of the 24 years, the market price was no more than 1.5 times the economic value of farmland. This finding does not refute the existence of real options since evidence by Falk and Lee (1998) shows from time to time that short-run shocks can push land prices away from equilibrium, but eventually they will return. In 9 out of 24 years the results support the existence of a speculative bubble in farmland prices. A speculative bubble does not refute the existence of real options, since a bubble by definition is simply a real option gone awry. But the finding of bubbles does refute the implicit notion of rationality in real-option pricing, and, in this respect, we must reject the idea that real options can explain bubbles. The evidence in this research suggests that econometricians did indeed discover bubbles rather than real options.

A number of issues need further investigation. One direction of research would be to survey landowners and determine if they purposefully delay selling land in the hopes of land-price increases or if they attempt to extract a premium from the buyer as compensation for future capital gains. In other words, how persistent is the hysteresis argument used in the proposition? In addition, specific examinations need to clarify why bubbles from time to time persist and what causes real-option values to rise beyond reason. Behavioral models, such as those described in Shiller (2000), provide some foundation principles, but the behavioral theories of asset overvaluation have yet to consider option values explicitly. It is easy to examine Equation (6) and its roots to show that option values increase with growth, but by any definition of a random walk, growth is a long-term parameter. Still, one can imagine a short-run increase in growth that increases land and option values. In a model, such as Hong and Stein (1999), a cascading of growth expectations, initially real and later imagined, can explain the result.

We must also be cautious about some of the underlying assumptions of the chapter. The assumption of perfect correlation between cash flows and land values at each moment in time can be easily questioned. This assumption plays a critical role in theory because otherwise it would be difficult to argue that the buyer has no option. Indeed, researchers may want to examine option values under the weaker assumption of less-than-perfect correlation by defining uncorrelated Brownian motions between two or more random influences. For example, cash flow and inflation as independent or weakly correlated Brownian motions could be used to explain the inflation-driven land prices of the 1970s, and the collapse of the inflation option may explain the dramatic decline of land prices in the late 1980s. The existence of agricultural-stabilization programs and disaster relief may also create defensible option values. Finally, the whole notion of hysteresis needs to be explored. From a classical standpoint, the notion that a farmer might postpone a sale by giving up certainty today for uncertainty in the future, is as much a function of psychology as economics, since it requires certain beliefs that may not be justified by a purely economic model. Without this psychology the real options discussed in this chapter may not be transacted at all. However, with the assumption of hysteresis in place, this chapter has provided the first step in a new avenue for investigating the dynamics of agricultural land values.

APPENDIX 10.A

Following Dixit and Pindyck (1994), the solution to Equation (6) is obtained by assuming that the option price is described by

$$F(\pi) = A\pi^\beta, \tag{10.A1}$$

where A and β are parameters to be determined. Applying the appropriate calculus to Equation (10.A1) and substituting these into Equation (6) yields the quadratic equation

$$\tfrac{1}{2}\sigma^2\beta(\beta-1)+\beta\alpha-r=0. \tag{10.A2}$$

Solving for the positive root of β gives

$$\beta = \left\{ \frac{1}{2} - \frac{\alpha}{\sigma^2} + \left[\left(\frac{\alpha}{\sigma} - \frac{1}{2} \right)^2 + \frac{2r}{\sigma^2} \right]^{1/2} \right\} \tag{10.A3}$$

Solving the smooth-pasting condition of Equation (9) in terms of A and substituting A into the value matching the condition of (8) yields

$$\pi^* = \left\{ \left[\frac{\beta}{\beta-1} \right] V(\pi)(r-\alpha) \right\}. \tag{10.A4}$$

Equation (10.A4) gives the optimal level of cash flow π^* at which the option to sell should be exercised (e.g., \$167 in Figure 10.1). It is a function of interest rates, volatility, growth, and fundamental value as defined by $V(\pi) = \pi/(r-a)$. To maximize expected NPV the optimal-decision rule is to postpone the sale until π^* occurs. Finally

$$A = \frac{\pi^*}{(r-\alpha)\beta\pi^{*\beta}}, \text{ and} \tag{10.A5}$$

the value of the option can be solved by

$$F(\pi^*) = A\pi^{*\beta}. \tag{10.A6}$$

REFERENCES

Abel, A.B., A.K. Dixit, J.C. Eberly, and R.S. Pindyck. (1996) "Options, the Value of Capital, and Investment." *Quarterly Journal of Economics* 111: 753–77.

Amran, M. and N. Kulatilaka. (1999) *Real Options: Managing Strategic Investment in an Uncertain World.* Boston: Harvard Business School Press.

Ansic, D and G. Pugh. (1999) "An Experimental Test of Trade Hysteresis: Market Exit and Entry Decisions in the Presence of Sunk Costs and Exchange Rate Uncertainty." *Applied Economics* 31: 427–36.

Bailey, W. (1991) "Valuing Agricultural Firms: An Examination of the Contingent-Claims Approach to Pricing Real Assets." *Journal of Economics Dynamics and Control* 15: 771–91.

Baker T.G., E.H. Ketchabaw, and C.G. Turvey. (1991) "An Income Capitalization Model for Land Values with Provisions for Ordinary Income and Long-Term Capital Gains Taxation." *Canadian Journal of Agricultural Economics* 39(1): 69–82.

Berk, J., R.C. Green, and V. Naik. (1999) "Optimal Investment, Growth Options, and Security Returns." *Journal of Finance* 53(5): 1553–1607.

Black, F. and M.S. Scholes. (1973) "The Pricing of Options and Corporate Liabilities." *Journal of Political Economy* 81 (May-June): 637–54.

Campbell, J.Y. (2000) "Asset Pricing at the Millennium." *Journal of Finance* 55(4): 1515–67.

Capozza, D.R. and R.W. Helsley. (1990) "The Stochastic City." *Journal of Urban Economics* 28: 187–203.

Capozza, D.R. and G.A. Sick. (1994) "The Risk Structure of Land Markets." *Journal of Urban Economics* 35: 297–319.

Cecchetti, S.G., P. Lam, and N.C. Mark. (1998) "Asset Pricing with Distorted Beliefs." National Bureau of Economic Research (NBER) Working Paper 6354, Federal Reserve Bank of NY and NBER, Ohio State University.

Clark, J.S., M. Fulton, and J.T. Scott. (1993) "The Inconsistency of Land Values, Land rents, and Capitalization Formulas." *American Journal of Agricultural Economics* 75(1): 147–55.

Copeland, T.E. and V. Antikarov (2001) *Real Options: A Practitioner's Guide*. Texere Publishing, N.Y.

Cox, J.C., J.E. Ingersoll, and S.A. Ross. (1985) "An Intertemporal General Equilibrium Model of Asset Prices." *Econometrica* 53 (2): 363–384.

Dixit, A.K. (1992). "Investment and Hysteresis." *Journal of Economic Perspectives* 6(Winter): 107–32.

_____. (1989) "Entry and Exit Decisions Under Uncertainty." *Journal of Political Economy* 97(3): 620–38.

Dixit, A.K. and R.S. Pindyck. (1994) *Investment Under Uncertainty*. Princeton: Princeton University Press.

Falk, B. (1991) "Formally Testing the Present Value Model of Farmland Prices." *American Journal of Agricultural Economics* 73(1): 1–10.

Falk, B. and B. Lee. (1998). "Fads Versus Fundamentals in Farmland Prices." *American Journal of Agricultural Economics* 80(4): 696–707.

Falk, B, B. Lee, and R. Susmel. (2001) "Fads versus Fundamentals in Farmland Pricing: Reply." *American Journal of Agricultural Economics* 83(4): 1078–81.

Featherstone, A.M. and T.G. Baker. (1988) "The Effects of Reduced Price and Income Supports on Farmland Rent and Value." *North Central Journal of Agricultural Economics* 10(1): 177–90.

_____. (1987) "An Examination of Farm Sector Real Asset Dynamics." *American Journal of Agricultural Economics* 69(3): 532–46.

Fishman, M.J. (1989) "Pre-emptive Bidding and the Role of the Medium of Exchange." *Journal of Finance* 44: 41–57.

Gorman, M. (1976) "A General Theory of Asset Valuation Under Diffusion State Processes." Giannni Foundation Working Paper No. 50, Berkeley: University of California.

Hansen, R. (1987) "A Theory for the Choice of Exchange Medium in Mergers and Acquisitions." *Journal of Business* 60: 75–95

Hanson, S. and R.J. Myers. (1995) "Testing for a Time-Varying Risk Premium in the Returns to U.S. Farmland." *Journal of Empirical Finance* 2(3): 265–76.

Heaton, J. and D. Lucas. (1999) "Stock Prices and Fundamentals." *NBER Macroeconomic Annual*. Cambridge: MIT Press (213–42).

Hong, H. and J.S. Stein. (1999) "A Unified Theory of Under reaction, Momentum Trading, and Overreaction in Asset Markets." *Journal of Finance* 54(6): 2143–84.

Katzner, D.W. (1999) "Hysteresis and the Modeling of Economic Phenomenon." *Review of Political Economy* 11(2): 171–81.

Krugman, P. (1989) *Exchange Rate Instability*. Cambridge: MIT Press.

Lambrecht, B.M. (2001) "The Timing and Terms of Mergers, Stock Offers, and Cash Offers." Paper presented at the Real Options Conference. UCLA, Los Angeles (July).

Leland, H.E. and D.H. Pyle. (1977) "Informational Asymmetries, Financial Structure, and Financial Intermediation." *Journal of Finance* 32(2): 242-58.

Martin, K.J. (1996) "The Methods of Payment in Corporate Acquisitions, Investment Opportunities, and Management Ownership." *Journal of Finance* 51(4): 1227–46.

Martzoukos, S.H. (2001). "Hysteresis Models of Investment with Multiple Uncertainties and Exchange Rate Risk." *Review of Quantitative Finance and Accounting* 16: 251–67

McDonald, R. and D. Siegel. (1986) "The Value of Waiting to Invest." *Quarterly Journal of Economics* 101(November): 707–27.

Merton, R.C. (1973) "Theory of Rational Option Pricing." *Bell Journal of Economics and Management Science* (Spring): 141–83.

Myers, S.C. and N.S. Majluf. (1984) "Corporate Financing and Investment Decisions When Firm's Have Information That Investors Do Not Have." *Journal of Financial Economics* 13: 187–221.

Pindyck, R.S. (1991) "Irreversibility, Uncertainty, and Investment." *Journal of Economic Literature* 29(September): 1110–52.

Quigg, L. (1993) "Empirical Testing of Real Option-Pricing Models." *Journal of Finance* 48(2): 621–40.

Richards, T.J. and P. M. Patterson. (1998) "Hysteresis and the Shortage of Agricultural Labor." *American Journal of Agricultural Economics* 80(4): 683—95.

Roche, M.J. (2001) "Fads versus Fundamentals in Farmland Pricing: Comment." *American Journal of Agricultural Economics* 83(4): 1074–77.

Shiller, R.J. (2000) *Irrational Exuberance*. Princeton: Princeton University Press.

Stokes, J, W.I. Nayda, and B.C. English. (1997) "The Pricing of Revenue Assurance." *American Journal of Agricultural Economics* 79: 439–51.

Sundaresan, S.M. (2000). "Continuous-Time Methods in Finance: A Review and an Assessment." *Journal of Finance* 55(4): 1569–1622.

Titman, S. (1985) "Urban Land Prices Under Uncertainty." *American Economic Review* 75(3): 505–14.

Trigeorgis, L. (1993) "Real Options and Interactions with Financial Flexibility." *Financial Management* 22(Autumn): 202–24.

Turvey, C.G., T.G. Baker, and A. Weersink. (1992) "Farm Operating Risk and Cash Rent Determination." *Journal of Agricultural and Resource Economics* 17(July): 186–94.

Weersink, A.J., J.S. Clark, C.G. Turvey, and R. Sarkar. (1999) "The Effects of Agricultural Policy on Farmland Values." Land Economics 75 (August): 425–39.

ENDNOTES

[i] In the finance literature, an emerging set of research is focusing on what Campbell (2000) refers to as the equity puzzle. The equity puzzle refers to persistent departures of market returns on stock from fundamental value. Heaton and Lucas (1999) attempt to reconcile the departure by calculating the implied returns or growth rates that would

match fundamental value to market value. Campbell (2000) and Sundaresan (2000) review the equity puzzle in terms of uncertainty with Sundaresan (2000) focusing on dynamic risks. Cecchetti et al. (1998) build an economic model based on subjective and fluctuating beliefs about growth and use this to explain economic expansions and contractions. Hong and Stein (1999) note that the failure of asset-pricing models to explain short-run departures and long-run reversions suggest that behavioral dynamics can also influence asset values. Shiller (2000) provides a similar explanation. These models suggest a type of fundamental hysteresis that departures from value are perpetuated by reinforced herd mentality. The limit of this behavior is a speculative bubble.

[ii] To see this, suppose that the sale is postponed one year. In that year π is received in cash flow and the value of land one year hence is $\pi(1+g)/(r-g)$. The present value is

$$PV = \left[\pi(r-g) + \pi(1+g)/(r-g)(1+r) = \pi(1+r)/(r-g)(1+r) = \pi/(r-g)\right],$$

which equals Equation (1). Postponement has no impact on real wealth.

[iii] $dV(\pi) = dV(\pi, t+1) - dV(\pi, t)$, so using the expected value of Equation (4), $V(\pi, t+1) = V(\pi, t)(1+\alpha) = \pi(1+\alpha)/(r-\alpha)$, which is the common form of the bid-price growth formula (e.g., Baker et al. 1991) when π is an observation rather than an expectation as defined in Equation (1).

[iv] The subject of this chapter and the forgoing model is the option value of delaying a sale in order to achieve a higher possible capital gain when cash flows are uncertain. The option on the capital-gain components is actually a lower bound on the option value to postpone the sale since the intervening rents, which accrue to the landowner, are not explicitly considered. To see that the option on the gain is a lower bound, assume that a perfect market exists which allows for a replicating portfolio [Dixit and Pindyck (1996), Copeland and Antikarov (2001)] of marketable securities that matches the cash flow and capital value of the underlying land market. For simplicity, assume current cash flow is $100/acre, discounted at 10 percent. With zero drift, the instantaneous present-value bid price is $1,000/acre. Over a one-period time step, cash flows can increase to $150 or fall to $50 and at each state the instantaneous present-value of land could be $1,500 and $500 respectively. In the good state, the owner receives $150 in cash and a capital gain of $500 for a total benefit of $650, and in the bad state, he faces a capital loss of $500 and a cash flow loss of $50 for a net benefit of –$450. To calculate the option to postpone, we construct a replicating portfolio with a payoff of $MAX(500,0)$ when only capital gains are considered and a payoff of $MAX(650,0)$ when cash flow and capital gains are considered. With x representing the position in the replicating risky asset and B representing the amount of risk-free bonds earning a rate of 5 percent, the portfolios are constructed as

For the option on the capital gain,
a) $X(1,500) + B(1.05) = 500$ in the good state, and
b) $X(500) + B(1.05) = 0$ in the bad state.
For the option on the cash flow plus capital gain,
c) $X(500) + B(1.05) = 0$ in the good state, and
d) $X(500) + B(1.05) = 0$ in the bad state.

Solving a) and b) simultaneously gives $X = .5$ and $B = -238.1$, which means that a replicating portfolio with a payoff schedule of Max $[500,0]$ occurs if the farmer borrows -238.1 at the risk-free rate and buys $.5$ units of the spanning security. The present value of this portfolio will simply be $X(1,000) + B = .5(1,000) - 128.1 = \$261.90$. Simply put, this means that it is worth \$261.90 to wait one period and sell the land at that time. Solving c) and d) in the same way gives $x = .65$ and $B = -309.52$, which says to borrow more and buy more of the spanning asset. The option value is then $x^*(1,000) + B = .65^* - 309.52 = 340.48$. Because of the intervening cash flow, the option to postpone is higher than the option on the capital gain, with the latter being equivalent to selling idle or vacant land for its best alternative use. The implication of this result suggests that the option to postpone the sale as presented in this chapter may in fact understate the true real-option value.

^v The hysteresis argument has been used in other agricultural contexts. Richards and Patterson (1998) use the hysteresis argument to explain labor movement between urban and rural economies.

^{vi} Again, it is important to distinguish between the real option discussed here and a financial option such as Black and Scholes (1973). If the buyer and the seller were to agree to a contract whereby the buyer could postpone the purchase of the farmland until some future date, the value of that option would be priced using a formula similar to, but not exactly the same as, a conventional European call option on non-dividend paying stock. If at some future date T, the market value exceeds the current value I, the buyer can then buy the land for I, otherwise the buyer can (without obligation) buy the land at a lower price. The use of a Black-Scholes type model to price real options over a defined time horizon is discussed by Amran and Kulatilaka (1999) and in further notes to this chapter.

^{vii} In McDonald and Siegel (1986) and Dixit and Pindyck (1996) an intermediate option will exist if the correlated Brownian motion defined by dI and $dV(\pi)$ are not perfectly correlated. McDonald and Siegel (1986) define the boundary condition Max$\left[0, V/1 - C^*\right]$. When V and I are both Brownian motion, the decision is to invest when the ratio $V/1$ exceeds a critical value C^*. If $I = V$, then clearly C^* can only equal 1, and the decision is to invest immediately.

^{viii} Other conditions could result in a similar conclusion. For example, property rights provide a natural monopoly to landowners.

^{ix} If $F\left(\pi^*\right)$ is the option value to waiting for π^* to occur, then $F(\pi) = F\left(\pi|\pi^*\right) < F\left(\pi^*\right)$ represents the opportunity costs of selling immediately at π. If $\pi \geq \pi^*$, then $F(\pi) \geq F\left(\pi^*\right)$ and the land will be sold immediately at its present-value bid price.

^x It has also been argued in the finance literature that market signaling can give rise to real options in a number of different ways. For example, Leland and Pyle (1977) develop a model whereby movement by one party signals a forthcoming benefit or cost to the second party. Applied to the farmland problem, a potential buyer placing a bid on a parcel of land may signal to the seller an imminent increase in cash flow and land values. Such a signal may cause the landowner to postpone the sale, giving rise to the real-option value. Likewise, a tender offer for the shares of a corporation will signal to

the owners a higher, perhaps synergistic, future value—with a tender offer in place, the shareholder has the option to postpone acceptance until the true value to the acquirer is revealed. Fishman (1989) argues that an all-cash offer signals a high value for the target and this may deter other bidders. Myers and Majluf (1984), Hansen (1987), and Martin (1996) argue that if a firm is perceived to be undervalued, the tender offer is made on a cash basis, while a perceived overvaluation results in an equity-based tender offer. When information is asymmetric, such signaling can result in the emergence of real options (Lambrecht 2001). As a hypothesis, cash offers would signal larger capital gains in the future, so that the shareholders of the target firm have a higher-valued option to postpone the sale. An equity deal, on the other hand, signals a lower chance of future capital gains, so shareholders will likely accept the tender offer sooner than later. The models by Lambrecht (2001) are based on symmetric information. This chapter argues that the timing and terms of mergers, stock offers, and cash offers evolve from a real-options framework.

[xi] Note that Equation (6) with $F_t \neq 0$ is very similar to the partial differential equation in Black and Scholes (1973). The solution to the call-option price is

$$C(\pi,t) = \pi e^{(a-r)T} N(d_1) - X e^{-rT} N(d_2), \text{ where}$$

$$d_1 = \frac{1}{\sigma\sqrt{T}} \left[\ln\left(\frac{\pi}{X}\right) + \left(\alpha + \frac{1}{2}\sigma^2\right)T \right] \text{ and } d_2 = d_1 - \sigma\sqrt{T}.$$

This call option is priced relative to the diffusion of cash flow, not to the diffusion of land. Since cash flow (or land, for that matter) is not a traded asset, a strong assumption must be made about the (potential) existence of a spanning portfolio of traded assets that is perfectly correlated with the volatility of cash flow. This option can also be used to create a riskless hedge in the cash position. If a risk-taker's hedge can be constructed, then $\alpha = 0$, r equals the risk-free rate of return, and $C(\pi,t)$ is the Black-Scholes model. If a spanning asset does not exist then a risk-free hedge is impossible (Dixit and Pindyck 1996; Stokes et al. 1997) and the option-pricing model will have to include the market price of risk as in Gorman (1976) and Cox et al. (1985). In this case, one can still set r equal to the risk-free rate but the term a becomes $\alpha - \theta_\sigma$, where θ_σ is the market price of risk. Finally, while it is possible to write a real option on cash flow, the most likely real option is on land itself. For example, suppose an investor wanted to buy farmland 3 years hence but wanted protection against land-price increases. The current price of land is V_0 so $X = V_0$ is the strike price. If in 3 years $V_T > V_0$, the investor will buy the land at the prevailing price V_T but will receive from the call option a payoff of $V_T - V_0$. The total payoff is $-V_0 = \left[-V_T + (V_T - V_0) \right]$. If $V_T \leq V_0$, the option expires without value and the investor pays the lower price for the land. In essence, the option on the real asset with a payoff $E\left[\text{Max}(0, V_T - V_0) \right]$ results in an investment payoff of $E\left[\text{Min}(V_T, V_0) \right]$.

[xii] In Figure 10.1, the optimal target is at b, because any points to the right of b have no meaningful interpretation (Dixit 1992), since the option to postpone the sale increases at a rate greater than value of the land. Transactions valued to the right of b are considered to be speculative in nature and suggest the formation of a speculative bubble.

[xiii] The risk premium reflects the risk premium on a low β stock in the sense of the Capital Asset Pricing Model. A higher-risk premium will reduce the fundamental value and increase the option value while a lower-risk premium will increase the fundamental value and lower the real option value.

[xiv] Not shown in Table 10.1.

Chapter 11

The Certainty Equivalence of Farmland Values: 1910 to 2000

Charles B. Moss, J.S. Shonkwiler, and Andrew Schmitz
University of Florida
University of Nevada-Reno
University of Florida

INTRODUCTION

This chapter estimates the effect of uncertainty on farmland values in the United States using an option-pricing approach. From 1910 to 2000, returns and interest rates applied to U.S. agricultural farmland were used to impute the value of farmland per acre for each year. In this chapter, we use these imputed values and divide them by the observed 2000 market values of farmland to yield a certainty equivalent, which measures the amount that future returns on farmland are discounted as a result of uncertainty. We also use a structural model of latent-variables (Bollen 1989) to estimate the effect of risk, within both interest rates and agricultural returns, on this certainty equivalent. Other studies such as those by Barry (1980) and Bjornson and Innes (1992), examine the role of risk in the valuation of agricultural assets. Their parameterizations, however, focus on the risk in agriculture with respect to a market portfolio. In addition, their analyses assume that the relative risk of agricultural assets is static over time. One notable exception to the assumption of a constant risk premium is that of Bjornson (1995) who examines the effect of business cycles on farmland values. Our results show that risk is a significant determinant of agricultural-asset values. Further, we show that the relative risk adjustments (i.e., the certainty equivalent due to the risk in returns to farmland) vary over time.

The closing decade of the twentieth century and the opening decade of the twenty-first century have proven tumultuous for agricultural policy in the United States. During the mid-1990s, agriculture in the United States experienced high output prices that added to the momentum for a general change in the basic nature of U.S. agricultural policy. This movement culminated in the passage of the Federal Agricultural Improvement Reform Act (FAIR) of 1996 that replaced the deficiency payments and modified the loan rates found in previous farm bills. [Agricultural supports began with the Agricultural Adjustment Act of 1933; deficiency payments and loan rates were introduced with the Agricultural Adjustment Act of 1970 (Schmitz et al. 2002)]. In place of deficiency payments, FAIR introduced production flexibility contract payments (PFCPs) that promised producers a gentle transition to a more market-oriented agricultural policy. However, the period of higher agricultural prices that led to the passage of FAIR was short-lived. As farm prices declined, the U.S. Congress resorted to emergency farm programs to augment the market loan-rate provisions that were embedded in FAIR. In addition to emergency provisions, the low prices experienced in the late 1990s contributed to a reversal of the movement toward market-oriented policies. The passage of the Farm Security and Rural Investment Act (FSRI) of 2002 signaled the return to many of the same U.S. agricultural policies that had been eliminated under FAIR.

Each U.S. policy regime (the Agricultural Adjustment Act of 1970, FAIR, and FSRI) has implications for expected agricultural returns and also for the variance (or relative risk) of those returns. Some say that the pre-FAIR agricultural polices increased the rates of return of agricultural assets and simultaneously they reduced the variability of those returns. FAIR, as some conjecture, contained fewer provisions than the pre-FAIR policies did at stabilizing U.S. agriculture. Given that differences in agricultural policy imply changes in the expected rates of return to farmland and alterations to the amount of variability of those returns, policy differences can also have significant implications for farmland values in the United States. According to financial economic literature, changes in agricultural policies that imply higher risk result in lower farmland values. Changes in farmland values that are due to alterations in relative risk are important for the United States because of the significance of farmland values to the U.S. agricultural balance sheet. Thus any decline in farmland values will have dramatic implications for the financial solvency of the U.S. farm sector over time. Farmland values accounted for 70 percent of all agricultural assets from 1960 through 2001 (USDA/ERS 2002a).

DERIVATION OF THE IMPUTED PRESENT VALUE

A basic concept from financial economics is that the investor will put money into an asset if the present value of the cash flow arising from that asset is greater than the purchase price of that asset. By logical extension, the price of an asset in a competitive economy should be equal to the present value of the cash flow arising from that asset. The problem with a straightforward application of this principle is complicated by the uncertainty of future cash flows and by the variability of interest rates. Given a sequence of returns to agricultural assets and interest rates, we develop a basic model for the estimation of certainty equivalence over time.

In a certain world, the value of agricultural assets is the stream of future income discounted at a known interest rate. Assuming that interest rates and cash flows may change over time, this value becomes

$$V_t = \sum_{i=0}^{\infty} \frac{CF_{t+i}}{\prod_{j=0}^{i}(1+r_{t+j})},$$

(1)

where V_t is the value of agricultural assets in period t, CF_{t+i} is the cash flow arising from those assets in period $t+i$, and r_{t+j} is the interest rate in period $t+j$.

The formulation in Equation (1) inadequately represents the process that generates asset values in domestic agriculture because it assumes that all future cash flows and interest rates are certain. The first step, when developing a model of net present value under risk, is to introduce the expectation of future cash flows. Defining $E_t CF_{t+i}$ as the expectation of cash flows in period $t+i$ given information in period t, the value of assets becomes

$$V_t = \sum_{i=0}^{\infty} E_t \left[\frac{CF_{t+i}}{\prod_{j=0}^{i}(1+r_{t+j})} \right].$$

(2)

If the interest rate is certain, we have

$$V_t = \sum_{i=0}^{\infty} \frac{E_t CF_{t+i}}{\prod_{j=0}^{i}(1+r_{t+j})}.$$

(3)

The specification of land value in Equation (3) incorporates the fact that future cash flows are not certain; however, this specification of land value does not integrate the effects of risk on the asset-valuation process through either risk aversion or through irreversibility. We use two traditional approaches to adjust the present-value formulation for increased risk. The first approach we use is the discount rate, which is increased to adjust for increased risk. The second approach we use to account for risk is to derive a certainty equivalence to the observed risk cash flow by multiplying the projected cash flow by a certainty-equivalence factor. We use the latter approach in this chapter based on the irreversibility framework proposed by Dixit and Pindyck (1994). We formulate the risk-adjusted asset value as

$$V_t = \theta_t \sum_{i=0}^{\infty} \frac{E_t CF_{t+i}}{\prod_{j=0}^{i}(1+r_{t+j})},$$

(4)

where θ_t is a multiplier that accounts for the reaction of the market to uncertainty. Intuitively, we expect that $0 < \theta_t < 1$, with $\theta_t \to 1$ as the returns from the asset become less risky.

EMPIRICAL RESULTS

In this chapter, we determine the effect of a change in the perceived relative risk for asset values over time by observing the change in the value of θ_t. To derive an estimate of θ_t, we compute an imputed asset value based on observed returns to farmland. We take income to farmland based on USDA historical data as nominal-cash income from farmland. We add back interest paid on the real estate and payments to nonfarm landlords [the use of imputed returns in valuing farmland is supported by the findings of Erickson et al. (2003)]. The interest rate is the commercial-paper rate published by the Federal Reserve Board of Governors (2002). The imputed asset value in period t is then derived as

$$\tilde{V}_t = \sum_{i=0}^{N} \frac{CF_{t+i}}{\prod\limits_{j=0}^{i}(1+r_{t+j})} , \tag{5}$$

where $\tilde{V}_t$ is the imputed value of farmland over the observed planning horizon and CF_{t+i} is the observed cash flow to agricultural assets in period $t+i$. [The use of CF_{t+i} implies short-run errors from weather, including both pests and the long-run change in expectations (Featherstone and Moss 2003)]. This cash flow represents the present value over a finite increment in time. In order to transmute the present value to perpetuity, we then compute

$$\hat{V}_t = \frac{\tilde{V}_t}{1-\left(\dfrac{1}{1+\bar{r}_t}\right)^N} , \tag{6}$$

where $\hat{V}_t$ is the imputed value of farmland in the United States and $\bar{r}_t$ is the geometric average interest rate over the remaining planning period.[i] Given the imputed value of farmland into perpetuity defined by Equation (6), Equation (4) can be rewritten as

$$V_t = \theta_t \hat{V}_t . \tag{7}$$

Dividing both sides of Equation (7) by the imputed value defined in Equation (6) yields an empirical estimate of θ_t.

Next, we compare the U.S. historical farmland values (USDA/ERS 2002b) to our imputed present value of farmland. [The imputed value of farmland is generated using USDA income data from 1910 to 2001 (USDA/ERS 2002a) and the nominal interest rate on commercial paper from the Federal Reserve Board of Governors (2002)]. We depict graphically the gap between the imputed value and the market value of farmland from 1910 to 2001 (Figure 11.1). The data in Figure 11.1 indicates that the gap between the market value of farmland and the imputed value of farmland was quite large from 1910 to 1950; in 1931, the imputed value of farmland was as high as 9.5 times its market value. The gap narrows to the point where the imputed value of farmland is approximately 1.6 times the market value between 1950 and 1970. The 1970s were marked by an increase in farmland values to 2.2 times the imputed value in 1979 with a general correction to the same 1960s levels in the late 1980s.

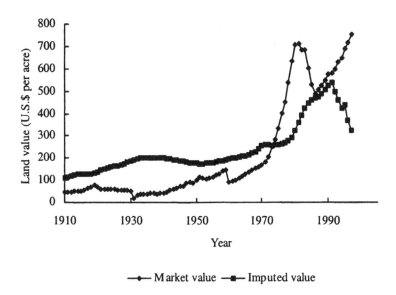

—♦— Market value —■— Imputed value

Figure 11.1 Market Value Versus Imputed Value Of U.S. Agricultural Assets (1910 to 2001)

Undoubtedly, the process generating the observed spread between the imputed value and the market value of farmland values is quite complex. Plotting the time path from 1911 to 1990 of the real interest rate and the rate of return on agricultural assets against the ratio of our imputed values yields insights into the valuation of agricultural assets (Figure 11.2). First, the initial gap (from 1915 to 1950) between the imputed value of agricultural assets and their market values occurred with a fairly volatile real return to capital. Second, between 1950 and 1975 the imputed and observed values narrowed as the volatility of real interest rates declined, but the real interest rate remained slightly positive. Finally, overvaluation of agricultural assets appeared when the real interest rate fell below zero in 1975 to 1985 and its volatility was fairly small. The empirical results in Figure 11.2 prompt us to search for explanations of the asset-valuation problem,

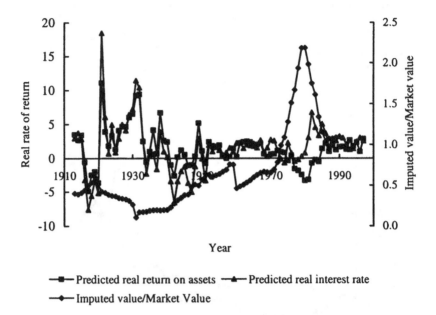

—■— Predicted real return on assets —▲— Predicted real interest rate
—◆— Imputed value/Market Value

**Figure 11.2 Ratio of Imputed Value to Market Value Versus Real Interest Rate and
Rate of Return on U.S. Agricultural Assets (1911 to 1990)**

including both the expectation and the volatility of future returns to capital. In
the next section, we present an explanation of the asset-valuation problem based
on the concepts of uncertainty and irreversibility.

THE EFFECT OF UNCERTAINTY
ON IRREVERSIBLE INVESTMENTS

Several models exist that relate volatility to asset values. The Lintner (1965) and
Sharpe (1964) classic capital-asset-price model (CAPM) develops a market explanation in which asset returns vary according to relative risk. An alternative
development of the same basic concept is the Ross (1976) arbitrage-pricing theory (APT); however, both of these theories describe changes in returns relative
to the market portfolio (or to the arbitrage portfolio as in the case of APT). The
rate of return on commercial paper that was discussed previously in this chapter
is a relatively risk-free interest rate. Thus, our discussion is more appropriately
focused on the effect of changes in the variability of the market portfolio over
time. An alternative framework is that of investment under uncertainty and irreversibility that develops the opportunity to invest as a financial option (Dixit and
Pindyck 1994).

The Dixit and Pindyck (1994) model of investment under uncertainty and irreversibility assumes that investors have the option to invest in a set of endeavors depending on their financial resources, including management expertise. The primary assumption of this model is that, once the assumption is made, the investor cannot easily reverse the investment. Thus traditional market-arbitrage models, such as CAPM and APT, are not adequate in thin asset markets such as farmland that have limited opportunities for arbitrage in the short-run (i.e., possible irreversibility in sales and purchases of farmland). Given this irreversibility, the decision to invest becomes one of exercising an option. The decision rule for investing in the option framework is to invest when the value of investing exceeds the value of waiting.

The decision is to invest if the annual return is greater than the hurdle rate,

$$H = \frac{\beta}{\beta - 1} \rho K, \tag{8}$$

where β is defined as

$$\beta = \frac{1}{2}\left[1 + \sqrt{1 + \frac{8\rho}{\sigma^2}}\right], \tag{9}$$

where ρ is the risk-adjusted rate of return on capital for a similar asset, K is the value of the asset, and σ^2 is the variance of the stochastic process determining the rate of return.[ii] Turning Equation 8 around yields

$$K = \frac{\beta - 1}{\beta} \frac{H}{\rho}. \tag{10}$$

The $(\beta - 1)/\beta$ term in this expression is one manifestation of the θ term in the original NPV model. Note that $\beta \to \infty \Rightarrow (\beta - 1)/\beta \to 1$, thus the investment rule under uncertainty and irreversibility approaches the traditional NPV decision rule. Further, we see from Equation (8) that $\beta \to \infty$ as $\sigma^2 \to 0$, or the investment becomes less risky. Thus, the gap between the imputed present value and the market value of agricultural assets is a function of the annual rate of return to agricultural assets, the risk-adjusted discount rate, and the variability of agricultural returns (Figure 11.1). In addition, the effect of variability is independent of any notion of risk aversion.

The effect of the various components determining θ can be observed over time. Differentiating the rearranged result in Equation (10) with respect to time yields

$$\frac{dK}{dt} = \frac{H}{\rho}\frac{1}{\beta}\frac{d\beta}{dt} + \frac{\beta - 1}{\beta}\frac{1}{\rho}\frac{dH}{dt} - \frac{\beta - 1}{\beta}\frac{H}{\rho^2}\frac{d\rho}{dt}, \tag{11}$$

where $d\beta$ includes changes in both variance and risk-adjusted interest rates. Similarly, modifications in the risk-adjusted discount rate may be composed of alterations in the both the risk-free rate and the relative-risk adjustment for that class of assets. Intuitively, Figure 11.2 depicts substantial shifts in the level of real interest rates with a small increase in volatility during the 1970s that had its origin in inflation.

The empirical model is derived from Equation (9) and Equation (10). In the current formulation, $H\!\!\!/\!\rho$ is the present value of the annual return required to pull the trigger on the investment or to kill the option to invest (Dixit 1992). In the context of Equation (6), $\hat{V}_t$ must equal $1\!\!\!/\!\theta_t$ times the purchase price of the asset for the investment to be made. Specifically

$$\frac{H}{\rho} \equiv \bar{V}_t = \frac{1}{\theta_t} V_t \equiv \frac{\beta}{\beta-1} K \qquad (12)$$

is the relationship under uncertainty implied Equations (7) and (10). V_t is the market value that is equal to K in Equation (9), and $\hat{V}_t$ is equal to the discounted value of returns sufficient to make the investment, which is $H\!\!\!/\!\rho$ or the present value of the hurdle return. $1\!\!\!/\!\theta_t = \beta\!\!\!/\!(\beta-1)$ is the risk adjustment incorporating the risk relationship. Equation (11) is then derived as a first-order Taylor series expansion (Chaing 1974: 268–74) of $\beta\!\!\!/\!(\beta-1)$

$$\frac{1}{\theta_t} = \frac{\beta(\rho,\sigma^2)}{\beta(\rho,\sigma^2)-1} \approx \frac{\beta(\rho_0,\sigma_0)}{\beta(\rho_0,\sigma_0)-1} -$$
$$\frac{1}{\left[\beta(\rho_0,\sigma_0)-1\right]^2}\left[\left.\frac{\partial\beta}{\partial\rho}\right|_{\rho_0,\sigma_0}(\rho-\rho_0)+\left.\frac{\partial\beta}{\partial\sigma}\right|_{\rho_0\sigma_0}(\sigma-\sigma_0)\right]. \qquad (13)$$

The effects of the risk-free rate and the volatility of returns on the certainty equivalent are determined by the effect of these variables on β. Casual inspection of the definition of β in Equation (9) implies that the derivative of β with respect to the risk-free interest rate is positive while the derivative of β with respect to volatility is negative. Thus Equation (13) implies that the inverse of the certainty equivalent increases with an increase in the volatility and declines with an increase in the risk-free rate.

Abstracting from the structural forms of the derivatives, we then specify an empirical model of θ_t

$$1/\theta_t = \gamma_0 + \gamma_1 r_t^a + \gamma_2 r_t^f + \gamma_3 \sigma_{a,t}^2 + \gamma_4 \sigma_{f,t}^2, \qquad (14)$$

where r_t^a is the real rate of return to agricultural assets in period t, r_t^f is the real risk-free interest rate in period t, $\sigma_{a,t}^2$ is the volatility of the real rate of return on agricultural assets, and $\sigma_{f,t}^2$ is the volatility in the real risk-free rate.

There are several difficulties to overcome when estimating this specification directly. First, the instantaneous volatility of the real risk-free interest rate and rate of return on agricultural assets is unobserved. Second, one could argue that the appropriate real rates of return are the ex ante rates. Third, a proxy measures the dependent variable. To incorporate these facets, we use a structural latent-variable approach (Bollen 1989). Specifically, an unobserved variable η that represents the true certainty equivalence is postulated as a function of four latent variables

$$\eta_t = \gamma_1 \xi_{1t} + \gamma_2 \xi_{2t} + \gamma_3 \xi_{3t} + \gamma_4 \xi_{4t} + \nu_5, \tag{15}$$

where ξ_{1t} is the latent expectation of the real rate of return on agricultural assets, ξ_{2t} is the latent expectation of the real risk-free interest rate, ξ_{3t} is the latent variance of the real rate of return on agricultural assets, ξ_{4t} is the latent variance of the real risk-free interest rate, and ν_t is the error in the latent certainty equivalence.

To quantify these latent independent variables we use a set of observable indicators

$$
\begin{bmatrix} x_{1t} \\ x_{2t} \\ x_{3t} \\ x_{4t} \\ x_{5t} \\ x_{6t} \\ x_{7t} \\ x_{8t} \\ x_{9t} \\ x_{10t} \end{bmatrix}
=
\begin{bmatrix}
1 & 0 & 0 & 0 \\
0 & 1 & 0 & 0 \\
0 & 0 & \beta_{33} & 0 \\
0 & 0 & \beta_{43} & 0 \\
0 & 0 & \beta_{53} & 0 \\
0 & 0 & \beta_{63} & 0 \\
0 & 0 & 0 & \beta_{74} \\
0 & 0 & 0 & \beta_{84} \\
0 & 0 & 0 & \beta_{94} \\
0 & 0 & 0 & \beta_{10,4}
\end{bmatrix}
\begin{bmatrix} \xi_{1t} \\ \xi_{2t} \\ \xi_{3t} \\ \xi_{4t} \end{bmatrix}
+
\begin{bmatrix} \delta_{1t} \\ \delta_{2t} \\ \delta_{3t} \\ \delta_{4t} \\ \delta_{5t} \\ \delta_{6t} \\ \delta_{7t} \\ \delta_{8t} \\ \delta_{9t} \\ \delta_{10,t} \end{bmatrix},
\tag{16}
$$

where x_{1t} is an autoregressive estimate of the real risk-free interest rate, x_{2t} is an autoregressive estimate of the real rate of return on agricultural assets, x_{3t}, x_{4t}, x_{5t}, and x_{6t} are t through $t-3$ lagged squared residuals from the autoregressive representation of real agricultural asset returns, x_{7t} through x_{10t} are similarly autoregressive models of risk-free interest rates, and δ_{it} are errors of measurement. We view this portion of the representation as a confirmatory factor analysis with, for example, the common portion of the variance between x_{3t}, x_{4t}, x_{5t}, and x_{6t} representing the current expectation of volatility of the real interest rate.

To complete the model, we add an imperfect observation for the endogenous latent variable

$$y_t = \eta_t + \varepsilon_t, \tag{17}$$

in which y_t is our imputed value of the certainty equivalence, $(1/\theta_t)$. In matrix form, we express the complete model as

$$\begin{aligned} \eta_t &= \Gamma \xi_t + v_t \\ y_t &= \eta_t + \varepsilon_t \\ x_t &= \beta \xi_t + \delta_t. \end{aligned} \tag{18}$$

To estimate this model we derive the implicit variance matrix, given a set of parameters,

$$\Sigma(\Omega) = \begin{bmatrix} \Gamma \Psi \Gamma' + \sigma_v^2 & \beta \Psi \Gamma' \\ \Gamma \Psi \beta' & \beta \Psi \beta' + \Theta \end{bmatrix}, \tag{19}$$

in which Ψ is the covariance matrix for the latent exogenous variables, σ_v^2 is the variance of the latent endogenous variable, and Θ, a diagonal matrix, is the noncommunal variation in the indicators of the latent variable. Estimation is then accomplished by minimizing

$$F = \ln|\Sigma(\Omega)| + tr\left[S\Sigma^{-1}(\Omega)\right], \tag{20}$$

where $tr[.]$ denotes the trace of the matrix and S is the covariance matrix of the observed data series, $[y \vdots x_1 \vdots x_2 \vdots \cdots \vdots x_{10}]$.

To complete the specification, we impose additional restrictions on Θ and Ψ matrices as well as on the σ_ε^2 coefficients. An inspection of Equation (18) indicates that both σ_ε^2 and σ_v^2 cannot be simultaneously parameterized. Thus σ_ε^2 will be set to zero. In addition, ξ_1 and ξ_2 are latent variables that symbolize the real rate of return on farmland and the real interest rate whose only indicators (i.e., observed proxies) are estimated expected real rates of return on farmland and real interest rates, respectively. For identification reasons, θ_{11} and θ_{22} are set to zero. Also, β_{33} and β_{74} are set to one in order to provide a scale of measurement to which the other $(\beta_{ij}\theta)$s may be compared.

ECONOMETRIC ESTIMATES

In the first step, we fit autoregressive models for real interest rates and real returns on agricultural assets. Choosing an arbitrary lag length of 3, these equations are estimated using ordinary least squares that yield $\bar{R}^2$ of 45.4 percent for the real interest rate and 32.1 percent for the real rate of return to agricultural assets. Further, the Ljung-Box statistics (Greene 2003: 622) for each equation are 21.97 and 23.51. Given that these statistics are distributed, $\chi^2_{(20)}$ fails to reject the hypothesis of white-noise residuals at the 5 percent confidence level.

The Phillips-Peron test (Greene 2003: 644–45) of nonstationarity yields statistics of –29.60 for the real risk-free interest rate and –37.58 for the real-returns series, so we conclude that these series are stationary.

The predictions from the autoregressive models are assumed to provide the true ex ante estimates of the real interest rate $(x_1 = \xi_1)$ and the real rates of return to the agricultural-asset $(x_2 = \xi_2)$ series. Also, the estimated squared errors for the current and three previous periods derived from the autoregressive models are assumed to be indicators of the corresponding unobserved risk measures $(\xi_3 \text{ and } \xi_4)$. We estimate the system represented in Equation (18) by centering the data on their means and maximizing the likelihood function in Equation (20). Convergence could not be obtained for the entire data set (i.e., 1920 through 1997) because of several outliers (from 1920 to 1923). After dropping these outliers, the data set contains only 81 observations. We present the estimates for the remaining years between 1913 and 1997 (Table 11.1).

The results of our analysis reveal a statistically significant autoregressive process for squared variance with both the first-order autoregressive parameters being significant at the 0.05 level of confidence. The estimated effect of latent risk on real returns to farmland γ_3 is negative. Specifically, an increase in the estimated variance in the real rate of return to farmland causes the ratio of the market value to its imputed value to decline. This estimated impact is statistically insignificant at the 0.10 level of confidence. However, this effect is conditioned on the statistical properties of risk. Simultaneously testing the autoregressive parameters and the effect of risk yields a test statistic of 14,361 that is distributed χ^2_4 and can be rejected at any standard confidence level. Thus, changes in risk affect the value of farmland over time.

CONCLUSIONS

In this chapter, we examined whether or not the changes in relative risk impacted the value of agricultural assets in the United States from 1910 to 2000. The statistical results indicated that the certainty-equivalence parameter θ varies over time. Increased volatility of both real interest rates and real returns to agriculture, given the same expected returns on agricultural assets, decreases the certainty equivalence of agricultural assets and results in lower asset values. This result appears to be consistent with risk aversion but does not require the assumption of risk aversion within the development of this model.

The result that the volatility of asset returns reduces the value of agricultural assets apart from any effect on the mean is particularly relevant to the current agricultural policy debate. Falk (1991), Alston (1986), Burt (1986), and Featherstone and Baker (1987) conclude that increases in returns to farmland yield higher

Table 11.1 Maximum Likelihood Estimates for Latent Variable Model.

Parameter	Estimate	Standard Deviation
$\beta_{4,3}$	0.070^{**}	0.039
$\beta_{5,3}$	0.002	0.040
$\beta_{6,3}$	0.004	0.038
$\beta_{8,4}$	0.050^{**}	0.024
$\beta_{9,4}$	-0.002	0.024
$\beta_{10,4}$	0.014	0.024
$\theta_{3,3} = \theta_{4,4} = \theta_{5,5} = \theta_{6,6}$	2.766^{***}	0.228
$\theta_{7,7} = \theta_{8,8} = \theta_{9,9} = \theta_{10,10}$	2.197^{***}	0.193
$\psi_{1,1}$	6.576^{***}	1.053
$\psi_{2,1}$	6.297^{***}	0.878
$\psi_{2,2}$	10.915^{***}	3.030
$\psi_{3,1}$	-1.717	1.529
$\psi_{3,2}$	-3.268^{*}	1.994
$\psi_{3,3}$	24.939^{***}	4.642
$\psi_{4,1}$	-3.537^{*}	2.164
$\psi_{4,2}$	-5.359^{**}	2.815
$\psi_{4,3}$	36.028^{***}	5.993
$\psi_{4,4}$	52.246^{***}	8.712
γ_1	-2.010	4.889
γ_2	1.400	1.575
γ_3	-1.475	29.995
γ_4	0.930	20.855
σ_v^2	15.465^{***}	2.479

[*]Denotes statistical significance at the 0.10 level of confidence.
[**]Denotes statistical significance at the 0.05 level of confidence.
[***]Denotes statistical significance at the 0.01 level of confidence.
Source: Authors' computation.

farmland values. Based on the results of Falk, Alston, Burt, and Featherstone and Baker, and assuming that farm programs increase the net return to farmland, the net effect of the elimination of government programs is the reduction of farmland values. Our results add another dimension to the debate. Specifically, the risk-reducing impact of government programs increases the value of agricultural assets even if commodity programs do not bring about an increase in mean returns. Further, the fact that certainty equivalence changes over time may strengthen the results of previous findings by attributing noise, which was previously included in the standard error of the mean effect, to changes in certainty equivalence. The variation in risk may also accentuate the relationship between returns and asset values; thus failure to address the volatility in the certainty equivalence may overstate the standard errors of the effect of farm returns on asset values.

Hence, the results of this study support a secondary effect of agricultural policies on farmland values in the United States. Specifically, agricultural policies that reduce the overall risk to agriculture may increase farmland values (and the wealth of the farm sector) apart from the average effect of farm program payments. Thus programs designed to be budget neutral (e.g., crop insurance programs) may increase farmland values. In addition, shifts in the overall philosophy of agricultural policies (e.g. the movement toward market-oriented agricultural policies implicit in the passage of FAIR) will affect farmland values if they are perceived to offer less protection from market fluctuations. Further, the reintroduction of countercyclical components in FSRI may increase farmland values by reducing the perceived risk of returns to farmland.

REFERENCES

Alston, J.M. (1986) "An Analysis of Growth of U.S. Farmland Prices, 1963–82" *American Journal of Agricultural Economics* 68(1): 1–9.

Barry, P.J. (1980) "Capital Asset Pricing and Farm Real Estate." *American Journal of Agricultural Economics* 62(3): 549–53.

Bjornson, B. (1995) "The Impacts of Business Cycles on Returns to Farmland Investments." *American Journal of Agricultural Economics* 77(3): 566–77.

Bjornson, B. and R. Innes. (1992) "Another Look at Returns to Agricultural and Nonagricultural Assets." *American Journal of Agricultural Economics* 74(1): 109–20.

Bollen, K.A. (1989) *Structural Equations with Latent Variables.* New York: John Wiley & Sons.

Burt, O.R. (1986) "Econometric Modeling of the Capitalization Formula for Farmland Prices." *American Journal of Agricultural Economics* 68(1): 10–26.

Chaing, A.C. (1974) *Fundamental Methods of Mathematical Economics* Second Edition. New York: McGraw-Hill Book Co.

Dixit, A. (1992) "Investment and Hysteresis." *Journal of Economics Perspectives,* 6(1): 107–32.

_____. (1993) *The Art of Smooth Pasting.* Chur: Harwood Academic Publishers.

Dixit, A. and R. Pindyck. (1994) *Investment under Uncertainty.* Princeton: Princeton University Press.

Erickson, K., A.K. Mishra, and C.B. Moss. (2003) See Chapter 12, this volume.

Falk, B. (1991) "Formally Testing the Present Value Model of Farmland Prices." *American Journal of Agricultural Economics* 73(1): 1–10.

Featherstone, A.M. and T.G. Baker. (1987) "An Examination of Farm Sector Real Estate Dynamics." *American Journal of Agricultural Economics* 69(3): 532–46.

Featherstone, A.M. and C.B. Moss. (2003) See Chapter 9, this volume.

Federal Reserve Board of Governors. (2002) "H.15 Selected Interest Rates." Internet Website:www.federalreserve.gov/releases/h15/data.htm.

Greene, W.H. (2003) *Econometric Analysis* Fifth Edition. Upper Saddle River: Prentice-Hall .

Harl, N. (1990) *The Farm Debt Crisis of the 1980s.* Ames: Iowa State University Press.

Lintner, J. (1965) "The Valuation of Risk Assets and the Selection of Risky Investments in Stock Portfolios and Capital Budgets." *Review of Economics and Statistics* 47(1): 13–37.

Moss, C.B., J.S. Shonkwiler, and J. Reynolds. (1989) "Government Payments to Farmers and Real Agricultural-Asset Values in the 1980s." *Southern Journal of Agricultural Economics* 21(2): 139–51.

Purvis, A., W.G. Boggess, C.B. Moss, and J. Holt. (1995) "Technology Adoption Decisions under Irreversibility and Uncertainty: An Ex Ante Approach." *American Journal of Agricultural Economics* 77(3): 541–51.

Ross, S.A. (1976) "The Arbitrage Theory of Capital Asset Pricing." *Journal of Economic Theory* 13(3): 341–60.

Schmitz, A. (1995) "Boom/Bust Cycles and Ricardian Rent." *American Journal of Agricultural Economics* 77(5): 1110–25.

Schmitz, A., W.H. Furtan, and K. Baylis. (2002) *Agricutural Policy, Agribusiness, and Rent-Seeking Behaviour.* Toronto: University of Toronto Press.

Sharpe, W.F. (1964) "Capital Asset Prices: A Theory of Market Equilibrium Under Conditions of Risk." *Journal of Finance* 19(3): 425–42.

USDA/ERS (U.S. Department of Agriculture, Economic Research Service). (2002a) Farm Balance Sheet Data. Internet Website: www.ers.usda.gov/Data/FarmBalanceSheet/
_____. (2002b) Farm Income Data. Internet Website: www.ers.usda.gov/Data/FarmIncome/.

ENDNOTES

[i] The geometric average of future interest rates is computed as

$$\overline{r}_t = \left[\prod_{s=t}^{N} (1 + r_s) \right]^{1/N-t}.$$

[ii] A complete discussion of the Dixit and Pindyck (1994) model as applied to an agricultural investment problem is presented in Purvis et al. (1995); however, the derivation of β and the hurdle return can be derived from either traditional option-pricing or dynamic-programming approaches. The assumption is that the rate of return follows the Brownian motion stochastic process (Dixit 1993), with the trigger being determined when the value of investing exceeds the initial cost of the investment plus the value of the option of investing in the future. The most concise explanation of the formulation of the model is found in Dixit (1992).

Chapter 12

Cash Rents, Imputed Returns, and the Valuation of Farmland Revisited

Kenneth Erickson, Ashok K. Mishra[*], and Charles B. Moss
U.S. Department of Agriculture, Economic Research Service
U.S. Department of Agriculture, Economic Research Service
University of Florida

INTRODUCTION

Most studies of farmland values are based (either directly or indirectly) on the notion of Ricardian rent, which is defined as the residual return to a fixed factor of production (e.g., farmland) after all other factors (e.g., fertilizer, fuel, and labor) have been paid their marginal product (Schmitz 1995). In this framework, the value of farmland is the present or discounted value of Ricardian rent into the future (Hirshleifer 1970: 99–132). While there is little disagreement about the basic tenets of Ricardian rent, its empirical definition has been the subject of contention in the farmland valuation literature.

Our study examines the U.S. Department of Agriculture, Economic Research Service (USDA/ERS), Moss (1997), and Phipps (1982, 1984) methodologies, which are three alternative empirical approaches for imputing the Ricardian (or economic) rents to farmland using accounting data. Starting with some measure of agricultural income (e.g., gross agricultural profit per acre), economic rents to farmland are derived (or imputed) by subtracting payments to exclude factors not directly included in operating costs (e.g., unpaid operator or family labor). However, this adjustment may also include additions to economic rents for cash expenses not directly associated with production. For example, following the Modigliani-Miller theory (Hirshleifer 1970: 261–64; Ingersoll 1987: 410–15) the choice of financing for an asset—debt or equity—does not affect the value of the asset, hence the interest paid on real-estate debt is added to impute economic rents. We include a fourth procedure, the cash-rental-rate

(CRR) method, which provides a market measure of economic rents to farmland. In this chapter we determine which of these four measures of economic rents best predicts changes in farmland values over time.

In our study, we test whether farmland values are nonstationary time-series data,[i] (Falk 1991; Moss 2002). The nonstationarity of farmland values suggests that the estimated relationship between farmland values and returns to farmland may be spurious regression results (Granger and Newbold 1974). We analyze relationships among nonstationary time-series data by determining whether or not a stationary relationship can be estimated such that the time-series data do not diverge over time. Four closely related approaches are used to analyze relationships among nonstationary time-series data: the error-correction model (Engle and Granger 1987), the common stochastic-trend model (Stock and Watson 1988), fully modified-least squares regression analysis (Phillips and Hansen 1990), and the maximum-likelihood formulation of the error-correction-model approach (Johansen and Juselius 1990). In this chapter, we apply the maximum-likelihood formulation of the error-correction-model approach proposed by Johansen and Juselius (1990) to estimate the relationship between each of the measures of economic rents and farmland prices over time.

IMPUTED RETURNS AND MODELS OF FARMLAND VALUES

The present-value relationship between expected returns to agriculture and farmland values can be used to derive a mathematical model for changes in farmland values over time (Schmitz 1995)

$$\Delta V_t = -\frac{E_{t-1}CF_t}{1+r_t} + \frac{r_t}{1+r_t}V_t + \gamma_t,$$ (1)

in which V_t is the value of farmland in period t, ΔV_t denotes the first difference in farmland values ($\Delta V_t = V_t - V_{t-1}$), $E_{t-1}CF_t$ is the expected cash flow (i.e., the expected return to farmland or Ricardian rent) from farmland in period t based on information available in period $t-1$, r_t is the expected discount rate in period t, and γ_t is the change in expected future cash flows due to the new information that becomes available in period t. Substituting observed cash flows in period t for expected cash flows yields

$$\Delta V_t = \frac{CF_t + v_t}{1+r_t} + \frac{r_t}{1+r_t}V_t + \gamma_t.$$ (2)

The new error term v_t in Equation (2) is the error introduced since we do not directly observe expected cash flows from farmland in Equation (1). Econometrically, the introduction of this error introduces inefficiency into the estimation process. Using the original formulation of Equation (1), Schmitz (1995) concludes that changes in farmland values in the long run conform to the standard present-value formulation of farmland values. However, the composed residuals (i.e., a combination of the error introduced by using observed cash flows in place of expected cash flows v_t, as presented in Equation (2), and the effect of changes in long-run expectations γ_t) are correlated or contain short-run information (Featherstone and Moss 2003).[ii] Therefore, while Schmitz finds that farmland values are appropriately valued in the long run, they exhibit short run boom-bust cycles. This conclusion is dependent on the properties of the measurement error (i.e., the properties of the composed error term) for expected cash flows.

Unfortunately, the error in measurement usually does not stop with the introduction of observed cash flows for expected cash flows; additional distortions are introduced through the measurement of returns to farmland. The definition of either cash flows or expected cash flows in Equations (1) or (2) should follow the notion of Ricardian rent. This notion of Ricardian rent is fundamental to economic theory, but additional support for the formulation is found in the mathematical application of Euler's theorem to production functions. Clark (1923) and Wicksteed (1933) use Euler's theorem to infer that distribution of factor returns generated by the market are economically efficient. (Demonstrating the economic efficiency of Euler's theorem is beyond the scope of this paper.) However, the basic mathematical Euler theorem result states that, under perfect competition and constant returns to scale, paying each factor of production the value of its marginal product equals the economic return on production. Thus, like the definition of Ricardian rent, the economic rents to farmland can be defined as the profit remaining after all other factors have been paid their value of marginal product.

Euler's theorem states that if a function is homogeneous of degree k

$$l^{k-1} f(x) = \sum_{i=1}^{N} \frac{\partial f(lx)}{\partial x_i} x_i,$$

(3)

and $f(x)$ is a production function. Given that the production function is homogeneous of degree one (i.e., $k = 1$), or that it exhibits constant returns to scale (CRTS), implies that

$$f(x) = \sum_{i=1}^{N} \frac{\partial f(x)}{\partial x_i} x_i.$$

(4)

Multiplying both sides of Equation (4) by the output price p yields

$$pf(x) = \sum_{i=1}^{N} p \frac{\partial f(x)}{\partial x_i} x_i.$$

(5)

Given the standard conditions for profit maximization $\left(w_i = p\,{\partial f(x)}\Big/{\partial x_i} \right)$, where w_i is the market price of input i, Equation (5) becomes

$$pf(x) = \sum_{i=1}^{N} w_i x_i. \tag{6}$$

Thus, letting the Nth input be farmland and restricting our analysis to per-acre returns, Euler's theorem implies that the market price for one acre of farmland in a competitive market would be

$$w_N = p\,f(x) - \sum_{i=1}^{N-1} w_i x_i, \tag{7}$$

which is consistent with the Ricardian rent to farmland.

The Euler theory formulation is subject to several caveats. Some of the most general caveats involve the assumption of CRTS and competitiveness within input and output markets. More nebulous is the implicit assumption that all other factors of production (e.g., labor and machinery) are variable. Specifically, a critical assumption in the derivation of the residual return to farmland using the imputed-return methodologies is that all other factors of production are paid the value of their marginal products (VMPs). One exception to factors of production earning the VMP is the case of the quasi-fixed input—an input whose level cannot be instantaneously varied. Thus if firms cannot instantaneously vary the level of an input, its shadow value—the constrained marginal value of the input in production $p\,{\partial f(x)}\Big/{\partial x_i}$—may differ from the market price. For example, if the fixed level of the input is constrained below the optimum level, the value of the fixed input in production will be above its market price. This potential divergence of shadow value and market price is examined within the context of asset fixity (Chambers and Vasavada 1983) in which asset fixity refers to the inability to instantaneously vary the level of an input. Implicitly, farmland is one of the most significant examples of quasi-fixity for most farms. However, the measurement of asset fixity is complicated if more than one asset is quasi-fixed. Mishra, et al. (2000) demonstrate how the presence of multiple quasi-fixed assets (labor as well as land) may affect the estimation of the farmland-pricing models. Specifically, if labor is temporarily captured and unable to shift from agriculture, the imputed return to farmland (assuming that labor is paid its VMP computed at prevailing wages) understates the true Ricardian rent (i.e., the VMP of labor at the market wage is higher than the true VMP of labor, thus subtracting the market-valued VMP for labor produces a lower return to farmland). In this situation, observed farmland prices do not properly reflect their opportunity costs or shadow prices.

The derivation of the notion of Ricardian rent within the Euler-theorem formulation reveals a basic difficulty when imputing the rate of return to farmland. A hidden assumption in Equation (7) is that all factors of production are variable. Dividing the inputs into variable and quasi-fixed inputs, the Euler result from Equation (7) becomes

$$w_N = p f(x) - \sum_{i=1}^{M} w_i x_i - \sum_{i=1}^{N-M-1} \lambda_i z_i, \tag{8}$$

where z_i is quasi-fixed inputs (e.g., operator labor and machinery) and λ_i is the shadow value of the quasi-fixed inputs in production. The formulation presented in Equation (8) demonstrates how the returns to farmland can be misstated if the quasi-fixed inputs are not in equilibrium. Specifically, by letting z_1 denote operator labor, if labor is temporarily captured in agriculture then the shadow value of labor in production λ_1 will be less than the prevailing market wage (mathematically $\lambda_1 < w_{M+1}$, where w_{M+1} is the prevailing market wage). Thus computing the imputed rate of return to farmland w_N based on the prevailing market wage understates the true Ricardian rent, which can be seen when we substitute w_{M+1} for λ_1 in Equation (8).

To examine the relationship between cash rents, imputed returns, and farmland values, we use alternative estimates of cash flows (alternative imputed returns and net-cash rents). We assume that imputed per-acre returns and per-acre net cash rents are highly correlated, and that both approaches will explain a large share of the long-run variation in farmland values. Because of measurement errors inherent in these imputations (i.e., the assumption of CRTS and competitiveness of input and factor markets, and the presence of one or more quasi-fixed factors), we compare several different methods of imputing the economic returns to land. We seek the one method of representing per-acre cash rent that is more consistent with changes in farmland value than the others.

First, we impute a return to labor and to management, letting returns to farm assets be the claimant of residual income (USDA/ERS method). Second, we add interest payments and net rent to nonoperator landlords to the return to farm assets (Moss method). Third, we estimate a residual return to land by subtracting the imputed income shares of non-land factors of production from the USDA/ERS measure of residual income (Phipps 1982; Phipps 1984), which reflects the residual return to land when other factors (e.g., non-land assets and operator time) are valued at their estimated opportunity costs. In addition, we also examine the relationship between cash rents and farmland values. We calculate cash rents by dividing the total cash rents paid by the number of acres cash rented. We then take the quotient of this division calculation and subtract from it the property taxes and depreciation on farm buildings.

Given the four methods of estimating Ricardian rent (i.e., USDA/ERS, Moss, Phipps, and CRR), we must find one method of representing per-acre cash rent that is more consistent with changes in farmland value than the other methods are. Given that farmland prices may be nonstationary, we estimate separate error-correction models for farmland values using the procedure outlined in Johansen (1988) and Johansen and Juselius (1990). This avoids the spurious-regression problem introduced by nonstationarity in the time-series data. The maximum likelihood specification of the error-correction model is

$$\Delta x_t = \Gamma_1 \Delta x_{t-1} + \cdots \Gamma_{k-1} \Delta x_{t-k+1} + \Pi x_{t-1} + \mu + \Psi D_t + \varepsilon_t, \tag{9}$$

where x_t is the vector of prices, Δx_t is the change in x_t, D_t is a vector of exogenous variables, $\Gamma_1, \ldots \Gamma_{k-1}, \Pi, \mu,$ and Ψ are estimated parameters, and ε_t is the error term. The system is cointegrated if Π is a singular matrix ($\Pi = \alpha' \beta$), in which the rank (or number of independent α vectors) is the number of significant eigenvalues in Π. The test for the number of significant eigenvalues is

$$2\ln Q\left[H_1(r) \mid H_1(p)\right] = -T \sum_{i=r+1}^{p} \ln\left(1 - \hat{\lambda}_i\right), \tag{10}$$

where $\hat{\lambda}_i$ represents the ith eigenvalue of Π. The statistic presented in Equation (10) tests for the significance of r-cointegrating vectors $H_1(r)$ against the hypothesis of p-cointegrating vectors $H_1(p)$.

To apply the error-correction formulation in Equation (9) to the problem of farmland values, we start from the simple capitalization formula

$$V_t = \frac{R_t}{r_t}. \tag{11}$$

We then take the natural logarithm of Equation (11), which yields the cointegrating relationship

$$\ln(V_t) - \ln(R_t) + \ln(r_t) = 0. \tag{12}$$

We expect at least one cointegrating vector to exist among the three time series. However, neither the farmland-valuation formulation in Equation (12) nor the maximum-likelihood formulation of the error-correction model Equation (9) precludes a second cointegrating vector that relates returns to farmland to interest rates.

To estimate per-acre net-cash rents, we use USDA/ERS (Jones 1997) and U.S. Department of Agriculture, National Agricultural Statistical Service (USDA/NASS 2002) estimates of gross per-acre cash rents. Following the Alston (1986) analysis of the relationship between cash rents and land values, and using data from the U.S. Census of Agriculture (USDA/NASS 2002), the Agricultural Economics Land Ownership Survey (USDA/NASS 2001), the ARMS Agricultural Resource Management Study [USDA/NASS/ARMS (USDA/ERS 2002a)], and the USDA/ERS income accounts for the farm sector (USDA/ERS 2002b), we estimate per-acre net cash rents as

$$\frac{\text{Gross cash rents per acre} - \left(\text{Property taxes per acre} + \text{Depreciation on farm buildings}\right)}{\text{Net cash rent per acre}} \tag{13}$$

Since state-level data for gross per-acre cash rents do not exist for all 50 States, we use data for the U.S. Corn Belt (Illinois, Indiana, Iowa, Missouri, and Ohio) from the USDA's Balance Sheet (USDA/ERS 2002c) and Income Statement data (USDA/ERS 2002b). To develop estimates of net per-acre cash rents in the Corn Belt, we use the number of cash-rented acres of land in each state, the amount of property taxes paid, and the depreciation expenses associated with those rented acres. (We estimate this information because the USDA income-account data used in this study provides property tax and depreciation information for all land in farms, not just for the farms that have cash-rented acres.)

RESULTS OF TESTING THE EMPIRICAL COINTEGRATION MODEL

Following the standard cointegration approach developed by Engle and Granger (1987), we first test each individual time-series datum of our imputed values for nonstationarity. We present the Dickey-Fuller statistics (Greene 2003: 637–46) for stationarity given the presence of a time-trend (Table 12.1). The results indicate that the per-acre value of farmland, the USDA/ERS method, and the Moss method for estimating the time series (with the exception of interest rates which are stationary at the 0.05 level of confidence) are nonstationary at any conventional confidence level. We find interest rates to be nonstationary only at the 0.01 level of confidence. Differencing the time-series data and reapplying the Dickey-Fuller test indicates that the remaining farmland values and the return-time series data are trend stationary in differences. Taken together, these results

Table 12.1 Results of the Dickey-Fuller Test for Unit Roots

Variable	Definitions	Trend levels	Trend differenced
$\ln(VALUE_t)$	Value of farmland per acre	−0.89	−4.11[***]
$\ln(INTEREST_t)$	Interest rate	−3.51[**]	−6.84[***]
$\ln(IMPRETNR1_t)$	Returns to farm assets 1 (limited adjustments, USDA/ERS method)	−2.07	−8.10[**]
$\ln(IMPRETNR2_t)$	Returns to farm assets 2 (includes net rents to nonoperator landlords and interest) Moss method	−2.01	−3.20[**]
$\ln(CASHRENT)$	Cash rent per acre in the Corn Belt region	0.69	−12.24[***]
Critical value (95%)		−3.19	−3.19

[**]Denotes statistical significance at the 0.05 level of confidence.
[***]Denotes statistical significance at the 0.01 level of confidence.
Source: Authors' computations using Eviews (2001).

indicate that both farmland values and returns time-series data are $I(1)$, that is, they are integrated of order 1. Thus a long-run equilibrium (or cointegrating relationship) may exist between farmland values and one or more of the individual time series used to represent the rates of return to farmland. In this study, any of the imputed-return series data may be used to estimate long-run changes in farmland values. In addition, while the evidence to support the nonstationarity of interest rates is weaker, it is included in our cointegration analysis (the present value formulation for asset price presented in Equation (12) suggests that the logarithm on the interest rate should be included in the time-series formulation; excluding it could potentially lead to a specification error).

Following the results of the Dickey-Fuller tests, the next step in the cointegration approach is to determine if one or more of the cointegrating relationships occur in the multivariate time series. Following the approach developed by Johansen (1988) and Johansen and Juselius (1990), in Table 12.2 we estimate the test for r-cointegrating relationships versus the p-cointegrating relationships that were presented in Equation (10) to test for the existence of long-run (cointegrating) relationships between farmland values and each imputed-return series. These results suggest the existence of a single cointegrating relationship (i.e., only the first eigenvalue is statistically significant at the 0.05 level of confidence). For each imputed-return series, a single long-run relationship exists between farmland values, returns to farmland, and interest rates for each method of imputing the rates of return to farmland.

Given the existence of a single cointegrating vector, we compute an error-correction model based on the eigenvectors associated with the eigenvalues of Equation (10) using each alternative specification of imputed returns and cash rents. Each cointegrating vector can then be compared with the implicit restrictions on the long-run relationship between farmland values, returns to farmland, and interest rates from Equation (12). Specifically, the theoretical model of farmland values presented in Equation (12) suggests that the coefficient on the returns to farmland in the cointegrating relationship should be 1.0 while the coefficient on the interest rate should be −1.0. First we estimate an error-correction model using the definition of imputed returns proposed by the USDA/ERS procedure. The coefficient of imputed returns (IMPRETNR1: USDA/ERS method) is positive and statistically significant at the 0.01 level of confidence (Table 12.3). Results indicate that farmland values increase in response to increases in

Table 12.2. Number of Cointegrating Vectors, All Endogenous

Null	Alternative	Maximum eigenvalue	Critical values	
			5 percent	1 percent
$r=0$[a]	$R \geq 1$	34.99[**]	31.46	36.65
$r=1$	$R \geq 2$	23.68	25.54	30.34
$r=2$	$r \geq 3$	13.87	18.96	23.65
$r=3$	$R \geq 4$	6.41	12.25	16.26

[a]r is the rank matrix.
[**]Denotes statistical significance at the 0.05 level of confidence.
Source: Authors' computations using Eviews (2001).

returns to farmland when those returns to farmland are defined using the USDA/ERS procedure. These results are consistent with the asset-pricing formula. The estimated coefficient for the return to farmland in the cointegrating relationship is 1.268, which, given a standard deviation of 0.199, is not statistically different than 1.0. Further, the estimated coefficient in the cointegrating relationship for interest rates is –0.846 which again, given a standard deviation of 0.746, is not statistically different than –1.0. Thus the USDA/ERS procedure for imputing the return to farmland produces a long-run (cointegrating) relationship that is consistent with the PVM for farmland values.

Table 12.3 Parameter Estimates for Value of U.S. Farmland, 1960 to 2000

Variable	$\ln(VALUE_t)$ (value of farmland per acre)	$\ln(VALUE_t)$ (value of farmland per acre)
$\ln(INTEREST_t)$	–0.846	–0.666
	(0.746)	(0.529)
$\ln(IMPRETNR1_t)$	1.268***	na
	(0.199)	
$\ln(IMPRETNR2_t)$	na	1.106***
		(0.131)
Intercept	4.181	3.600

***Denotes statistical significance at the 0.01 confidence level.
Source: Authors' computations using Eviews (2001).

Next, we present the error-correction model for the second imputation method (IMPRETNR2, Moss method). Like the results for the USDA/ERS procedure, these estimated results are consistent with the asset-pricing formula presented in Equation (12). The coefficient of imputed returns (IMPRETNR2, Moss method) is positive and statistically significant at the 1 percent level (Table 12.3). Like the estimated coefficient using the USDA/ERS returns, the estimated coefficient is close to 1.0. Specifically, the estimated coefficient is 1.106, with a standard deviation of 0.131, which is not statistically different than 1.0. In addition, the estimated coefficient on interest rates in this formulation are negative and not statistically different than –1.0 at any conventional level of significance. When using the Phipps procedure, our study finds that none of the parameters in the cointegrating relationship is consistent with asset-pricing fundamentals as presented in Equation (12). This is partly because of the difficulties of imputing returns to non-land factors of production. In fact, largely for this reason, the USDA/ERS does not follow this procedure.

Following the approach of Alston (1986) and Falk (1991), we use cash rents to measure the returns to farmland. Because of limitations in cash-rental data on the national level and of aggregation bias, we perform the analysis on cash rents only on the U.S. Corn Belt region.[iii] These results are similar to the results when using measured imputed returns (i.e., the USDA/ERS and Moss methods). The parameter, 1.977, is larger than the estimated coefficients for either the USDA/ERS or Moss data. Further, given a standard deviation of 0.113 (Table 12.4), the estimated coefficient is significantly different than 1.0 which is implied by Equation (12). On the other hand, the coefficient of interest rates is negative and statistically significant at the 0.10 level of significance. Results suggest that a 1 percent increase in interest rates decreases farmland value by 0.38 percent (Table 12.4). The cointegrating relationship is consistent with the present-value formulation of farmland. Thus increases in cash rents lead to long-run increases in the value of farmland, while increases in the interest rates lead to long-run declines in farmland values. However, the magnitude of these impacts is not consistent with the present-value formulation. The long-run increase in farmland values following an increase in cash rents is larger than anticipated, while the long-run effect of increases in interest rates is smaller.

Table 12.4. Parameter Estimates for Value of Farmland, Corn Belt Region, 1960 to 2000

Variable	$\ln(\text{VALUE}_t)$ (value of farmland per acre)
$\ln(\text{INTEREST}_t)$	-0.381^*
	(0.226)
$\ln(\text{CASHRENT}_t)$	1.977^{***}
	(0.113)
Intercept	0.489

*Denotes statistical significance at the 0.10 level of confidence.
***Denotes statistical significance at the 0.01 level of confidence.
Source: Authors' computations using Eviews 4.1.

DISCUSSION AND IMPLICATIONS

The imputed-return (USDA/ERS and Moss) methodologies outperformed the CRR methodology when explaining long-run changes in farmland values (and perform about as well or better as a predictor of farmland values in the long-run than cash rents do). However, all three (the USDA/ERS, Moss and cash rent procedures) outperform the Phipps methodologies by producing the theoretically anticipated signs for the estimated coefficients. The dominance of the imputed-returns series over the CRR methodology was contrary to our expectations, since we expected that imputed per-acre returns would be highly correlated with net per-acre cash rents. The discrepancies we found among the results for the imputed-return methodologies are also informative. In a sense, the USDA/ERS, Moss, and Phipps imputation methods can be ranked in terms of complications (or in terms of additional information required) to adjust for unpaid factors of production. The Moss method of estimating imputed farmland values of returns only adjusts for interest paid on real estate. The USDA/ERS procedure makes adjustments for additional factors, including farm labor and capital consumption. The Phipps technique is still more inclusive because it adjusts for such factors as inventory costs. Following the derivation in Equation (8), we expect that greater precision when adjusting for changes in returns to factors of production will yield a more precise measure of net returns to farmland. However, this conjecture assumes that each adjustment is made without error. Using market prices for quasi-fixed factors of production in place of shadow values (e.g., using wages to compute the value of farmer labor) may add more noise (i.e., provide less transparency) than information in the derivation of returns to farmland. Thus our results support the use of a simple imputation, such as the Moss procedure, for the estimation of the returns to farmland.

The empirical dominance of imputed returns over cash rents in the estimation of long-run changes in farmland prices is somewhat more problematic. Previous studies maintain that the productive value of farmland is most readily apparent in the rental market. The price of farmland in the rental market is free of factors that include anticipated capital gains and boom-bust cycles. However, the rental market may be subject to alternative distortions. One possible distortion could result from the assumption of homogeneous productivity (i.e., smaller parcels of land that are more costly to farm may reflect rental markets that are dominated by retired farmers). Alternatively, the rental market may represent a segmented market in which farmers, who cannot obtain long-term credit, place higher bids in the rental market in part because of credit constraints. Under either scenario, the cash-rental market may not measure the Ricardian rents accruing to the average tract of farmland while imputed returns, based on average income statements and balance sheets, may be more appropriate. Finally, we recognize that farmland values are increasingly (over time) influenced by nonfarm factors such as urbanization and spatial factors and these must therefore also be considered.

REFERENCES

Alston, J.M. (1986) "An Analysis of Growth of U.S. Farmland Prices, 1963–82." *American Journal of Agricultural Economics* 68(1): 1–9.

Chambers, R.G. and U. Vasavada. (1983) "Testing Asset Fixity for U.S. Agriculture." *American Journal of Agricultural Economics* 65(1): 761–9.

Clark, J.M. (1923) *Studies in the Economics of Overhead Costs.* Chicago: University of Chicago.

Engle, R.F. and C.W.J. Granger. (1987) "Co-Integrating and Error Correction: Representation, Estimation, and Testing." *Econometrica* 55(2): 251–76.

Eviews (Eviews User's Guide). (2001) *Eviews 4.1* Irvine: Quantitative Micro Software.

Falk, B. (1991) "Formally Testing the Present-Value Model of Farmland Prices." *American Journal of Agricultural Economics* 73(1): 1–10.

Featherstone, A.M. and C.B. Moss. (2003). See Chapter 9, this volume.

Granger, C.W.J. and P. Newbold. (1974) "Spurious Regressions in Econometrics." *Journal of Econometrics* 2(2): 111–20.

Greene, W.H. (2003) *Econometric Analysis,* Fifth Edition. Upper Saddle River: Prentice-Hall, Inc.

Hirshleifer, J. (1970) *Investment, Interest, and Capital.* Englewoods Cliffs: Prentice-Hall, Inc.

Ingersoll, J.E. (1987) *Theory of Financial Decision Making,* Totowa: Rowan & Littlefield Publishing Co.

Johansen, S. (1988) "Statistical Analysis of Cointegration Vectors." *Journal of Economic Dynamics and Control* 12(2-3): 231–54.

Johansen, S. and K. Juselius. (1990) "The Full Information Maximum Likelihood Procedure for Inference on Cointegration with Applications to the Demand for Money." *Oxford Bulletin of Economics and Statistics* 52(2): 169–210.

Jones, J. (1997) CASH RENTS FOR U.S. FARMLAND [computer file]. #90025.Washington, DC: USDA/ERS.

Mishra, A.K., C.B. Moss, and K.W. Erickson. (2000) "Effect of Debt Servicing, Returns, and Government Payments on Changes in Farmland Values: A Fixed and Random-Effects Model." Proceedings of the Regional Committee NC-221. Minneapolis (October 2–3).

Moss, C.B. (1997) "Returns, Interest Rates, and Inflation: How They Explain Changes in Farmland Values." *American Journal of Agricultural Economics* 79(4): 1311–8.

Moss, C.B. (2002) "Estimating Western Farmland Values: The Effect of Returns on Assets, Productivity Growth, and Urbanization over Time." Paper Presented at the Western Agricultural Economics Association Meetings, Long Beach, California, (July).

Phillips, P.C.B. and B.E. Hansen. (1990) "Statistical Inference in Instrumental Variables Regression with I(1) Processes. *Review of Economic Studies* 57(1): 99–125.

Phipps, T.T. (1982) "The Determination of Price in the U.S. Agricultural Land Market." Unpublished Ph.D. dissertation, University of California-Davis.

Phipps, T.T. (1984) "Land Prices and Farm-Based Returns." *American Journal of Agricultural Economics* 66(4): 422–29.

Schmitz, A. (1995) "Boom-Bust Cycles and Ricardian Rents." *American Journal of Agricultural Economics* 77(5): 1110–25.

Stock, J.H. and M.W. Watson. (1988) "Testing for Common Trends." *Journal of the American Statistical Association* 83(404): 1097–1107.

USDA/ERS (U.S. Department of Agriculture). (2002a) "Agricultural Resource Management Survey (ARMS): Detailed documentation." Internet Website: www.ers.usda.gov/briefing/ARMS/detaileddocumentation.htm.

____. (2002b) "Farm Income Data." Internet Website: www.ers.usda.gov/Data/FarmIncome/.

____. (2002c) "Farm Balance Sheet Data." Internet Website: www.ers.usda.gov/Data/FarmBalanceSheet/.

USDA/NASS (U.S. Department of Agriculture). (2001) "Agricultural Cash Rents" Sp Sy 3(02), July.

____. (2002) Census of Agriculture 1997. Internet Website: www.nass.usda.gov/Census.

U.S. DOC (U.S. Department of Commerce, Bureau of the Census). (1990) *Census of Agriculture, Agricultural Economics and Landownership Survey* 3(2): U.S. Department of Commerce.

Wicksteed, P.H. (1933) *The Common Sense of Political Economy.* London: G. Routledge & Sons.

ENDNOTES

* The opinions expressed here are those of the authors and not necessarily those of the U.S. Department of Agriculture.

[i] Nonstationarity in time-series data are defined unit roots in the autoregressive representation. Following Greene (2003: 611–14), the autoregressive time-series formulation can be expressed as

$$x_t = \rho x_{t-1} + \varepsilon_t,$$

where x_t is the observed time-series variable, ρ is the autoregressive parameter, and ε_t is an unobserved error. The time-series process is stationary (or has a finite variance over time) if $-1 < \rho < 1$. If $\rho = 1$ the time-series process has a unit root and is nonstationary. In addition, the variance of the time-series data is unbounded as the sample size grows.

[ii] Additional errors are introduced by the substitution of the observed discount rate for the expected discount rate. However, one could argue that fixed interest rates on mortgages reduce this uncertainty.

[iii] The Corn Belt region of the United States includes Illinois, Indiana, Iowa, Missouri, and Ohio.

Section IV:
Transaction Costs and Farmland Values

Chapter 13

On the Dynamics of Land Markets under Transaction Costs

Jean-Paul Chavas
University of Maryland

INTRODUCTION

Much research has focused on the determinants of land prices. One basic argument is that land prices reflect the discounted expected value of the stream of future farm income or rent (Melichar 1979; Burt 1986; Alston 1986). Yet, this capitalization formula has been found to be too simple to capture the dynamics of land markets (Featherstone and Baker 1987; Falk 1991; Clark et al. 1993; Just and Miranowski 1993), which has stimulated the search for other factors that affect land prices. First, agricultural risk has a negative effect on land prices due to risk aversion (Just and Miranowski 1993; Barry et al. 1996). Second, there is evidence that transaction costs do affect land markets (Just and Miranowski 1993; Shiha and Chavas 1995; Chavas and Thomas 1999). For example, Shiha and Chavas show that transaction costs create barriers to the flow of external equity capital into farm real-estate markets. Third, inter-temporal preferences seem to play some role in asset valuation (Epstein and Zin 1991; Barry et al. 1996; Chavas and Thomas, 1999). Both inter-temporal preferences and transaction costs can help to explain why the standard capital-asset pricing model (CAPM) fails to capture some aspects of land pricing (Bjornson and Innes, 1992; Hanson and Myers, 1995; Barry et al. 1996).

This chapter evaluates the effect of transaction costs on the dynamics of land prices by developing a general household model of asset allocation. The model includes production, time allocation, investment, and consumption decisions, and incorporates the effects of risk, transaction costs, and inter-temporal preferences. The model is then used to investigate the implications for asset pricing. Implications for the functioning of land markets are also discussed. Special attention is given to the effects of risk and transaction costs, since both are particularly relevant in the analysis of land prices. First, weather risk, unanticipated pest damage, and price risk all contribute to uncertain agricultural income and thus to uncertain returns to land. Second, realtor fees, heterogeneity in land quality, and the spatial characteristics of land all contribute to transaction and information costs faced by participants in the land markets. Our analysis helps shed some light on the recent evolution of land markets under risk and transaction costs.

A HOUSEHOLD MODEL

Consider a household making production, consumption, and investment decisions over time. Let y_t denote consumption at time t; let $a_t = (a_{0t}, a_{1t}, \ldots, a_{Kt})'$ be a $(K+1) \times 1$ vector of assets held at time t; and let $x_t = (x_{1t}, \ldots, x_{Mt})$ be a $(M \times 1)$ vector of netputs (where outputs are positive and inputs are negative) used in the production process. Let a_{0t} be a riskless asset (such as a government bond) having a unit price and yielding a sure rate of return r_{t+1} at time $t+1$. Let $(a_{1t}, \ldots, a_{Kt})'$ be a $(K \times 1)$ vector of risky assets (that is, assets with uncertain rates of returns). Denote by $m_t = (m_{0t}, m_{1t}, \ldots, m_{Kt})'$ a $(K+1) \times 1$ vector of investments in which m_{Kt} is investment (disinvestment if negative) in the kth asset at time $t, k = 0, 1, \ldots, K$. At each time period the household allocates its total time available T among leisure L_t, work off the farm W_t, and work on the farm (within the household) H_t. Household decisions must satisfy the time constraint

$$L_t + W_t + H_t = T, \tag{1a}$$

as well as technological feasibility

$$(x_t, H_t, a_{1,t-1}, \ldots, a_{K,t-1}) \in \Omega_t, \tag{1b}$$

where Ω_t represents the household-production set.

The state equation representing asset dynamics at time t is

$$a_{kt} = a_{k,t-1} + m_{kt}, k = 0, 1, \ldots, K. \tag{2}$$

Define $q_{0t} > 0$ as the market price of the consumption good y_t, w_t is the wage rate, $q_t = (q_{1t}, \ldots, q_{Mt})'$ is the market price for x_t, and $p_t = (p_{1t}, \ldots, p_{Kt})$ is the market price for a_t at time t. The household-budget constraint at time t is

$$r_t a_{0,t-1} + w_t W_t + q_t' x_t = q_{0t} y_t + m_{0t} + \sum_{k \geq 1} p_{kt} m_{kt}. \tag{3a}$$

This simply states that income from the riskless asset $r_t a_{0,t-1}$, plus labor income $w_t W_t$, plus profit $q_t' x_t$ must be allocated between consumption expenditures $q_{0t} y_t$ and investment $\left(m_{0t} + \sum_{k \geq 1} p_{kt} m_{kt}\right)$. Defining household wealth at time t by $A_t \equiv a_{0t} + \sum_{k \geq 1} p_{kt} a_{kt}$, the budget constraint in Equation (3a) can be alternatively expressed as

$$A_t = R_t A_{t-1} - q_{0t} y_t, \tag{3b}$$

where $R_t \equiv \left[(1 + r_t) a_{0,t-1} + \sum_{k \geq 1} p_{kt} a_{k,t-1} + w_t W_t + q_t' x_t\right] / A_{t-1}$ is the rate of return on wealth at time t.

Assume that the household has dynamic preferences represented by the recursive utility function

$$U_t = u\left(y_t, L_t; y_{t+1}; L_{t+1}; \ldots\right) = U\left[(y_t, L_t, M_t(U_{t+1}))\right],$$

where $M_t(U_{t+1})$ is an aggregator function representing the certainty equivalent of the stream of future consumption given the information available at time t. Future uncertainty can involve production uncertainty (e.g., due to weather effects) as well as price uncertainty in the labor, commodity, and/or asset markets. Throughout this chapter we assume that the utility function $U(y_t, L_t, M_t)$ is differentiable and quasi-concave. We also assume non-satiation in y_t.

A special case of interest is the specification proposed by Epstein and Zin (1989 and 1991), which involves the aggregator function $M_t(U_{t+1}) = \left(E_t U_{t+1}^\alpha\right)^{1/\alpha}$. If $\alpha > 0$, then $(1 - \alpha)$ is the Arrow-Pratt coefficient of relative risk aversion, where $\alpha = 1$ corresponds to risk neutrality and $\alpha < 1$ corresponds to risk aversion. The Epstein-Zin specification also involves the constant elasticity of substitution (CES) utility function $U(y_t, L_t, M_t) = \left\{(1 - \beta)\left[u(y_t, L_t)^\rho + \beta M_t^\rho\right]\right\}^{1/\rho}$, where $\rho < 1$ is an inter-temporal substitution parameter. When $\alpha = \rho$, this gives as a special case the time-additive utility function $U_t = \left[(1 - \beta) E_t \sum_{j \geq 0} \beta^j u(y_t, L_t)^\alpha\right]^{1/\alpha}$. This shows that the traditional time-additive utility specification is rather restrictive: it implies exogenous discounting of future utility (with β as a fixed discount factor). This fails to distinguish between inter-temporal substitution (as measured by parameter ρ) and risk aversion (as reflected by the parameter α). However, when $\alpha \neq \rho$, this allows for non-additive time preferences. This more general specification distinguishes between the effects of inter-temporal substitution (as measured by ρ) from the effects of risk aversion (as reflected by α). It also allows for endogenous discounting, whereby the relative importance of the future may depend on current income and wealth. This may be particularly relevant in low-income situations, in which survival concerns can affect household incentives to invest.

Given recursive preferences $U[y_t, L_t, M_t(U_{t+1})]$, household decisions are consistent with the maximization problem

$$V_t(a_{t-1}) = \text{Max} \left\{ U[y_t, L_t, M_t V_{t+1}(a_t)] \right\}: \text{Equations (1a), (1b), (2) and (3b)}$$

$$= \text{Max}_{a_t, x_t, H_t, L_t} \left\{ U\left[\left(R_t A_{t-1} - a_{0t} - \sum_{k \geq 1} p_{kt} a_{kt} \right) / q_{0t}, L_t, M_t V_{t+1}(a_t) \right] \right\}$$

$$: \left[(x_t, H_t, a_{1,t-1}, \ldots, a_{k,t-1}) \in \Omega_t \right], \tag{4}$$

where $V_t(a_{t-1})$ is the indirect utility function (or value function) at time t,

$$A_t \equiv a_{0t} + \sum_{k \geq 1} p_{kt} a_{kt}, \text{ and}$$

$$R_t \equiv \left[(1 + r_t) a_{0,t-1} + \sum_{k \geq 1} p_{kt} a_{k,t-1} + w_t (T - H_t - L_t) + q_t' x_t \right] / A_{t-1},$$

assuming $W_t > 0$.

Next, we introduce transaction costs in the markets for risky assets. Under transaction costs, the net asset price paid (or received) by each household can vary depending on its market position. For the kth asset at time t, we consider price specification

$$p_{kt}(m_{kt}) = p_{kt}^+ \equiv p_{kt}^0 + v_{kt}^+, \text{ if } m_{kt} \equiv a_{kt} - a_{k,t-1} > 0,$$

$$= p_{kt}^- \equiv p_{kt}^0 - v_{kt}^- \text{ if } m_{kt} \equiv a_{kt} - a_{k,t-1} < 0, \tag{5}$$

$k = 1, \ldots, K$ where p_{kt}^0 is the observed transaction price, $v_{kt}^+ \geq 0$ is the unit transaction cost paid when $m_{kt} > 0$, p_{kt}^+ is the net buying price paid for a_{kt}, $v_{kt}^- \geq 0$ is the unit transaction cost paid when $m_{kt} < 0$, and p_{kt}^- is the net selling price for a_{kt}. In general, $p_{kt}^+ \geq p_{kt}^-, k = 1, \ldots, K$. When $p_{kt}^+ = p_{kt}^-$, there are no transaction costs ($v_{kt}^+ = v_{kt}^- = 0$) and all buyers and sellers face the same net price p_{kt}^0. However, when $p_{kt}^+ > p_{kt}^-$, the net purchase price paid by a buyer becomes larger than the net selling price received by a seller. The difference, $(p_{kt}^+ - p_{kt}^-) \equiv v_{kt}^+ + v_{kt}^- \geq 0$, is a price wedge that represents unit transaction costs in the kth asset market. These costs are also sunk costs in the sense that a buyer cannot change his or her mind and sell what was just bought without a loss. In general, the difference $(p_{kt}^+ - p_{kt}^-) \equiv v_{kt}^+ + v_{kt}^-$ reflects transportation costs, information costs, and brokerage fees (among other costs) involved in trading the kth asset. In the presence of transaction costs, note that the net-revenue function $[p_{kt}(m_{kt}) m_{kt}]$ in (3a) is continuous but has a kink at $m_{kt} = 0$. Indeed, from (5), its marginal value is p_{kt}^+ when $m_{kt} > 0$, but it is p_{kt}^- when $m_{kt} < 0$. As a result, when $p_{kt}^+ > p_{kt}^-$, the net revenue function $[p_{kt}(m_{kt})m_{kt}]$ in (3a) is not differentiable at $m_{kt} = 0$.

Under non-satiation in y_t, the optimization problem of Equation (4) implies that x_t and H_t are chosen so as to maximize profit,

$$\pi_t\left(q_t, w_t, a_{t-1}\right) = \text{Max}_{x_t, H_t}\left[q_t' x_t - w_t H_t : \left(x_t, H_t, a_{t-1}\right) \in \Omega_t\right],$$

where $\pi_t\left(q_t, w_t\right)$ is the indirect profit function. It also implies that the consumption and leisure decisions $\left(y_t, L_t\right)$ solve the utility maximization problem

$$\text{Max}_{y_t, L_t}\left[U\left(y_t, L_t, M_t\right) : q_{0t} y_t + w_t L_t \leq w_t T + \pi_t\left(q_t, w_t, a_{t-1}\right)\right.$$
$$\left. + (1 + r_t) a_{0,t-1} - a_{0t} + \sum_{k \geq 1} P_{kt} a_{k,t-1} - \sum_{k \geq 1} P_{kt} a_{kt}\right\}.$$

In this formulation the budget constraint defines household income as being the sum of three parts: full labor income $\left(w_t T\right)$, profit $\pi_t\left(q_t, w_t, a_{t-1}\right)$, and net income from investments $\left[(1 + r_t) a_{0,t-1} - a_{0t} + \sum_{k \geq 1} P_{kt} a_{k,t-1} - \sum_{k \geq 1} P_{kt} a_{kt}\right]$. In this context, the wage rate w_t has three effects: (1) it measures the opportunity cost of leisure L_t; (2) it has a positive effect on full labor income $\left(w_t T\right)$; and (3) it has a negative effect on profit $\pi_t\left(q_t, w_t, a_{t-1}\right)$, since $\pi_t\left(q_t, w_t, a_{t-1}\right)$ is a decreasing function of w_t from Hotelling's lemma. The first effect is a price effect, while the second and third effects are income effects. This generates the well-known result that the net effect of a change in the wage rate w_t on consumption-time and household-time allocation is ambiguous depending on the relative magnitude of these three effects.

Finally, under transaction costs, the first-order necessary conditions for an interior solution for a_t in Equation (4) are

$$a_{0t} : (1 + r_t) = q_{0t}\left[(\partial U / \partial M_t)(\partial M_t / \partial a_{0t}) / (\partial U / \partial y_t)\right], \qquad (6)$$

$$a_{kt} : p_{kt}^- = q_{0t}\left[(\partial U / \partial M_t)(\partial M_t / \partial a_{kt}) / (\partial U / \partial y_t)\right], \text{ if } m_{kt} < 0, \qquad (7a)$$

$$p_{kt}^- \leq q_{0t}\left[(\partial U / \partial M_t)(\partial M_t / \partial a_{kt}) / (\partial U / \partial y_t)\right] \leq p_{kt}^+, \text{ if } m_{kt} = 0, \qquad (7b)$$

$$p_{kt}^+ = q_{0t}\left[(\partial U / \partial M_t)(\partial M_t / \partial a_{kt}) / (\partial U / \partial y_t)\right], \text{ if } m_{kt} > 0, \qquad (7c)$$

for $k = 1, \ldots, K$. Equation (6) states that at the optimum for the riskless asset a_{0t}, marginal cost $(1 + r_t)$ equals marginal benefit

$$q_{0t}\left[(\partial U/\partial M_t)(\partial M_t/\partial a_{0t})/(\partial U/\partial y_t)\right].$$

Equations (7a) and (7c) state similar results for the kth risky asset. They state that marginal cost p_{kt}^- in Equation (7a) under a disinvesting scenario, or p_{kt}^+ in Equation (7c) under an investment scenario) equals marginal benefit $\left[q_{0t}(\partial U/\partial M_t)(\partial M_t/\partial a_{kt})/(\partial U/\partial y_t)\right]$. In addition, Equation (7b) identifies optimal behavior when the marginal benefit $\left[q_{0t}(\partial U/\partial M_t)(\partial M_t/\partial a_{kt})/(\partial U/\partial y_t)\right]$ is larger than p_{kt}^- but smaller than p_{kt}^+. This is the zone of asset fixity, where the household has no incentive to invest or to disinvest in the kth asset. This generates the following result:

Proposition 1: Transaction Costs Provide a Disincentive to Either Invest or to Disinvest

Illustrated in Figure 13.1, Equations (7a), (7b), and (7c) identify three distinct regimes: (a) the regime of disinvestment when the marginal benefit is relatively low, (b) the regime of asset fixity, and (c) the regime of investment when the marginal benefit is relatively high. Note that this generates a situation of hysteresis. To illustrate, consider a situation in which there is a transitory economic change

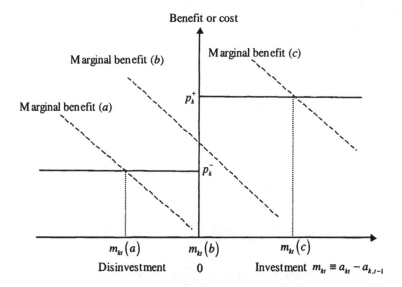

Figure 13.1 Investment under Transaction Costs

from (b) to (c) then back to (b) (e.g., due to temporary changes in output price). The first change from (a) to (b) creates an incentive to invest. However, the second change in regime back to (b) does not create an incentive to disinvest. Thus while the household faces exactly the same economic conditions before and after the change, its asset holding will differ. Under stated Equation (2), this difference will persist forever. A similar result would be obtained under a transitory change in regime from (b) to (a) then back to (b) (e.g., due to a temporary decrease in output price). In this case, the household would first disinvest but then would fail to reinvest after the change. This gives the following result:

Proposition 2: Under Transaction Costs, Transitory Changes in Economic Conditions can have Long-Term Effects on the Path of Asset Accumulation

Denote the optimal decision for a_t in (4) by a_t^*. Consider the decision rule for the kth asset: $a_{kt}^*\left(a_{k,t-1}, \cdot\right), k = 1, \ldots, K$. Equations (7a), (7b) and (7c) imply the existence of three regimes. Define $d_{kt} = \inf_{\alpha \geq 0}\left[\alpha : a_{kt}^*(\alpha, \cdot) - \alpha < 0\right]$ and $g_{kt} = \sup_{\alpha \geq 0}\left[\alpha : a_{kt}^*(\alpha, \cdot) - \alpha > 0\right]$. The term d_{kt} is the smallest amount of capital $a_{k,t-1}$ that would induce disinvestment in the kth asset at time t. The term g_{kt} is the largest amount of capital $a_{k,t-1}$ that would induce investment in the kth asset at time t. The decision rule $a_{kt}^*\left(a_{k,t-1}, \cdot\right)$ then satisfies

$$a_{kt} = a_{kt}^* < a_{k,t-1}, \text{ if } a_{k,t-1} > d_{kt}, \quad \text{(disinvestment),} \qquad (8a)$$

$$= a_{k,t-1}, \text{ if } g_{kt} \leq a_{k,t-1} \leq d_{kt}, \quad \text{(asset fixity),} \qquad (8b)$$

$$= a_{kt}^* \geq a_{k,t-1}, \text{ if } a_{k,t-1} < g_{kt}, \quad \text{(investment),} \qquad (8c)$$

which is illustrated in Figure 13.2 and shows that any point in the zone of asset fixity $\left(g_{kt} \leq a_{k,t-1} \leq d_{kt}\right)$ implies that the household does not participate in the kth asset market at time t. Figure 13.1 also shows that investment takes place when $a_{k,t-1}$ is relatively small with $a_{k,t-1} < g_{kt}$. Alternatively, disinvestment takes place when $a_{k,t-1}$ is relatively large with $a_{k,t-1} > d_{kt}$.

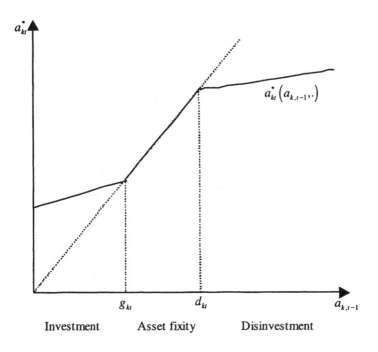

Figure 13.2 Optimal Capital Accumulation under Transaction Costs

Market Equilibrium

Consider a market comprising a set of households. We allow heterogeneity among market participants by classifying households into different types. Assume that households are identical within a type, but that they can differ across types. Let N denote the set of household types, and let $f(i)$ be the number of households in the ith type, $i \in N$. Given $a_{i,t-1} = (a_{1i,t-1}, \ldots, a_{Ki,t-1})$ and $p_t^0 = (p_{1t}^0, \ldots, p_{Kt}^0)$, denote by $a_{kit}^*(a_{i,t-1}, p_t^0)$ the optimal decision rule for the kth asset in the ith household type at time t. Then, $\left[a_{kit}^*(a_{i,t-1}, p_t^0) - a_{ki,t-1} \right]$ is the excess demand for the kth asset in the ith household type. The aggregate excess demand function for the kth asset at time t is $\sum_{i \in N} \left[a_{kit}^*(a_{i,t-1}, p_t^0) - a_{ki,t-1} \right] f(i)$.

Define the market equilibrium condition for the K risky assets as the prices p_t^0 that satisfy a zero aggregate excess demand function

$$\sum_{i \in N} \left[a_{kit}^*(a_{i,t-1}, p_t^0) - a_{ki,t-1} \right] f(i) = 0, \tag{9}$$

where $k = 1,\ldots,K$. Equation (9) represents market-clearing conditions. It is a system of K equations in K unknown variables $p_t^0 = \left(p_{1t}^0,\ldots,p_{Kt}^0 \right)$. This shows that finding the market equilibrium prices for the K assets involves finding the value $p_t^* = \left(p_{1t}^*,\ldots,p_{Kt}^* \right)$ that solves Equation (9).

To illustrate, assume there is a single risky asset a_t (with $K = 1$), and that the marginal benefit function in Equation (7) for the ith household type takes the form

$$\left[q_{0t} \left(\partial U / \partial M_t \right)\left(\partial M_t / \partial a_t \right) / \left(\partial U / \partial y_t \right) \right] \equiv \alpha_{i1t} + \alpha_{2t}a_{it} + \alpha_{i3t}a_{i,t-1},$$

where $\alpha_{2t} < 0$ and $\alpha_{2t} + \alpha_{i3t} < 0, i \in N$. For simplicity, we assume that p_t^0 and α_{2t} are the same for all households. Then, for the ith household type at time t,

$$d_{it} = \left(p_t^0 - v_{it}^- - \alpha_{i1t} \right)/\left(\alpha_{2t} + \alpha_{i3t} \right), \quad g_{it} = \left(p_t^0 + v_{it}^+ - \alpha_{i1t} \right)/\left(\alpha_{2t} + \alpha_{i3t} \right),$$

and the optimal decision in Equations (8a), (8b), and (8c) becomes

$$a_{it}^* = \left(p_t^0 - v_{it}^- - \alpha_{i1t} - \alpha_{i3t}a_{i,t-1} \right)/\alpha_{2t}, \text{ if } a_{i,t-1} > d_{it}, \text{ (disinvestment),} \quad (10a)$$

$$= a_{i,t-1}, \text{ if } g_{it} \leq a_{i,t-1} \leq g_{it}, \text{ (asset fixity),} \quad (10b)$$

$$= \left(p_t^0 + v_{it}^+ - \alpha_{i1t} - \alpha_{i3t}a_{i,t-1} \right)/\alpha_{2t}, \text{ if } a_{i,t-1} < g_{it}, \text{ (investment),} \quad (10c)$$

where household heterogeneity is represented by $v_{it}^-, v_{it}^+, \alpha_{i1t}, a_{i3t}$, and $a_{i,t-1}$, which can vary across household types, $i \in N$. For the ith household type, let

$$\beta_{i1t} \equiv -a_{i,t-1} + \left(-v_{it}^- - \alpha_{i1t} - \alpha_{i3t}a_{i,t-1} \right)/\alpha_{2t}, \text{ and}$$

$$\beta_{i2t} \equiv -a_{i,t-1} + \left[v_{it}^+ - \alpha_{i1t} - \alpha_{i3t}a_{i,t-1} \right]/\alpha_{2t}, i \in N.$$

At time t, denote the distribution function for β_{i1t} and β_{i2t} across households by $F_{1t}(\cdot)$ and $F_{2t}(\cdot)$, respectively. Given $m_{it}^* \equiv a_{it}^* - a_{i,t-1}$, Equations (10a), (10b), and (10c) can be written as

$$m_{it}^* = \beta_{i1t} + \beta_{3t}p_t^0, \text{ if } \beta_{i1t} + \beta_{3t}p_t^0 < 0, \text{ (disinvestment),} \quad (11a)$$

$$= 0, \text{ if } \beta_{i2t} + \beta_{3t}p_t^0 \leq 0 \leq \beta_{i1t} + \beta_{3t}p_t^0, \text{ (asset fixity),} \quad (11b)$$

$$= \beta_{i2t} + \beta_{3t}p_t^0, \text{ if } \beta_{i2t} + \beta_{3t}p_t^0 > 0, \text{ (investment),} \quad (11c)$$

where $\beta_{3t} = 1/\alpha_{2t} < 0$ and $\beta_{ilt} \geq \beta_{i2t}, i \in N$. The market-equilibrium condition of Equation (9) becomes $E_{it}(m_{it}^{*}) = 0$ (expressed on a per-household basis), when E_{it} is the expectation operator over all households at time t, and can be written as

$$G_t\left(p_t^0, \gamma\right)$$
$$\equiv \int_{-\infty}^{-\beta_3 p_t^0}\left(\beta_{ilt} + \beta_{3t}p_t^0\right)dF_{1t}\left(\beta_{ilt}\right) + \int_{-\beta_3 p_t^0}^{\infty}\left(\beta_{i2t} + \beta_{3t}p_t^0\right)dF_{2t}\left(\beta_{i2t}\right) = 0, (12)$$

where γ is a vector of parameters. Then, the market equilibrium price p_t^* is the solution of $G_t\left(p_t^0, \cdot\right) = 0$ for p_t^0.

Using the Leibnitz rule, differentiating $G_t\left(p_t^0, \cdot\right)$ at points where the function is differentiable with respect to p_t^0 gives

$$\partial G_t / \partial p_t^0 = \beta_{3t}\left[F_{1t}\left(-\beta_3 p_t^0\right) + 1 - F_{2t}\left(-\beta_3 p_t^0\right)\right]. \tag{13}$$

In general, $0 \leq \left[F_{1t}\left(-\beta_3 p_t^0\right) + 1 - F_{2t}\left(-\beta_3 p_t^0\right)\right] \leq 1$. It includes two extreme special cases: (1) zero transaction costs and (2) infinite transaction costs. First, consider the absence of transaction costs in which $v_{it}^- = v_{it}^+ = 0$, and $\beta_{ilt} = \beta_{i2t}, i = \in N$. Then, $\left[F_{1t}\left(-\beta_3 p_t^0\right) + 1 - F_{2t}\left(-\beta_3 p_t^0\right)\right] = 1$, and Equation (12) becomes $G_t\left(p_t^0, \gamma\right) \equiv \beta_{0t} + \beta_{3t}p_t^0 = 0$, where β_{0t} is the average value of β_{ilt} and β_{i2t} across all households. It follows from Equation (13) that $\partial G_t / \partial p_t^0 = \beta_{3t} < 0$. Thus in the absence of transaction costs, β_{3t} is the slope of the excess demand function $G_t\left(p_t^0, \cdot\right)$.

Second, consider the case in which transaction costs v_{it}^- and v_{it}^+ are very large for all households $i \in N$, which means that the zone of asset fixity is very large (i.e., no household has an incentive to participate in the asset market). In this extreme case, $F_{1t}\left(-\beta_3 p_t^0\right) > 0$ only for very high prices p_t^0, and $F_{2t}\left(-\beta_3 p_t^0\right) < 1$ only for very low prices p_t^0. For intermediate prices $p_t^0, F_{1t}\left(-\beta_3 p_t^0\right) = 0$, $F_{2t}\left(-\beta_3 p_t^0\right) = 1$, and there is neither buyer nor seller in the asset market. From Equation (13), this also implies that $\partial G_t / \partial p_t^0 = 0$, because price p_t^0 fails to influence the quantity traded. Thus, when transaction costs become excessive, the asset market becomes inactive. The market equilibrium condition $G_t\left(p_t^0, \cdot\right) = 0$ holds only for price p_t^0, in which there is no trade. At such prices, the price elasticity of the aggregate excess-demand function is zero. In intermediate situations, transaction costs may be positive but not large enough to make the market completely inactive. This reflects the situation found in most asset markets, in which we have $0 < \left[F_{1t}\left(-\beta_3 p_t^0\right) + 1 - F_{2t}\left(-\beta_3 p_t^0\right) \right] < 1$ for some market-equilibrium prices p_t^0. From Equation (13), this implies that any increase in price p_t^0 tends to decrease the aggregate quantity demanded $\partial G_t / \partial p_t^0 < 0$. This gives the intuitive result that the aggregate excess-demand function $G_t(p_t^0, \cdot)$ is downward sloping. It also means that Equation (12) has a unique market equilibrium price p_t^*. Assuming differentiability is present, applying the implicit function theorem to Equation (12) yields

$$\partial p_t^* / \partial \gamma = -\left(\partial G_t / \partial \gamma\right) \big/ \left(\partial G_t / \partial p_t^0\right) = \text{sign}\left(\partial G_t / \partial \gamma\right). \tag{14}$$

Equation (14) implies that any factor γ that shifts up (down), the excess demand function $G_t(p_t^0, \gamma)$ tends to increase (decrease) the market equilibrium price p_t^*. This is intuitive, given that the aggregate excess-demand function is downward sloping. In addition, when $0 < \left[F_{1t}\left(-\beta_3 p_t^0\right) + 1 - F_{2t}\left(-\beta_3 p_t^0\right) \right] < 1$, Equation (13) implies that $\left| \partial G_t / \partial p_t^0 \right| < \left| \beta_{3t} \right|$. Since we have seen that β_3 is the slope of the excess demand function in the absence of transaction costs, this generates the following result:

Proposition 3: Transaction Costs Tend to Reduce the Price Elasticity of the Aggregate Excess-Demand Function

Proposition 3 states that transaction costs tend to decrease the aggregate-quantity response to a price change. Intuitively, this is due to a reduction in the incentive to participate in market exchange, which suggests that any particular shift in the excess demand function will have a larger impact on price p_t^0 as transaction costs increase. When the shifts are due to unanticipated shocks, it follows that asset prices are expected to become more volatile in the presence of transaction costs. Alternatively, a reduction in transaction costs would increase the elasticity of the aggregate excess-demand function and would reduce the effects of unanticipated shocks on asset prices.

The effect of transaction costs on price determination is illustrated in Figure 13.3 in the context of a two-household market. Figure 13.3a shows the market equilibrium under moderate transaction costs. In this case, there is a price p_t^0 where the market clears (with $m_{1t}^* + m_{2t}^* = 0$) with active trading (and household 1 sells to household 2). In contrast, Figure 13.3b shows a situation of high transactions costs, where market clearing takes place only in the absence of trade. Figure 13.3a and 13.3b also illustrates that the effects of price on the aggregate excess-demand function become larger (smaller) when the number of market participants increases (decreases).

a. Market equilibrium under moderate transaction cost

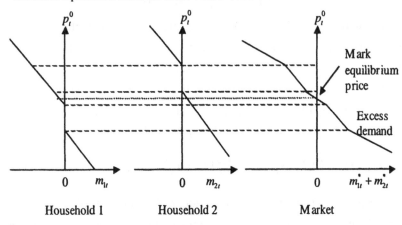

b. Market equilibrium under high transaction cost

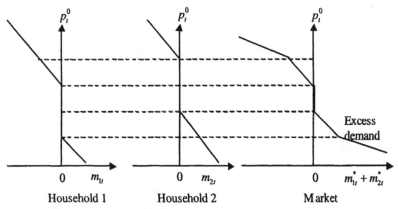

Figure 13.3 Price Determination in the Presence of Transaction Costs

IMPLICATIONS

Our analysis has a number of implications for the functioning of land markets. It provides some insights on the dynamic process of creating and improving market performance. Note the role played by household heterogeneity in Equation (9). If all households were the same, there would be no incentive to trade. Thus exchange is motivated by heterogeneity among households, and by low transaction costs. This suggests that the key factors in the proper functioning of asset markets are: (1) to identify the potential buyers and sellers, and (2) to facilitate the exchange process with low information and transaction costs.

In this chapter, we focused our attention on the role of risk and the effects of transaction costs on land markets. As noted in the introduction, production risks as well as price risks are significant in agriculture and contribute to uncertain returns to land. Further, realtor fees, heterogeneity in land quality, and the spatial characteristics of land all contribute to transaction and information costs faced by participants in the land markets.

First, consider the effects of risk on land prices. The move toward agricultural-market liberalization in the 1990s contributed to lower farm prices and to greater farm price instability. Both put downward pressure on farmland prices. However, recent increases in government payments to farmers have had the opposite effect and have put upward pressure on land prices.

In the context of the Epstein-Zin model (Epstein and Zin, 1989 and 1991), the value of the parameters α and ρ shed some light on the impact of risk on land prices (Chavas and Thomas 1999; Lence 2000). When $\alpha = 0$, the model generates as a special case—CAPM. In CAPM, the risk associated with an asset is measured by the covariance between its return and the overall return. Alternatively, when $\alpha = \rho \neq 0$ the Epstein-Zin model generates the inter-temporal consumption-based, capital-asset pricing model (CCAPM) whereby the risk associated with an asset is measured by the covariance of its return with the growth rate of consumption. In this context, risk aversion is closely related to consumption-smoothing activities. More generally, under alternative values α and ρ, the Epstein-Zin model gives a general formulation in which both income smoothing and consumption smoothing can be relevant in risk management and asset pricing. In their analysis of U.S. land prices using the Epstein-Zin approach, Chavas and Thomas (1999) find evidence in favor of the CCAPM (and against the static CAPM). This means that the risk associated with land holding must be evaluated compared to the uncertainty in the growth rate of consumption. In other words, risk aversion involves consumption smoothing (as opposed to income smoothing). In this regard, it is worth emphasizing the role of nonfarm income. Indeed, even in the presence of significant farm-income risk (due to both price and production uncertainty), nonfarm income can play two important roles: (1) it tends to be less risky than farm income and (2) it helps stabilize household consumption. As such, it reduces the effect of risk on asset prices. This suggests that the growth in nonfarm income observed over the last few decades has contributed to reducing the effects of risk on U.S. land prices.

Second, consider the effect of transaction costs on land prices. We have seen that transaction costs create a zone of asset fixity, which reduces the number of households participating in the market (Proposition 1). This means that transaction costs contribute to the existence of thin markets. In addition, transaction costs reduce the elasticity of the aggregate excess demand (Proposition 3). In thin markets, prices become heavily influenced by a few household-specific shocks. In this context, the market price is expected to be volatile for two reasons: (1) the market price will depend on the few shocks affecting the participating households and (2) the inelasticity of aggregate excess demand will translate these shocks into potentially large price swings. This shows that price instability is a basic characteristic of thin markets.

In addition to contributing to thin markets, transaction costs also affect market dynamics. For example, consider a thin market in which few potential buyers search among few potential sellers. When the number of market participants is small, search costs can increase. To the extent that search costs contribute to transaction costs, there may actually be an increase in transaction costs as a market becomes thinner. In this case, we obtain a reverse causal effect: Thin markets contribute to increasing transaction costs. This suggests that market exchange may become difficult below some minimum threshold in the number of market participants. An extreme situation is one with inactive markets, wherein transaction costs are large and there is no market-equilibrium price that can generate market participation (Figure 13.3b). Alternatively, an initial reduction in transaction costs would provide incentives to trade. Increasing the number of market participants may actually reduce search costs, which would further stimulate market participation in the buying and selling of farmland.

As noted by Shiha and Chavas (1995), transaction costs in land markets can generate market segmentation. By reducing trade incentives, transaction costs create barriers to the flow of equity capital into farm real-estate markets. By reducing the number of market participants, they also contribute to the creation of local land markets that fail to be well integrated into global land and capital markets. This segmentation lowers the benefits from market exchange and reduces the efficiency of the allocation of capital markets. This situation also implies that land prices are particularly sensitive to local shocks that fail to be arbitraged over space. Given a lower elasticity of demand, land prices are particularly sensitive to local shocks.

Our analysis provides a basis for evaluating the effects of recent structural changes in land markets. The growing use of the internet in real-estate markets has contributed to reducing information costs and lowering realtor fees. As a result of new information technology over the last decade, transaction and information costs have declined for participants in U.S. land markets. This provides some useful insights on the recent changes in these markets. With lower transaction costs, the zone of asset fixity shrinks, the size of the market expands, the aggregate excess-demand function becomes more elastic. This improves market integration, over both space and across equity markets. Lower transaction costs can also contribute to reducing the volatility of land prices. First, by increasing the number of market participants, they stimulate price arbitrage. Second, they increase the price elasticity of the aggregate demand for land, which reduces the sensitivity of land prices to unanticipated shocks. As a result, lower transaction costs help stimulate the flow of equity capital toward the agricultural sector, they increase the gains from market exchange, and they raise the efficiency of the allocation of capital in agriculture and in the economy.

REFERENCES

Alston, J.M. (1986) "An Analysis of Growth of U.S. Farmland Prices, 1963–82." *American Journal of Agricultural Economics* 68(1): 1–9.

Barry, P.J., L.J. Robison, and G.V. Neartea. (1996) "Changing Time Attitudes in Inter-Temporal Analysis." *American Journal of Agricultural Economics* 78: 972–81.

Bjornson, B. and R. Innes. (1992) "Another Look at Returns to Agricultural and Nonagricultural Assets." *American Journal of Agricultural Economics* 74: 109–19.

Burt, O.R. (1986) "Econometric Modeling of the Capitalization Formula for Farmland Prices." *American Journal of Agricultural Economics* 68: 10–26.

Chavas, J.P. and A. Thomas. (1999) "A Dynamic Analysis of Land Prices." *American Journal of Agricultural Economics* 81(1): 772–84.

Clark, J.S., M. Fulton, and J.T. Scott, Jr. (1993) "The Inconsistency of Land Values, Land Rents, and Capitalization Formulas." *American Journal of Agricultural Economics* 75: 147–55.

Epstein, L.G. and S.E. Zin. (1991) "Substitution, Risk Aversion, and the Temporal Behavior of Consumption and Asset returns: An Empirical Analysis." *Journal of Political Economy* 99: 263–86.

———. (1989) "Substitution, Risk Aversion, and the Temporal Behavior of Consumption and Asset Returns: A Theoretical Framework." *Econometrica* 57: 937–969.

Falk, B. (1991) "Formally Testing the Present Value Model of Farmland Prices." *American Journal of Agricultural Economics* 73(1): 1–10.

Featherstone, A.M. and T.G. Baker. (1987) "An Examination of Farm Sector Real Asset Dynamics: 1910–85." *American Journal of Agricultural Economics* 69(3): 532–46.

Hanson, L.P. and R.J. Myers. (1995) "Testing for a Time-Varying Risk Premium in the Returns to U.S. Farmland." *Journal of Empirical Finance* 2(3): 265–76.

Just, R.E. and J.A. Miranowski. (1993) "Understanding Farmland Price Changes." *American Journal of Agricultural Economics* 75(3): 156–68.

Lence, S.H. (2000) "Using Consumption and Asset Return Data to Estimate Farmers' Time Preferences and Risk Attitudes." *American Journal of Agricultural Economics* 82: 934–47.

Melichar, E. (1979) "Capital Gains versus Current Income in the Farm Sector." *American Journal of Agricultural Economics* 61(5): 1085–92.

Shiha, A.N. and J.P. Chavas. (1995) "Capital Market Segmentation and U.S. Farm Real-Estate Pricing." *American Journal of Agricultural Economics* 77(2): 397–407.

Chapter 14

Spectral Evidence on the Investment Horizon and Transaction Costs for Present-Value Models of Iowa Farmland Prices

Douglas J. Miller
Purdue University

INTRODUCTION

One of the most widely used asset-pricing models, especially for farmland values, is the constant discount rate, present-value model (CDR-PVM). The model represents the current asset price as $P_t = \delta E_t \left(P_{t+1} + D_{t+1} \right)$ where P_t is the farmland price, δ is a constant risk-adjusted discount factor, D_t is the dividend earned from the asset at time t, and E_t is the conditional expectation operator (based on information at time t). By iterative substitution of this expression, we can solve for an alternative version of the CDR-PVM

$$P_t = \sum_{s=1}^{\infty} \delta^s E_t \left(D_{t+s} \right) \tag{1}$$

that expresses the current asset price solely in terms of future dividends. The CDR-PVM in Equation (1) implies that the current asset price is the sum of discounted conditional expected values of future dividends $E_t \left(D_{t+s} \right)$. The price represents an equilibrium outcome in the asset market because higher (lower) values of P_t provide an incentive to sell (or buy) land. The equilibrium condition is also based on the implicit assumption that it is costless to transfer land between buyers and sellers.

As noted by Lence and Miller (1999), the CDR-PVM has been rejected by several earlier studies of U.S. farmland prices. One plausible explanation for such broad rejection of the CDR-PVM is the unrealistic assumption of zero transaction costs. If buyers and sellers must pay for the services of brokers and other intermediaries involved in the transaction, market participants may not act if the transaction costs exceed the expected gains from the purchase or sale of land. Consequently, the equilibrium asset price may not be exactly equal to the sum of discounted expected dividends, and the asset-pricing equilibrium condition will include a range of inactivity (examples are provided below). Given that the transaction cost for land transfers is relatively large (Wunderlich [1989] reports a range of buying and selling costs with midpoint 7.5 percent) compared to other types of assets, the presence of non-zero transaction costs may be an important factor in farmland markets that is ignored by the stated CDR-PVM.

Lence and Miller (1999) evaluate the empirical evidence in support of the CDR-PVM based on observed annual Iowa farmland prices from 1900 to 1994. They derive the equilibrium conditions for the present value model with transaction costs and with one-period and infinite investment horizons. The equilibrium conditions for both cases are evaluated with formal bootstrap tests, which fail to reject the one-period-horizon model and reject the infinite-horizon model. As noted in Lence and Miller, if we believe the one-period-horizon model is appropriate, the CDR-PVM with transaction costs provides an adequate model of farmland prices (at least for this sample of Iowa data). Further, Lence (2001) notes that the results imply the existence of a finite investment-horizon boundary such that the CDR-PVM with transaction costs holds for all shorter horizon lengths. Evidence regarding the existence and magnitude of this horizon boundary is potentially important, as it would not only help to develop a deeper understanding of farmland prices but would also provide useful information regarding government policies that influence farmland values (e.g., agricultural land-use regulations, direct- or indirect-payment programs, conservation-reserve programs and environmental programs, and tax policies).

The purpose of this chapter is to evaluate additional evidence regarding the existence and magnitude of a finite investment horizon based on observed Iowa farmland prices. The method used to gather the evidence is based on the premise that if there exists a finite investment horizon such that the CDR-PVM with transaction costs holds for all shorter horizon lengths, then short-term or transitory movements in farmland prices that occur within the investment horizon should satisfy the required equilibrium conditions. Conversely, long-term movements in farmland prices that evolve beyond the investment horizon may not satisfy the equilibrium conditions. To examine the behavior of Iowa farmland investors in response to short-term and long-term components of the expected discounted excess returns, I use a spectral or frequency-specific version of the excess-volatility test. The empirical results from the spectral procedure may confirm the existence of a finite investment horizon under the CDR-PVM with transaction costs or indicate other types of CDR-PVM violations. The results may also provide evidence regarding the magnitude of the finite horizon boundary. The data used for this study are the Iowa farmland returns examined by Lence and Miller (1999). I review their findings in the next section.

THE LENCE AND MILLER (1999) FINDINGS

Let T_S and T_B represent the transaction costs of selling and buying farmland as a percentage of the asset price. Under a one-period investment horizon, an agent will sell (buy) land in period t if the expected discounted return from holding land until period $t+1$ is less (more) than the current price net of transaction costs. The agent's one-period-horizon transaction decision is

$$\text{Buy if } (1+T_B)P_t < \delta E_t \left[(1-T_S)P_{t+1} + D_{t+1} \right]; \tag{2}$$

$$\text{Sell if } (1-T_S)P_t > \delta E_t \left[(1+T_B)P_{t+1} + D_{t+1} \right]. \tag{3}$$

Otherwise agents will not act because the transaction costs exceed the expected gain from a sale or purchase.

The range of inactivity may be characterized by reversing the inequalities in Equations (2) and (3) to form the simultaneous conditions

$$(1+T_B)P_t \geq \delta E_t \left[(1-T_S)P_{t+1} + D_{t+1} \right], \text{ and} \tag{4}$$

$$(1-T_S)P_t \leq \delta E_t \left[(1+T_B)P_{t+1} + D_{t+1} \right]. \tag{5}$$

The range of inactivity may be restated to form an equilibrium condition for the discounted excess returns, $h_t \equiv \delta(P_{t+1} + D_{t+1})/P_t - 1$. Under the one-period investment horizon, the expected discounted excess returns must satisfy

$$-\left(\frac{T_B + T_S}{1+T_B} \right) \leq E_t(h_t) \leq \left(\frac{T_B + T_S}{1-T_S} \right). \tag{6}$$

If transaction costs are zero $(T_S = T_B = 0)$, the expected discounted excess return from holding land for one period should be zero, $E_t(h_t) = 0$, for all t. Otherwise, we have reason to reject the CDR-PVM. If transaction costs are non-zero $(T_S > 0 \text{ or } T_B > 0)$, the expected discounted excess returns from holding land for one period may not be zero, $E_t(h_t) \neq 0$. We should only reject the CDR-PVM if at least one of the $E_t(h_t)$ is adequately large such that

$$-\left(\frac{T_B + T_S}{1+T_B} \right) > E_t(h_t) \text{ or } E_t(h_t) > \left(\frac{T_B + T_S}{1-T_S} \right). \tag{7}$$

for at least one t.

The other case considered by Lence and Miller (1999) is the purchase or sale decision facing agents with an infinite investment horizon. Given the transaction costs of purchase or sale, the agent's decision may be expressed as

$$\text{Buy if } (1+T_B)P_t < \sum_{s=1}^{\infty} \delta^s E_t (D_{t+s});\tag{8}$$

$$\text{Sell if } (1-T_S)P_t > \sum_{s=1}^{\infty} \delta^s E_t (D_{t+s}).\tag{9}$$

The range of inactivity may be derived by reversing the inequalities (as above), and the condition may be restated in terms of the infinite-horizon expected discounted excess returns

$$-T_S \le E_t (g_t) \le T_B,\tag{10}$$

where $g_t \equiv \left(P_t^{-1} \sum_{s=1}^{\infty} \delta^s D_{t+s} \right) - 1$ is the discounted excess return from holding farmland indefinitely. Although g_t is not observable, Lence and Miller (1999) develop a proxy variable based on a finite approximation to the infinite sum. If transaction costs are zero $(T_S = T_B = 0)$, the expected discounted excess return from holding land indefinitely should be zero, $E_t (g_t) = 0$, for all t. Otherwise, we have reason to reject the CDR-PVM. If transaction costs are non-zero $(T_S > 0 \text{ or } T_B > 0)$, the expected discounted excess returns from holding land indefinitely may not be zero, $E_t (g_t) \ne 0$. Further, we should only reject the CDR-PVM if at least one of the $E_t (g_t)$ is adequately large such that

$$-T_S > E_t (g_t) \text{ or } E_t (g_t) > T_B.\tag{11}$$

Lence and Miller (1999) use annual observations of Iowa farmland prices and returns for 1900 to 1994 to formally test the null hypotheses associated with Equations (6) and (10). Due to the observed non-normality of the excess return series, they use a bootstrap test to evaluate the null hypotheses. Further, the rejection conditions [Equations (7) and (11)] depend on the magnitude of the transaction costs. To assist the reader in evaluating the test results, Lence and Miller (1999) conduct the bootstrap-test procedure at a variety of Type I error probabilities (α) and sums of the transaction costs $(T_S + T_B)$. They use the bootstrap test results to form a test-rejection frontier that represents the combined values of test size and transaction costs that lead to rejection of the null hypotheses. In summary, they cannot reject the CDR-PVM for Iowa farmland prices in the one-period-horizon case at any reasonable combination of test size and transaction-cost level. In contrast, they strongly reject the CDR-PVM based on an infinite investment horizon.

Again, one of the key implications of these results is that there exists some finite investment horizon boundary for the observed Iowa farmland prices such that the CDR-PVM with transaction costs holds for all shorter horizons. In the following section I outline a spectral or frequency-specific approach that may be used to diagnose violations of the CDR-PVM and to gather evidence regarding the existence and magnitude of the investment-horizon boundary.

SPECTRAL ANALYSIS OF THE EXPECTED DISCOUNTED EXCESS RETURNS

One of the earliest hypothesis test procedures used to evaluate asset-pricing behavior is the excess volatility test introduced by Shiller (1981) and by LeRoy and Porter (1981). Although the original excess volatility tests were criticized (Flavin 1983) and other methods such as regression-based test procedures (e.g., Tegene and Kuchler 1993) may be more powerful, excess-volatility tests based on excess-returns data can avoid some of the cited problems. For present purposes, I extend the excess-volatility concept to develop a test of the CDR-PVM under short-term (high-frequency) and long-term (low-frequency) components of the expected discounted excess returns. As noted above, the empirical results may provide evidence regarding the existence and magnitude of a finite investment horizon under the CDR-PVM with transaction costs.

To introduce the excess-volatility concept, note that the equilibrium conditions used as the null hypotheses for the bootstrap tests of the CDR-PVM with transaction costs, Equations (6) and (10), may be stated in terms of the unconditional variance of excess returns. In particular, Equation (6) for the one-period-horizon case may be restated as

$$\left[\hat{E}_t\left(h_t\right)\right]^2 \leq \left(\frac{T_B+T_S}{1-T_S}\right)^2, \tag{12}$$

which is a slightly weaker equilibrium condition than Equation (6) because $\left(T_B+T_S\right)/\left(1+T_B\right) < \left(T_B+T_S\right)/\left(1-T_S\right)$. We also know that Equation (12) must hold for the unconditional expectation, $E\left\{\left[\hat{E}_t\left(h_t\right)\right]^2\right\} = \text{var}\left[\hat{E}_t\left(h_t\right)\right] = \sigma^2$. We should reject the one-period-horizon CDR-PVM with transaction costs if the unconditional variance exceeds the upper bound stated in Equation (12). For the Iowa data, the estimate of σ^2 satisfies Equation (12), and we do not reject the CDR-PVM in the one-period-horizon case.

Accordingly, we can use the variation in the infinite-horizon excess returns to gather evidence regarding the finite investment horizon. In particular, we can divide the unconditional variance σ^2 into variance components associated with low-frequency and high-frequency movements in the expected discounted excess returns. If the variance components associated with low-frequency (long-term) movements in the excess returns exhibit excess volatility relative to the upper bound in Equation (12), we have evidence that is consistent with the existence of a finite investment horizon. If the frequency-specific variances exhibit other patterns of excess volatility, we may have evidence of model-specification error or the presence of other unobserved components in the data.

The frequency-specific variance components of the excess returns data may be computed by converting the observations from the time domain to the spectral or frequency domain. Under Cramer's spectral representation (Hamilton 1994: 157; Harvey 1993: 177), any covariance stationary mean-zero stochastic process may be represented in the frequency domain as

$$y_t = \int_0^\pi u(\lambda) \cos(t\lambda) d\lambda + \int_0^\pi v(\lambda) \sin(t\lambda) d\lambda, \tag{13}$$

where $u(\lambda)$ and $v(\lambda)$ are random functions continuous in $\lambda \in (0, \pi)$. The spectral representation implies that the stochastic process may be constructed from cyclic components with low-frequency (slow moving, λ near zero) and high-frequency (fast moving, λ near π) character. The empirical results reported by Lence and Miller (1999) indicate that the estimated expected discounted excess return series for the one-period- and infinite-horizon cases, $\hat{E}(h_t)$ and $\hat{E}(g_t)$, are covariance stationary processes that have spectral representations.

For any covariance stationary mean-zero stochastic process (i.e., y_t from Equation (13) above), the assumed properties of the $u(\lambda)$ and $v(\lambda)$ functions (Harvey, 1993) imply that we can represent the unconditional variance of y_t as Equation (14)

$$\sigma^2 = E\left[y_t^2\right] = E\left\{\left[\int_0^\pi u(\lambda) \cos(t\lambda) d\lambda + \int_0^\pi v(\lambda) \sin(t\lambda) d\lambda\right]^2\right\} = 2\int_0^\pi f(\lambda) d\lambda, \tag{14}$$

where $f(\lambda)$ is the power spectrum for the stochastic process. In other words, $f(\lambda)$ is proportional to the variation in y_t contributed by cycles with frequency near λ. Typically, much of the variation in economic time series (including asset prices and returns) is due to low-frequency events that evolve over a long period, and the power spectrum declines as λ increases from zero.

Following Hamilton (1994), we can decompose the unconditional variance var(y_t) as the sum of variance components from a finite partition of the interval $[0, \pi]$, $0 < \lambda_1 < \lambda_2 < ... < \lambda_{J-1} < \pi$,

$$\sigma^2 = 2\int_0^{\lambda_1} f(\lambda) d\lambda + ... + 2\int_{\lambda_{J-1}}^\pi f(\lambda) d\lambda = 2\sum_{j=1}^J \sigma_j^2. \tag{15}$$

For the one-period-horizon case, we can use this to restate Equation (12) as

$$2\sum_{j=1}^{J}\sigma_{j}^{2} \le \left(\frac{T_{B}+T_{S}}{1-T_{S}}\right)^{2}. \tag{16}$$

Note that this condition is weaker than the original result stated in Equation (6) due to our use of squared expected discounted excess returns and the unconditional expectation. However, it does provide a plausible means for evaluating consistency of the data with the CDR-PVM across the frequency domain. If any subset of the variance components exceeds the bound in Equation (16), the associated cyclic components of the expected discounted excess returns are not consistent with the observed asset prices under the CDR-PVM with transaction costs.

To illustrate, we begin with the expected discounted excess returns for the one-period-horizon case. Although there are a number of ways to evaluate the frequency-specific variation in the $\hat{E}_{t}(h_{t})$ outcomes, I use a simple transformation of the data from the time domain to the frequency domain based on the real Fourier transformation (Harvey, 1993). The key advantage of the real Fourier transformation over other Fourier transformations is that the frequency-specific outcomes are not complex-valued and may be used to form plots or compute summary statistics. The frequency-specific outcomes for the one-period case are shown in Figure 14.1. The domain of the plot is measured in units $\omega = \lambda/2\pi$, and the reciprocal of ω is the length of the cycle in years. For example, $\omega = 0.2$ indicates a cycle length of five years.

Given that h_{t} series is a stationary mean-zero $AR(1)$ process (Lence and Miller 1999: Table 1), the estimated expected discounted excess returns $\hat{E}_{t}(h_{t})$ are also stationary $AR(1)$. Rather than compute the frequency-specific variance components σ_{j}^{2}, we can use the known spectral properties of stationary $AR(1)$ processes to benchmark the frequency-specific variation in the $\hat{E}_{t}(h_{t})$ outcomes plotted in Figure 14.1. The power spectrum of a stationary $AR(1)$ process with coefficient ρ and variance σ^{2} is

$$f(\lambda) = \frac{\sigma^{2}}{\left(1+\rho^{2}-\rho\cos[\lambda]\right)}. \tag{17}$$

The curves appearing above and below the horizontal axis in Figure 14.1 are the 90-percent-error bands, $\pm 1.645\sqrt{f(\lambda)}$, and are proportional to the frequency-specific standard deviation σ. Note that a larger share of the variation for the $AR(1)$ process is due to low-frequency events, which is consistent with the typical behavior of $f(\lambda)$ for economic variables. Also, the observed spectral data are consistent with the properties of the $AR(1)$ process at all frequencies, and the results are consistent with the outcome of the time-domain test of the CDR-PVM conducted by Lence and Miller (1999).

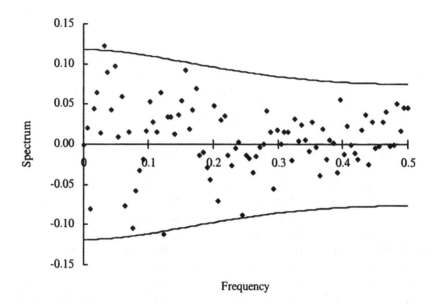

Figure 14.1 Expected Excess Returns In The Spectral Domain (Finite-Horizon Case)

Now I consider the spectral evidence regarding the infinite-horizon CDR-PVM with transaction costs, which was rejected by Lence and Miller (1999). Following the preceding analysis, a weaker version of the equilibrium condition in Equation (10) may be stated as

$$\left[\hat{E}_t\left(g_t\right)\right]^2 \le \max\left(T_B, T_S\right)^2,\tag{18}$$

and the upper bound must also hold for the unconditional variance of the estimated expected discounted excess returns, $E\left\{\left[\hat{E}_t\left(g_t\right)\right]^2\right\} = \operatorname{var}\left[\hat{E}_t\left(g_t\right)\right]$. The unconditional variance may be stated as the finite sum of frequency-specific variance components. Finally, if subsets of the components exceed the upper bound in Equation (18), the observed farmland prices do not represent outcomes of the CDR-PVM with transaction costs at these frequencies.

To examine the frequency-specific variance components for the infinite-horizon case, the observed estimates $\hat{E}_t\left(g_t\right)$ are converted to the spectral domain with the real Fourier transform and the outcomes are plotted in Figure 14.2. As before, the 90-percent-error bands, $\pm 1.645\sqrt{f\left(\lambda\right)}$, are indicated by the lines appearing above and below the horizontal axis and are based on the fitted $AR(1)$ model of the $\hat{E}_t\left(g_t\right)$ series. The spectral outcomes exhibit variation within the $AR(1)$ error bands for all frequencies greater than 0.1 (i.e., cycles in $\hat{E}_t(g_t)$ lasting from 2 years to 10 years in length; right of the vertical line at frequency 0.1). However, the variation in the estimated excess returns for the low-frequency outcomes (i.e., cycles in $\hat{E}_t(g_t)$ longer than 10 years; left of the vertical line at frequency 0.1) increases sharply, and nearly half of the spectral outcomes fall outside the $AR(1)$ error bands. Consequently, these long-term variance components of the expected discounted excess returns are not consistent with the equilibrium conditions for the infinite-horizon version of the CDR-PVM with transaction costs.

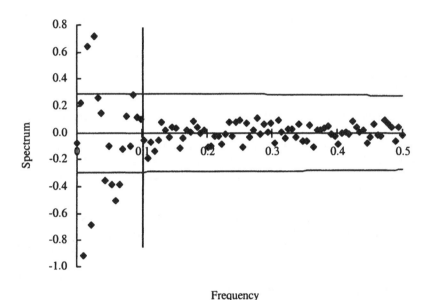

Frequency

Figure 14.2 Expected Excess Returns in the Spectral Domain (Infinite-Horizon Case)

CONCLUSIONS AND IMPLICATIONS

Although the findings presented in the preceding section are not outcomes of formal hypothesis tests, the results do provide useful information about observed investor behavior in the Iowa farmland market. First, the findings for the one-period-horizon version of the CDR-PVM with transaction costs are consistent with those reported by Lence and Miller (1999). There does not appear to be any reason to reject this model based on the spectral evidence. More importantly, the rejection of the infinite-horizon CDR-PVM appears to be due to excess variation in the excess-returns data associated with cycles lasting longer than 10 years. This pattern is consistent with the existence of a finite investment horizon under the CDR-PVM with transaction costs.

Further, the results provide empirical support for the CDR-PVM with transaction costs under finite investment horizons that may be roughly five to ten years in length. Participants in the Iowa farmland market do not appear to account for variation in the expected excess returns when making investment decisions beyond this horizon length. Admittedly, the spectral procedures used in this chapter may have relatively low power, and additional research is required to verify the findings. In particular, formal hypothesis tests regarding the magnitude of the frequency-specific variance components may be derived and conducted. Also, it may be possible to derive more powerful tests based on the equilibrium conditions for present-value models with transaction costs and specific multi-year investment horizons.

REFERENCES

Duffie, D. (2001) *Dynamic Asset Pricing Theory*, 3rd ed., Princeton: Princeton University Press.

Flavin, M. (1983) "Excess Volatility in the Financial Markets: A Reassessment of the Empirical Evidence." *Journal of Political Economy* 91(December): 929–56.

Hamilton, J. (1994) *Time Series Analysis*. Princeton: Princeton University Press.

Harvey, A. (1993) *Times Series Models*, 2nd Edition. Cambridge: MIT Press.

Lence, S. (2001) "Farmland Prices in the Presence of Transaction Costs: A Cautionary Note." *American Journal of Agricultural Economics* 83(November): 985–92.

Lence, S., and D. Miller. (1999) "Transaction Costs and the Present-Value Model of Farmland: Iowa, 1900 to 94." *American Journal of Agricultural Economics* 81(2): 257–72.

LeRoy, S., and R. Porter. (1981) "Stock Price Volatility: Tests Based on Implied Variance Bounds." *Econometrica* 49(May): 97–113.

Shiller, R. (1981) "Do Stock Prices Move Too Much to be Justified by Subsequent Changes in Dividends?" *American Economic Review* 71(3): 421–36.

Tegene, A. and F. Kuchler. (1993) "A Regression Test of the Present Value Model of U.S. Farmland Prices." *Journal of Agricultural Economics* 44(2): 135–43.

Wunderlich, G. (1989) "Transaction Costs and the Transfer of Rural Land." *Journal of the American Society of Farm Managers and Rural Appraisers* 53(April): 13–16.

Chapter 15

Using Threshold Autoregressions to Model Farmland Prices under Transaction Costs and Variable Discount Rates

Sergio H. Lence
Iowa State University

INTRODUCTION

The value of U.S. farmland was estimated at U.S. $593 billion on 31 December 1994, or roughly 9 percent of U.S. gross domestic product for that year. Farmland is the single largest asset in the U.S. farm sector, accounting for about two-thirds of the value of all farm assets. These figures clearly indicate that farmland is a non-negligible asset for the economy as a whole, and that it is of central importance to the U.S. agricultural sector. Interestingly, farmland prices have fluctuated notoriously through cycles of boom and bust periods.

Figures 15.1a and 15.1b show the behavior of Iowa farmland prices during the two price cycles that occurred in the twentieth century.[i] In the first cycle, Iowa land prices increased in every year of the first two decades of the century. By the peak of the cycle in 1920, prices had increased almost fivefold since 1900. During the subsequent downturn, prices fell in each of the following 13 years, ending up in 1933 at only 25 percent of the price level in 1920. Farmland prices in the second cycle behaved similarly. Prices increased steadily from 1960 through 1981 for a total gain of 720 percent. Immediately thereafter, prices dropped in every year from 1981 through 1986 for a total loss of over 60 percent.

Because farmland real estate has been by far the dominant asset in the balance sheet of the U.S. farm sector, the boom-bust cycles in farmland prices have produced great changes in the sector's financial position (Schmitz 1995). The

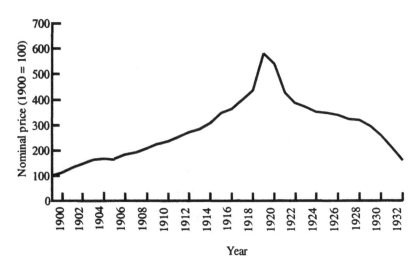

Figure 15.1a Iowa Farmland Prices, 1900 to 1933 (1900 = 100)

adjustments that occurred during the downturns have been extremely painful, and can be properly characterized as agricultural crises. In addition, large drops in land values have typically been accompanied by substantial reductions in the availability of credit to the sector. The reason for this credit problem is that land has been a major source of collateral in agricultural lending. A myriad of studies have attempted to explain the determinants of farmland values. Pope et al. (1979) critically assess land-pricing models that were available prior to 1979. Robison and Koenig (1992) discuss farmland-price modeling up to 1990. Weersink et al. (1996) review the literature on research performed on the determinants of farmland values up to 1996. Two major conclusions emerge from the research that has been

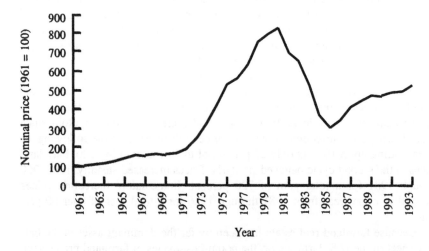

Figure 15.1b Iowa Farmland Prices, 1961 to 1994 (1961 = 100)

performed on farmland prices. First, there is a notorious lack of consensus regarding the determinants of U.S. farmland pricing behavior (Pope et al. 1979; Robison and Koenig 1992). Second, recent studies are based on the constant discount-rate present-value model (CDR-PVM) of farmland prices, and they typically reject it. (Falk 1991; Clark et al. 1993; Tegene and Kuchler 1993).

An important subject in many recent studies has been the impact of transaction costs on farmland prices (de Fontnouvelle and Lence 2002; Lence 2001; Chavas and Thomas 1999; Lence and Miller 1999; Just and Miranowski 1993). Attention has been drawn to this subject because of the magnitude of transaction costs involved in the transfer of farmland ownership. The U.S. Department of Agriculture, Economic Research Service (USDA/ERS) (USDA/ERS 1964), Moyer and Daugherty (1982), Thompson and Whiteside (1987), and Wunderlich (1989) also have studies on the impact of transaction costs on farmland prices.

According to a U.S, Department of Agriculture (USDA) study on the costs of transferring ownership of farm real estate (USDA/ERS 1964), the most popular method of land sale is through brokers, which accounts for about half of voluntary sales. In this instance, the most common sales commission charged for land sales is 5 percent, 6 percent, and 10 percent (60 percent, 12 percent, and 15 percent of the time, respectively). Wunderlich (1989) reports that transaction costs exclusive of sales commission averages 3 percent of the value of land transferred within the United States. One overall consensus, reached by these studies, is that 7.5 percent represents a minimum estimate of the total transaction costs involved in the transfer of farmland ownership.

Based on the Lence (2001) critical review of the literature that analyzes the effect of transaction costs on farmland prices, it may be concluded that transaction costs provide at least a partial explanation for the empirical rejection of the CDR-PVM in earlier studies. Both Lence and Miller (1999), and de Fontnouvelle and Lence (2002) find support for the CDR-PVM under transaction costs and a one-period holding horizon. However, Lence and Miller (1999) reject the CDR-PVM under transaction costs for the case of an infinite holding horizon.

The concern for transaction costs in farmland markets has echoed the considerable effort devoted to the study of frictions in asset markets in general by financial economists (e.g., He and Modest 1995; Luttmer 1996; Aiyagari 1993). The main conclusion from the literature on asset-market frictions is that frictions may explain some of the most common asset-pricing puzzles. It is well known that in frictionless equilibrium, asset prices must satisfy the conditional moment-equality restriction that conditional expected-excess returns are zero. In the presence of frictions, however, it is possible for the market to be in equilibrium even if conditional expected-excess returns deviate from zero. This assertion is true as long as frictions prevent agents from exploiting such opportunities to profit. Hence the necessary condition for asset-pricing equilibrium in the presence of frictions is a conditional moment-inequality restriction (e.g., He and Modest 1995).

As a result, frictions are likely to lead to autocorrelated returns. However, standard autoregression models seem ill equipped to represent the behavior of returns in equilibrium because such models may yield conditional expected returns that violate the necessary inequality restriction. A model that exhibits autocorrelation, but may also yield conditional expectations that always satisfy the necessary inequality restriction, is the threshold autoregression (TAR) that was first proposed by Tong (1978). Intuitively, TARs generalize the standard autoregressive models by allowing for autocorrelated variables to depend nonlinearly on their lagged values.

Given our knowledge about land values, an important line of inquiry is the investigation of the pricing implications of allowing for variable discount rates and market frictions. Accordingly, this chapter makes the following contributions to the literature. First, I use the canonical asset-pricing paradigm to show that TAR represents the necessary conditions for asset-market equilibrium in the presence of rational expectations and transaction costs. Second, I perform the first application of TARs to model asset pricing with data on Iowa farmland from 1900 to 1994. The long period analyzed is an important characteristic of my study because the data span the two farmland price cycles observed in this century (Figures 15.1a and 15.1b), and the quality of time-series analysis is heavily dependent on the number of observations in the time domain, even more so for data-intensive techniques such as TARs. Third, I carry out a comprehensive time-series analysis of farmland returns as a by-product of the TAR modeling. In particular, I conduct a formal test of structural change in farmland markets because of the long period studied. Finally, I use the results from the TAR model to analyze the consistency of farmland prices with the necessary conditions for market equilibrium in the presence of variable discount rates and typical transaction costs.

The findings presented in this chapter are summarized as follows. First, there is no evidence of structural change in the market for Iowa farmland from 1900 to 1994. Second, TAR provides a better representation of farmland-pricing behavior than does standard autoregressive models. Third, farmland prices are inconsistent with variable discount rates and frictionless market equilibrium. Fourth and most important, the data are consistent with the necessary conditions for asset-market equilibrium in the presence of variable discount rates and the transaction costs typically observed in farmland markets.

ASSET PRICING IN THE PRESENCE OF TRANSACTION COSTS

One of the most important results in finance theory is the fundamental equation of asset pricing (e.g., Dybvig and Ross 1992)

$$P_t = E_t \left[\delta_{t+1} \left(P_{t+1} + D_{t+1} \right) \right], \tag{1}$$

where P_t is the ex-dividend price of an asset at time t, δ_{t+1} is the stochastic discount factor,[ii] $E_t(\cdot)$ is the expectation operator conditional on information at time t, and D_{t+1} is the dividend paid by the asset at time $t+1$. According to the fundamental equation of asset pricing, the current price of an asset is equal to the expectation of the next period's asset payoffs (i.e., price plus dividends), in which the latter are discounted by a stochastic factor. This is a necessary condition for equilibrium in a frictionless-asset market. If the asset's current price were smaller (greater) than the right-hand side of Equation (1), agents would find it attractive to buy (sell) the asset because doing so would yield an expected return above the required return (after accounting for risk); therefore, such a situation is inconsistent with equilibrium.

The fundamental Equation (1) of asset pricing nests all of the standard asset-pricing models. By suitable specifications of the stochastic discount factor δ_{t+1}, Equation (1) yields Sharpe's (1964) capital asset pricing model (CAPM) Breeden's (1979) Consumption CAPM, and Ross' (1976) arbitrage pricing theory (APT) (Dybvig and Ross 1992; Ingersoll 1987). The simplest form of Equation (1) is attained when δ_{t+1} is constant ($\delta_{t+1} = \delta$, for all t). In this instance, Equation (1) yields the CDR-PVM of asset pricing used in the farmland-pricing literature (Falk 1991):

$$P_t = \delta \sum_{s=0}^{\infty} \delta^s E_t \left(D_{t+1+s} \right). \tag{2}$$

Expression (2) is the stable forward solution of Equation (1), obtained by recursive application of the latter.

Although often overlooked, a crucial assumption implicit in Equation (1) is that asset markets are frictionless. When there are frictions, Condition (1) is sufficient but not necessary for market equilibrium. Assuming a one-period holding horizon, conditions for asset-market equilibrium in the presence of transaction costs may be derived as follows.[iii] Let T_Π and T_Σ be the transaction costs that must be paid on the purchase and sale of the asset, respectively, expressed as a proportion of the asset price. In the presence of such costs, agents will buy (sell) the asset as long as expression (3) [expression (4)] holds

$$\left(1+T_\Pi\right)P_t < E_t\left\{\delta_{t+1}\left[(1-T_\Sigma)P_{t+1} + D_{t+1}\right]\right\}, \tag{3}$$

$$\left(1-T_\Sigma\right)P_t > E_t\left\{\delta_{t+1}\left[(1+T_\Pi)P_{t+1} + D_{t+1}\right]\right\}. \tag{4}$$

Therefore, in equilibrium it must be true that Equations (5) and (6) hold simultaneously

$$\left(1+T_\Pi\right)P_t \geq E_t\left\{\delta_{t+1}\left[(1-T_\Sigma)P_{t+1} + D_{t+1}\right]\right\}, \tag{5}$$

$$\left(1-T_\Sigma\right)P_t \leq E_t\left\{\delta_{t+1}\left[(1+T_\Pi)P_{t+1} + D_{t+1}\right]\right\}. \tag{6}$$

Using the fact that

$$E_t\left[\delta_{t+1}D_{t+1}/(1-T_\Sigma)\right] \geq E_t\left(\delta_{t+1}D_{t+1}\right) \geq E_t\left[\delta_{t+1}D_{t+1}/(1+T_\Pi)\right],$$

rearrangement of Equations (5) and (6) yields the following equilibrium condition

$$\lambda^L \le E_t\left(h_{t+1}\right) \le \lambda^U \tag{7}$$

for all t, where $h_{t+1} \equiv \delta_{t+1}\left(P_{t+1} + D_{t+1}\right)/P_t - 1$, $\lambda^L \equiv -\left(T_\Pi + T_\Sigma\right)/\left(1 + T_\Pi\right)$, and $\lambda^U \equiv \left(T_\Pi + T_\Sigma\right)/\left(1 - T_\Sigma\right)$.

Variable h_{t+1} is a stochastic discounted excess return. In general $h_{t+1} \ne 0$, but its conditional expectation $E_t(h_{t+1})$ must satisfy Equation (7) for the market to be in equilibrium. Comparing Equations (1) and (7) reveals that the former expression is a special case of the latter. In the absence of transaction costs, $\left(T_\Pi = T_\Sigma = 0\right)$, $\lambda^L = \lambda^U = 0$, so that expression (7) collapses to Equation (1). In the presence of transaction costs $\left(T_\Pi, T_\Sigma > 0\right)$, however, Equation (7) indicates that there is a band of inaction given by $\left[\lambda^L, \lambda^U\right]$, inside which agents do not react to new information. If agents did react to new information within the band $\left[\lambda^L, \lambda^U\right]$, the expected gains from doing so would be more than offset by the losses associated with the costs of transacting the asset.

ESTIMATION METHODS

The present empirical analysis is based on the fact that the fundamental asset-pricing Equation (7) cannot hold if the h_t series contains either deterministic or stochastic trends. This is true because $E_t\left(h_{t+1}\right)$ would eventually be smaller than λ^L or greater than λ^U if the h_t series were to exhibit either deterministic or stochastic trends, thus violating Equation (7). Hence, testing for the presence of trends in the h_t series is an obvious way to determine whether the fundamental asset-pricing Equation (7) holds. If such trends exist, then the model in Equation (7) must be rejected.

If the null hypothesis of no trend in the h_t series were not rejected, we would estimate the following autoregressive model of order p $\left[AR(p)\right]$

$$h_{t+1} = \phi_0 + \phi_1 h_t + \ldots + \phi_p h_{t+1-p} + e_{t+1}, \tag{8}$$

where fixed coefficients are denoted by ϕ and the identically and independently distributed (iid) error term is expressed as e_{t+1}. Estimation of Equation (8) is useful because the condition $\phi_0 = \phi_1 = \ldots = \phi_p = 0$ must hold to be consistent with the frictionless version of Equation (7) (i.e., $\lambda^L = E_t\left(h_{t+1}\right) = \lambda^U = 0$). Otherwise, in general

$$E_t(h_{t+1}) = \phi_0 + \phi_1 h_t + \ldots + \phi_p h_{t+1-p} \ne 0,$$

if at least one of the ϕ coefficients is different from zero. This suggests that a test of whether the frictionless version of the fundamental asset-pricing equation holds consists of estimating Equation (8) and testing the null hypothesis

$$H_0 : \phi_0 = \phi_1 = \ldots = \phi_p = 0$$

against the alternative that the null hypothesis is not true (Lence and Miller 1999). One may conclude that there is strong evidence against the frictionless fundamental asset-pricing equation if the null hypothesis is rejected.

Importantly, it must be noted that rejection of the null hypothesis

$$H_0 : \phi_0 = \phi_1 = \ldots = \phi_p = 0$$

does not imply rejection of the fundamental asset-pricing model of Equation (7) in the presence of transaction costs $\left(\lambda^L < 0 < \lambda^U \right)$. To reject model (7) under transaction costs, at least one of the $E_t \left(h_{t+1} \right)$s must be either significantly smaller than λ^L or significantly greater than λ^U. To perform the latter analysis, however, it is important to have as well-specified a time-series model as is possible. In this regard, there are four major concerns for model adequacy; structural change, nonlinearities, time-varying conditional variance, and normality of the residuals.

Investigating the existence of structural changes is important because the time series under study spans almost one century. Following Lence and Miller (1999), Andrews' (1993) Wald-like statistic is used to test for structural change without specifying the precise time of the likely structural change.[iv] Andrews' (1993) statistic is designed to test for one-time changes, but it has power against more general forms of parameter instability.

Another likely source of model misspecification is nonlinearities. To see how these may arise in the asset-pricing model with transaction costs in Equation (7), note that according to Equation (7) the conditional expectations must stay inside the interval $\left[\lambda^L, \lambda^U \right]$. In theory, however, standard $AR(p)$ models like Equation (8) will yield conditional expectations outside the interval $\left[\lambda^L, \lambda^U \right]$ for large enough absolute values of h_{t-1}. Unlike the standard $AR(p)$ in Equation (8), a TAR model like Equation (9) does not present such a problem

$$
h_{t+1} = \begin{cases}
\phi_0^{(1)} + \phi_1^{(1)} h_t + \ldots + \phi_p^{(1)} h_{t+1-p} + e_{t+1}^{(1)}, \\
\quad \text{if } \lambda^L < \kappa^L \leq \phi_0^{(1)} + \phi_1^{(1)} h_t + \ldots + \phi_p^{(1)} h_{t+1-p} < \kappa^U \leq \lambda^U, \\
\phi_0^{(2)} + e_{t+1}^{(2)}, \lambda^L \leq \phi_0^{(2)} \leq \kappa^L, \\
\quad \text{if } \kappa^L > \phi_0^{(1)} + \phi_1^{(1)} h_t + \ldots \phi_p^{(1)} h_{t+1-p}, \\
\phi_0^{(3)} + e_{t+1}^{(3)}, \kappa^U \leq \phi_0^{(3)} \leq \lambda^U, \\
\quad \text{if } \phi_0^{(1)} + \phi_1^{(1)} h_t + \ldots + \phi_p^{(1)} h_{t+1-p} > \kappa^U.
\end{cases} \tag{9}
$$

In Equation (9), superscripts within parentheses denote regimes, and κ^L and κ^U are constants that define such regimes. In region (1), h_{t+1} behaves as a standard $AR(p)$ random variable. However, if regime (1) yields a conditional expectation below κ^L (above κ^U), h_{t+1} switches to regime (2) [(3)]; that is, h_{t+1} then behaves as a constant plus an error term. In this way, $E_t(h_{t+1})$ always stays within the band of inaction defined by Equation (7).

The TAR process defined by Equation (9) allows for time-varying conditional heteroscedasticity, as well. This assertion is true because error variances $\left(\sigma_{e_{(i)}}^2 s \right)$ may be different across regimes. This property of TARs is important because it is a stylized fact that prices and other financial series tend to exhibit time-varying conditional variances.

Given the aforementioned characteristics of the TAR process of Equation (9), it seems that such a process provides a very plausible time-series representation of the asset-pricing model in the presence of transaction costs in Equation (7). Hence, the presence of TAR processes in general is tested by means of the $\hat{F}$- statistic developed by Tsay (1989).

As noted above, time-varying conditional variances may be another cause of model misspecification. Typically, time-varying conditional variances are modeled via autoregressive conditionally heteroscedastic (ARCH) processes (Bollerslev et al. 1992). Furthermore, Hanson and Myers (1995) recently find evidence of ARCH behavior in aggregate U.S. farmland prices. If h_t follows an ARCH process, then ordinary-least-squares estimates of Equation (8) are still best linear unbiased but inefficient (Harvey, 1991). Here, the Lagrange-multiplier test developed by Engle (1982) is used to detect ARCH behavior.

For hypothesis-testing purposes, however, the most important assumption is that of normally distributed disturbances. Although typically not tested for, normality of residuals is crucial for most tests of hypotheses involving $AR(p)$ models. Hence, in the present study error normality is checked for by means of the Lagrangian multiplier test advocated by Jarque and Bera (1980).

After having developed an adequate model to represent the behavior of h_t, such a model may be used to make inferences about $E_t(h_{t+1})$. In the interest of space, such inferences are discussed only in the context of TAR in Equation (9), which is shown later to be the model that best fits the data. According to TAR in Equation (9), $\phi_0^{(2)}$ and $\phi_0^{(3)}$ yield the smallest and the largest conditional expectation for h $\left[E_t(h_{t+1}) \right]$, respectively. Therefore, Equation (7) may be stated alternatively as the condition that

$$\lambda^U \equiv \frac{T_\Pi + T_\Sigma}{1 - T_\Sigma} \geq \text{largest } E_t\left(h_{t+1}\right) = \phi_0^{(3)}, \tag{11}$$

must hold simultaneously.

Inequality Equations (10) and (11) suggest two strategies to make inferences regarding $E_t\left(h_{t+1}\right)$. One strategy consists of selecting particular levels of T_Π and T_Σ, e.g., T_Π° and T_Σ°, and then calculating the p-value of the null hypothesis that Equations (10) and (11) hold simultaneously for $T_\Pi = T_\Pi^\circ$ and $T_\Sigma = T_\Sigma^\circ$. If the p-value thus obtained exceeds the conventional α percent significance level, the null hypothesis that Equation (7) holds for $T_\Pi = T_\Pi^\circ$ and $T_\Sigma = T_\Sigma^\circ$ should be rejected. An alternative but equivalent strategy consists of calculating simultaneously the $(1-\alpha)$-percent confidence intervals (CIs) for $\phi_0^{(2)}$ and $\phi_0^{(3)}$. The upper bound of the $(1-\alpha)$-percent CI for $\phi_0^{(2)}$ and the lower bound of the $(1-\alpha)$-percent CI for $\phi_0^{(3)}$ may then be plugged into the definitions of λ^L and λ^U, respectively, to solve for the smallest purchase and sale transaction costs consistent with Equation (7) at the α-percent significance level.

The present problem involves simultaneous inferences; therefore the significance levels of the individual tests must be adjusted to reflect the desired level of significance of the joint test (Bickel and Doksum 1977). The significance level of the individual tests and CIs that reflects a α-percent level of significance for the simultaneous tests and CIs equals $\left[1 - (1-\alpha)^{1/2}\right]$ because residuals $e_{t+1}^{(2)}$ and $e_{t+1}^{(3)}$ are independent (Godfrey 1988). Thus, the simultaneous $(1-\alpha)$-percent CIs for $\phi_0^{(2)}$ and $\phi_0^{(3)}$ are calculated as

$$\left[\hat{\phi}_0^{(i)} - \sqrt{F_{1,n^{(i)}-1,1-(1-\alpha)^{1/2}}} \frac{\hat{\sigma}_{e^{(i)}}}{\sqrt{n^{(i)}}}, \hat{\phi}_0^{(i)} + \sqrt{F_{1,n^{(i)}-1,1-(1-\alpha)^{1/2}}} \frac{\hat{\sigma}_{e^{(i)}}}{\sqrt{n^{(i)}}}\right], \tag{12}$$

$$i = 2, 3,$$

where hats denote sample estimates, $F_{1,n^{(i)}-1,1-(1-\alpha)^{1/2}}$ is the $[(1-\alpha)^{1/2}]^{th}$ quantile of the F distribution with 1 and $(n^{(i)}-1)$ degrees of freedom, and $n^{(i)}$ is the number of observations in the ith regime.

DATA

As discussed in the previous section, the empirical analysis is based on the variable $h_{t+1} \equiv \delta_{t+1}\left(P_{t+1} + D_{t+1}\right)/P_t - 1$, where P is the per-acre value of cash-rented Iowa farm real estate, D is the corresponding gross cash rents minus property taxes, and δ is the inverse of one plus the interest rate charged on new land mortgage loans made by the Farm Credit System. In the present case, price and cash-rent definitions are the same as in Falk (1991); that is, the sources for both

Iowa farm real estate, D is the corresponding gross cash rents minus property taxes, and δ is the inverse of one plus the interest rate charged on new land mortgage loans made by the Farm Credit System. In the present case, price and cash-rent definitions are the same as in Falk (1991); that is, the sources for both series are the USDA[v] and the Iowa State University Extension Service. Also, the U.S. Department of Agriculture reported taxes per U.S. $100 of full market value levied on Iowa agricultural real estate (DeBraal and Jones 1993).

The period analyzed in this chapter is from 1900 to 1994, which is more than 40-percent longer than the 1921 to 1986 period used by Falk (1991). In addition to updating the data to cover the most recent years, the present study spans the 1900 to 1920 period. The data from 1900 to 1920 are unique in that the USDA published them a long time ago and Iowa is the only state for which they are available (Johnson 1948).[vi] To my knowledge, these are the longest high-quality annual series of farm real estate prices with respective cash rents for the United States.

For present purposes, the importance of the long series is twofold. First, the series provides an adequate coverage of the two major land-price cycles that occurred in the twentieth century (Figure 15.1 a & b). Second, the reliability of time-series analysis is highly dependent on the length of the series used (Box and Jenkins 1976). The length of the time series is even more crucial with regards to testing and modeling TARs.

Data on Farm Credit System interest rates are from the USDA's *Agricultural Statistics* (USDA/NASS 1936 to 1994). The Farm Credit System has historically been the institution holding the single largest share of farm real estate loans, with around 20 percent to 35 percent of the total U.S. farm real-estate debt. Unfortunately, the Farm Credit System was created in 1916, and therefore the respective interest rates are available only since 1917. Because of the lack of any suitable interest-rate series covering previous years, the corresponding discount factor is computed using a surrogate land-mortgage interest rate (LMIR) series. The latter is calculated by means of the equation[vii]

$$LMIR_t = 1.75 + 0.85YCB_t, \tag{13}$$

where YCB is the percentage yield on corporate bonds with 30 years to maturity reported by the U.S. Department of Commerce (USDOC 1976). Equation (13) is obtained from the corresponding maximum-likelihood regression fitted over the 1917 to 1970 period allowing for first-order autocorrelation in the errors.[viii]

The calculated h-series is depicted in Figure 15.2. The sample average of h is 0.04, the sample standard deviation is 0.11, and the minimum and maximum values are -0.31 and 0.31, respectively.

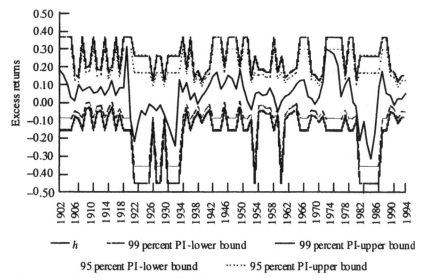

Figure 15.2 Predictive Intervals (PI) for h_{t+1}

RESULTS AND DISCUSSION

Results for the trend tests are as follows. The autoregressive model

$$\Delta h_t \equiv h_t - h_{t-1} = \varphi_0 + \varphi_1 h_{t-1} + \varphi_2 time + \varphi_{2+1}\Delta h_{t-1} + \ldots + \varphi_{2+p}\Delta h_{t-p} + e_t \ (14)$$

fitted to the data has a lag length of $p = 0$, selected by using the Schwarz-Bayes Information Criterion (Greene 2003). The augmented Dickey-Fuller (1981) test based on Equation (14) yields a $\hat{\tau}_\tau$-statistic of -4.52, leading to rejection of the null hypothesis that $\varphi_1 = 0$ since the critical value at the 1-percent-significance level equals -4.06. Thus, the null hypothesis of a stochastic trend is strongly rejected.

The test corresponding to the null hypothesis of no deterministic trend is based on the model

$$h_t = \beta_0 + \beta_1 time + \beta_2 h_{t-1} + \ldots + \beta_{1+p} h_{t-p} + e_t. \tag{15}$$

In this instance, the Schwarz-Bayes Information Criterion leads to the selection of a lag length of $p = 1$. The Student t-statistic corresponding to the estimate of β_1 equals -0.04, which does not allow us to reject the null hypothesis that $\beta_1 = 0$ at any reasonable significance level. In short, it can be safely concluded that the h-series contains neither deterministic nor stochastic trends.

Based on the results from the unit-root tests, the best fit using standard autoregressive models is obtained with the following $AR(1)$

$$h_t = 0.0144 + 0.637 h_{t-1} + e_t \tag{16}$$

(0.0097) (0.080)

$R^2 = 0.411$, $\hat{\sigma}_e = 0.087$, Akaike Information Criterion $= -454.08$,
93 observations,

where numbers between parentheses below coefficient estimates denote the respective standard deviations. The residuals from Equation (16) do not exhibit autocorrelation problems, as the Ljung-Box (1978) modified Q-statistics for first- and second-order autocorrelation of the residuals equal 0.24 and 0.34, respectively. Such values are much smaller than the corresponding critical values at the 5-percent-significance level, which are $\chi^2_{1,0.05} = 3.84$ and $\chi^2_{2,0.05} = 5.99$, respectively.

As measured by R^2, the model's explanatory power is moderate. The Student t-statistic for the estimated coefficient associated with the lagged dependent variable equals 7.97, which is significantly different from zero at the 1-percent level. From an asset-pricing perspective, the latter finding is important because it implies a sound rejection of the frictionless asset-pricing model of Equation (1).

The null hypothesis of no structural change over the 1900 to 1994 period cannot be rejected, because Andrews' (1993) test for structural change yields a Wald-like t-statistic of 9.25 and the 5-percent-significance level critical value equals 10.85. In contrast, the null hypothesis of no TAR process is strongly rejected. Tsay's (1989) TAR test yields $\hat{F}=9.09$, which exceeds the critical value at the 1-percent level of significance ($F_{2,80,0.01} = 4.88$). Therefore, a TAR is better suited than a standard $AR(1)$ to model the behavior of h.

Given the results from the tests for structural change and TAR, a single TAR is fitted to the whole period. Tsay's (1989) advocated method yields the TAR model presented in Table 15.1. As expected, the selected TAR exhibits a better fit than the standard $AR(1)$ in Equation (16). The fitted TAR has an R^2 of 0.464 and an Akaike Information Criterion of -496.58, versus $R^2 = 0.411$ and Akaike Information Criterion $= -454.08$ for the $AR(1)$. As shown in the bottom half of Table 15.1, tests of fit on the standardized TAR residuals do not reveal misspecification problems due to autocorrelation or to ARCH effects. The estimated residuals are also distributed in a manner consistent with the assumption of normality. Although typically overlooked in empirical work, the normality assumption is crucial for the validity of most hypothesis tests.[ix] The good fit of the TAR model can be appreciated in Figure 15.2. This graph depicts the actual h series along with the 95-percent and the 99-percent predictive intervals calculated from the TAR model in Table 15.1.

Consistent with the specified levels of confidence, there are only 4 out of 93 observations outside the 95-percent (99-percent) predictive intervals, which correspond to the years 1920, 1921, 1950, and 1960. Other predictive intervals are not shown to avoid cluttering the picture, but there are 12 (17, 43) out of 93 observations outside the 90-percent (80-percent, 50-percent) predictive intervals. Hence by all of these accounts the fitted TAR is a very good representation of the actual behavior of h.

Table 15.1 Estimated Threshold Autoregressive Model for h_{t+1}

Model:

$$h_{t+1} = \begin{cases} 0.0152^* + h_t + e_{t+1}^{(1)}, \hat{\sigma}_{e^{(1)}} = 0.042, \text{ if } -0.006 \le 0.0152 + h_t \le 0.093, \\ (0.0067)^a \qquad\qquad\qquad\qquad\qquad\qquad 40 \text{ observations;} \\ -0.094^{**} + e_{t+1}^{(2)}, \hat{\sigma}_{e^{(2)}} = 0.12, \text{ if } -0.006 > 0.0152 + h_t, \\ (0.028) \qquad\qquad\qquad\qquad\qquad\qquad 19 \text{ observations;} \\ 0.108^{**} + e_{t-1}^{(3)}, \hat{\sigma}_{e^{(3)}} = 0.094, \text{ if } 0.0152 + h_t > 0.093, \\ (0.016) \qquad\qquad\qquad\qquad\qquad\qquad 34 \text{ observations;} \end{cases}$$

$R^2 = 0.464$, Akaike Information Criterion$= -496.58$

Tests on standardized residuals:

Autocorrelation:

$Q'(1)^b = 0.22 \ [0.64]^c$

$Q'(2) = 0.38 \ [0.83]$

ARCH:

$LMA(1)^d = 3.60 \ [0.06]$

$LMA(2) = 3.66 \ [0.16]$

Normality:

$LMN^e = 0.34 \ [0.84]$

---

*(**) Significantly different from zero at the 5- (1-) percent level of significance based on the two-tailed t-statistic.

^{a}The numbers between parentheses below coefficient estimates denote the respective standard deviations.

$^b Q'(i)$ is the Ljung-Box portmanteau test or modified Q-statistic for i-order autocorrelation (Ljung and Box 1978).

^{c}Numbers between brackets at the right-hand side of t-statistics denote the corresponding p-values for the individual tests. The significance level of the composite test lies between the maximum of the individual significance levels and their sum (Godfrey 1988).

$^d LMA(i)$ is the Lagrange multiplier test for i-order autoregressive conditional heteroscedasticity (Engle 1982).

$^e LMN$ is the Lagrange multiplier test for normality (Jarque and Bera 1980).

The structure of the fitted TAR is of interest because it can be readily interpreted in terms of the asset-pricing model in the presence of transaction costs represented by Equation (7). The fitted TAR exhibits three regimes. The normal regime [i.e., regime (1)] is characterized by a unit root with a positive drift. When the conditional expected-excess returns from regime (1) tend to be too low (high), excess returns switch to regime (3) so that the conditional expected-excess returns stay within the band of inaction postulated by Equation (7).

Also of interest is to note that, like most financial series, the fitted TAR exhibits time-varying conditional heteroscedasticity. This assertion is true because the standard F-test for the inequality of two variances indicates that the variance of regime (1) is significantly different from the variances of regimes (2) and (3). The F-statistics corresponding to the null hypotheses that $\sigma^2_{e(1)} = \sigma^2_{e(2)}$ and $\sigma^2_{e(1)} = \sigma^2_{e(3)}$ equal 8.12 and 4.93, respectively, which exceed the corresponding critical values at the 1-percent-significance level ($F_{18,39,0.005} = 2.68$ and $F_{33,39,0.005} = 2.38$, respectively). The null hypothesis that regimes (2) and (3) have the same variance ($\sigma^2_{e(2)} = \sigma^2_{e(3)}$) cannot be rejected by the data, as the F-statistic equals 1.65 and the critical value at the 5-percent level of significance is $F_{18,33,0.025} = 2.19$.

As important as testing whether the h-series behaves like the TAR implied by the asset-pricing model in the presence of transaction costs represented by Equation (7) is analyzing whether the conditional expectations are consistent with the level of transaction costs observed in actual farm real-estate markets. The p-values and CIs described in the section about estimation methods indicate that the data strongly reject the frictionless asset-pricing model (1) but are consistent with the transaction-costs model (7).

The p-value for zero transaction costs ($T_\Pi = T_\Sigma = 0$) is less than 10^{-5}, a value much smaller than standard levels of significance. In contrast, for an average-to-low estimate of actual transaction costs—e.g., 8 percent ($T_\Pi = T_\Sigma = 4$ percent) —the p-value equals 0.25. Similarly, assuming transaction costs are equally shared between purchases and sales, the smallest magnitude of transaction cost consistent with the asset-pricing model in Equation (7) at the 5-percent (1-percent, 10-percent) level of significance is 6.78 percent (5.77 percent, 7.26 percent).[x] Given that transaction costs of 7.5 percent may be considered a minimum estimate of actual farmland transaction costs, it can be concluded that the data do not provide enough evidence to reject the asset-pricing model in the presence of transaction costs represented by Equation (7).

CONCLUDING REMARKS

Historically, the importance of farmland for the financial health of the U.S. agricultural industry, coupled with the observed boom-bust cycles in farmland prices, has caused concern not only to the farm sector but to other sectors as well (e.g., financial intermediation). [xi] Research has echoed such concerns by devoting ample resources to exploring and understanding the behavior of land prices. Researchers have also started to study the implications of the large transaction costs involved in transfers of farmland ownership for farmland price behavior.

This chapter is the first study to advocate and apply TARs to model the behavior of asset prices in the presence of transaction costs. It is shown that TARs are able to accommodate the conditional moment inequality restrictions required for equilibrium-asset pricing under transaction costs. TARs also allow for time-varying conditional heteroscedasticity, a property that characterizes most financial time series.

Using the longest time series available on U.S. land prices and rents, this exhaustive analysis indicates that there is no evidence of structural change in the market for Iowa farmland over the 1900 to 1994 period. However, the null hypothesis of a standard autoregression model of land returns is rejected in favor of the alternative hypothesis of a TAR model. The fitted TAR has the structure that one would expect from the theoretical restrictions on conditional expected returns in the presence of transaction costs.

The fitted TAR is also used for hypothesis testing regarding equilibrium asset pricing. In agreement with recent studies of farmland prices that assume constant discount rates, the variable discount rate frictionless equilibrium model of land prices is soundly rejected. However, I found the data are consistent with the necessary conditions for market equilibrium in the presence of variable discount rates and the typical transaction costs involved in the transfer of farmland ownership. In addition, I found that returns are characterized by time-varying conditional heteroscedasticity.

The results from my study imply that land-price behavior is consistent with the necessary conditions for market equilibrium under rational expectations and the typical transaction costs observed in land markets. Although land markets may be considered inefficient because transaction costs prevent real-estate agents from reflecting in land prices all of the information available to them, rendering such markets more efficient is a nontrivial problem. Exploring the latter issue and devising stronger tests of land pricing in the presence of transaction costs seem important and promising avenues for future research.

REFERENCES

Aiyagari, S.R. (1993) "Explaining Financial Market Facts: The Importance of Incomplete Markets and Transaction Costs." *Federal Reserve Bank of Minneapolis Quarterly Review* (Winter): 17–31.

Andrews, D.W.K. (1993) "Tests for Parameter Instability and Structural Change with Unknown Change Point." *Econometrica* 61: 821–56.

Bickel, P.J. and K.A. Doksum. (1977) *Mathematical Statistics*. Oakland: Holden-Day.

Bollerslev, T., R.Y. Chou, and K.F. Kroner. (1992) "ARCH Modeling in Finance: A Review of the Theory and Empirical Evidence." *Journal of Econometrics* 52: 5–59.

Box, G.E.P. and G.M. Jenkins. (1976) *Time Series Analysis: Forecasting and Control, 2nd Edition*. San Francisco: Holden-Day.

Breeden, D. (1979) "An Inter-Temporal Asset-Pricing Model with Stochastic Consumption and Investment Opportunities." *Journal of Financial Economics* 7: 265–96.

Chavas, J.P. and A. Thomas. (1999) "A Dynamic Analysis of Land Prices." *American Journal of Agricultural Economics* 81(4): 772–84.

Clark, J.S., M. Fulton, and J.T. Scott. (1993) "The Inconsistency of Land Values, Land Rents, and Capitalization Formulas." *American Journal of Agricultural Economics* 75(1): 147–55.

DeBraal, J.P. and J. Jones. (1993) *Agricultural Real Estate Taxes*. U.S. Department of Agriculture, Economic Research Service, Computer File #92002. Washington DC: USDA/ERS.

de Fontnouvelle, P. and S.H. Lence. (2002) "Transaction Costs and the Present Value 'Puzzle' of Farmland Prices." *Southern Economic Journal* 68: 549–65.

Dickey, D.A. and W.A. Fuller. (1981) "Likelihood Ratio Statistics for Autoregressive Time Series with a Unit Root." *Econometrica* 49: 1057–72.

Dybvig, P.H. and S.A. Ross. (1992) "Arbitrage." In *The New Palgrave Dictionary of Money and Finance*, P. Newman, M. Milgate, and J. Eatwell, eds., Volume 1: 43–50. London: The Macmillan Press Limited.

Enders, W. (1995) *Applied Econometric Times Series*. New York: John Wiley & Sons, Inc.

Engle, R.F. (1982) "Autoregressive Conditional Heteroscedasticity with Estimates of the Variance of United Kingdom Inflation." *Econometrica* 50: 987–1007.

Falk, B. (1991) "Formally Testing the Present-Value Model of Farmland Prices." *American Journal of Agricultural Economics* 73: 1–10.

Godfrey, L.G. (1988) *Misspecification Tests in Econometrics*. New York: Cambridge University Press.

Greene, W.H. (2003) *Econometric Analysis, 5th Edition*. Upper Saddle River: Prentice Hall.

Hanson, S.D. and R.J. Myers. (1995) "Testing for a Time-Varying Risk Premium in the Returns to Farmland." *Journal of Empirical Finance* 2(3): 265–76.

Harvey, A.C. (1991) *The Econometric Analysis of Time Series, 2nd Edition*. Cambridge: MIT Press.

He, H. and D.M. Modest. (1995) "Market Frictions and Consumption-Based Asset Pricing." *Journal of Political Economy* 103: 94–117.

Ingersoll, J.E. (1987) *Theory of Financial Decision Making*. Savage: Rowman & Littlefield Publishers, Inc.

Jarque, C.M. and A.K. Bera. (1980) "Efficient Tests for Normality, Homoscedasticity, and Serial Independence of Regression Residuals." *Economics Letters* 6: 255–59.

Johnson, A.R. (1948) *The Farm Real Estate Situation 1946–47*. U.S. Department of Agriculture, Bureau of Agricultural Economics, Circular No. 780. Washington, DC: USDA/BAE.

Just, R.E. and J.A. Miranowski. (1993) "Understanding Farmland Price Changes." *American Journal of Agricultural Economics* 75(1): 156–68.

Lence, S.H. (2001) "Farmland Prices in the Presence of Transaction Costs: A Cautionary Note." *American Journal of Agricultural Economics* 83: 985–92.

Lence, S.H. and D.J. Miller. (1999) "Transaction Costs and the Present Value Model of Farmland: Iowa, 1900 to 1994." *American Journal of Agricultural Economics* 81(May): 257–72.

Ljung, G.M. and G.E.P. Box. (1978) "On a Measure of Lack of Fit in Time-Series Models." *Biomètrika* 65: 297–303.

Luttmer, E.G.J. (1996) "Asset Pricing in Economies with Frictions." *Econometrica* 64: 1439–67.

Moyer, D.D. and A.B. Daugherty. (1982) "Land Purchases and Acquisitions, 1975 to 1977—A Report on a Landownership Follow-On Survey." U.S. Department of Agriculture, Economic Research Service Staff Report No. AGES820407. Washington, DC: USDA/ERS.

Pope, R.D., R.A. Kramer, R.D. Green, and B.D. Gardner. (1979) "An Evaluation of Econometric Models of U.S. Farmland Prices." *Western Journal of Agricultural Economics* 4: 107–19.

Robison, L.J. and S.R. Koenig. (1992) "Market Value Versus Agricultural Use Value of Farmland." In *Costs and Returns for Agricultural Commodities—Advances in Concepts and Measurement*, M.C. Ahearn and U. Vasavada, eds.,. Boulder: Westview Press.

Ross, S.A. (1976) "The Arbitrage Theory of Capital Asset Pricing." *Journal of Economic Theory* 13: 341–60.

Schmitz, A. (1995) "Boom-Bust Cycles and Ricardian Rents." *American Journal of Agricultural Economics* 77: 1110–25.

Sharpe, W.F. (1964) "Capital Asset Prices: A Theory of Market Equilibrium under Conditions of Risk." *Journal of Finance* 19: 425–42.

Tegene, A. and F. Kuchler. (1993) "A Regression Test of the Present-Value Model of U.S. Farmland Prices." *Journal of Agricultural Economics* 44: 135–43.

Thompson, C.S. and W.S. Whiteside. (1987) "Effects of Market Channels on Prices of Farm Land in South Carolina." *Agricultural Finance Review* 47: 119–24.

Tong, H. (1978) "On A Threshold Model." In *Pattern Recognition and Signal Processing*, C.H. Chen, ed., Amsterdam: Sijthoff & Noordhoff.

Tsay, R.S. (1989) "Testing and Modeling Threshold Autoregressive Processes." *Journal of the American Statistical Association* 84: 231–40.

USDA/ERS (U.S. Department of Agriculture, Economic Research Service). (1964) "Costs of Transferring Ownership of Farm Real Estate." *Farm Real Estate Market Developments CD-66*, October: 29–38. Washington, DC: USDA/ERS.

USDA/NASS (U.S. Department of Agriculture, National Agricultural Statistics Service). (1936–94) *Agricultural Statistics.* Washington, DC: U.S. Government Printing Office.

USDOC (U.S. Department of Commerce, Economics and Statistics Administration, Bureau of the Census). (1976) *The Statistical History of the United States, from Colonial Times to the Present: Historical Statistics of the United States, Colonial Times to 1970.* New York: Basic Books.

Weersink, A., C. Turvey, S. Clark, and R. Sarkar. (1996) "The Effect of Agricultural Policy on Farmland Values." Paper presented at the *Annual Meeting of the American Agricultural Economics Association*, San Antonio, Texas (30 July).

Wunderlich, G. (1989) "Transaction Costs and the Transfer of Rural Land." *Journal of the American Society of Farm Managers and Rural Appraisers* 53: 13–6.

ENDNOTES

[i]Iowa is probably the premiere agricultural state in the United States. In addition, it has available to it the longest annual series of U.S. farmland prices than does any other state in the union.

[ii]δ_{t+1} is also called the state-price density, the pricing operator, the pricing kernel, or the inter-temporal marginal rate of substitution.

[iii]Attention is restricted to the special case of a one-period holding horizon because of the complexities associated with the most general situation of uncertain holding horizons (Lence 2001). Lence and Miller (1999) discuss the infinite holding horizon scenario.

[iv]Andrews' (1993) test for structural change is performed after the unit-root test because structural changes tend to bias the latter toward nonrejection of the null hypothesis of unit roots (Enders 1995).

[v]The kind assistance of John Jones at the USDA in providing data and related information is gratefully acknowledged.

[vi]For other states, farmland price and rent data go back to 1921 at the most.

[vii]In Equation (13), both LMIR and YCB are measured in percentages.

[viii]Such regression has an R^2 of 0.971, and yields *t*-statistics of 7.90 and 16.86 for the intercept and the slope estimates, respectively. The estimated coefficient of autocorrelation in the errors is 0.734 with a *t*-statistic of 7.59.

[ix]It is worth stressing that Table 15.1 reports p-values for each test individually. The significance level of the composite test lies between the maximum of the individual significance levels and their sum (Godfrey 1988), thus lending further support for the adequacy of the fitted TAR.

[x]That is, $T_\Pi = T_\Sigma = 3.39$ percent   (2.885 percent, 3.63 percent) at the 5-percent- (1-percent-, 10-percent-) significance level.

[xi]For example, at the bottom of the farmland price cycle in 1985, 1986, and 1987 (Figure 15.1b), 68, 65, and 69 agricultural banks failed, respectively. In contrast, only one agricultural bank failed in 1981.

Section V:
Urbanization, Environmental Quality, and Farmland Values

Chapter 16

Local Land Markets and Agricultural Preservation Programs

Lori Lynch and Sabrina J. Lovell[*]
University of Maryland
Environmental Protection Agency

INTRODUCTION

Since 1977, local and state governmental entities have implemented transfer of development rights (TDR) and purchase of development rights or purchase of agricultural conservation easements (PDR/PACE) programs to permanently preserve farmland throughout the United States (AFT 2001a; AFT 2001b; AFT 2001c). In each of these programs the sale of development rights results in an easement attached to the title of the land that restricts current and all future owners from converting the parcel to residential, commercial, or industrial uses. The value of the land in alternative uses can affect an owner's willingness to participate in these programs and in the program costs. Thus information on the value that the private market places on parcel characteristics is important when determining participation behavior and payment levels. In addition, knowledge of the marginal contributions of different parcel characteristics to both private-market prices and easement values can help program administrators decide which easement purchases can maximize society's benefit at the lowest cost.

Lynch and Lovell (2002) find that the agricultural land-preservation programs in the Calvert, Howard, and Carroll counties of Maryland pay higher per-acre easement values for farmland close to the nearest employment center, smaller farms, and farms with a high percentage of prime soils; they pay lower values for farms with a high percentage of cropland. Lynch and Lovell (2002) also find that the type of agricultural-preservation program (i.e., TDR or PDR/PACE) into which farmland is enrolled affects the value paid to certain land characteristics. In an analysis of whether or not easement restrictions affect the preserved parcels' market price, Nickerson and Lynch (2001) examine private market sales prices for 200 farmland parcels in the same Maryland (Calvert,

Howard, and Carroll) counties. They find that the private market pays higher per-acre prices for farmland close to the nearest employment center, for smaller farms, for non-forested parcels, and for those parcels in Calvert and Howard counties. They find that prime soils are not important when determining the parcel price in the private market. Comparing the results of these two studies, we find both similarities and differences in the effect of different characteristics on easement values and on private-market prices for agricultural land.

In this chapter, we explore these similarities and differences further by investigating if the private land market pays similar values for parcel characteristics. We also investigate to see whether or not the payment schemes of the preservation programs are market driven. Analyzing a spatially explicit dataset of 2,592 arm's-length transactions, we correct for possible spatial correlation that might occur due to the proximity of the observations to one another. We also include nonfarm parcels that are no longer in use agriculturally. By examining the local market for land, we can determine if the easement value (indicated by the supply curve of eligible land to be preserved based on the easement-program payments) is comparable to the prices received by recently sold local land.

Preservation programs can use this information to adjust their payment schemes to ensure the purchase of future acres, assuming that the underlying conditions remain the same. In addition, program administrators have been proposing the use of a point system, which assigns monetary values to different parcel characteristics, rather than expending limited program dollars and time using the more expensive appraisal process to determine the market value. To guarantee enrollment under such systems, the programs need to have point systems that result in easement values that satisfy a landowner's participation constraint. In addition, these point systems need to be justified to county commissioners and state authorities, especially if programs value characteristics that the market does not typically reward (such as wetlands or other resource features) but which are values that may maximize the welfare of society. A hedonic model that analyzes recent market transactions for both agricultural and nonagricultural parcels of at least one acre will illustrate the monetary values that have been attached to land characteristics.

Preservation programs preserve agricultural lands and woodlands to provide sources of agricultural products, to control urban expansion, and to protect open-space land (Maryland Agricultural Land Preservation Foundation 2001). Lynch and Musser (2001) translate these goals to preserving those farms most likely to be converted in the near future (close to the city or town); preserving productive farms (prime soils; growing crops); maximizing the number of acres (many farms; large farms); and preserving farms close to one another (large blocks of land). While researchers suggest that preserving farmland is not necessary for food-security purposes, agricultural-preservation programs can maintain environmental amenities that include wildlife habitat, groundwater recharge, and rural and scenic views; can contribute to curbing urban and suburban sprawl; and can sustain a viable local agricultural economy (Gardner 1977; Wolfram 1981; Fischel 1985; McConnell 1989; Bromley and Hodge 1990). Society values these public goods, according to contingent valuation analyses that have been conducted (Halstead 1984; Bergstrom et al. 1985; Beasly et al. 1986; Drake

1992; Pruckner 1995). Furthermore, according to the Land Trust Alliance (2000), U.S. citizens appear willing to finance these types of programs. Numerous ballot initiatives have been designed to preserve parks, open space, farmland, and other amenities. In 2000, U.S. \$7.4 billion in conservation funding was authorized; in 1999, U.S. \$1.8 billion; and in 1998, U.S. \$8.3 billion.

While these studies demonstrate that the general public supports agricultural and open-space preservation, society may desire to preserve land with different characteristics other than those that are highly valued in private-market transactions. The preservation of an individual farm with a given set of characteristics may achieve all the goals mentioned above. However, in many cases, program administrators will have to make trade offs between different characteristics, as most farms will have some but not all of those desired characteristics. Thus information on the existing land-market returns for various characteristics may assist in ensuring that the incentive structure for potential participants is well designed. For example, if the private market does not reward prime soils and yet this characteristic is preferred by the program's goals, an appraisal process based on market transactions to determine the easement value may not reflect the value society would place on preserving a particular parcel with a high percentage of prime soils. With information on the influence of the various parcel characteristics to the private market value of the land, even land which is already developed, program administrators of agricultural land-preservation programs may be better able to select among the farms offered to be preserved or may be better able to set up a scheme to compensate and attract landowners with farms having the desired characteristics.

METHODS

In this chapter, we use a hedonic approach, corrected for possible spatial correlation, to model the per-acre sales price of land in Calvert County, Howard County and Carroll County, Maryland. When deciding whether to place his or her land on the market, a landowner examines the relative returns to the parcel's characteristics in recent sales in the local land market. Similarly, the buyer evaluates the relative cost of the parcel's characteristics before deciding to purchase the property. In Equation (1), the per-acre price of land in real-estate transaction i is modeled as the net present value of the stream of agricultural rents A_i, as a function of the parcel's characteristics (X_i and time t), until an optimal date to develop the land t^* and of the stream of the residential rents R_i as a function of the parcel's characteristics (X_i and time t), the time when t is greater than t^* such that

$$P_i = \left[\int_{t=0}^{t^*} A_i(X_i,t)e^{-rt}dt + \int_{t^*}^{\infty} R_i(X_i,t)e^{-rt}dt \right]. \tag{1}$$

   Parcel characteristics include soil quality, land use, and geographic location, as well as other attributes. Parcels already in residential use $(t > t^*)$ and those that continue in agricultural use $(t < t^*)$ are included in the analysis.

   We estimate a hedonic model to explain how these characteristics are rewarded for a locus of equilibrium land prices in recent private-market transactions. The empirical form of the land-value model can be specified as

$$P = X\beta + \varepsilon, \tag{2}$$

where P is a vector of the natural log of the per-acre private market-sales price, X is a matrix of exogenous parcel characteristics influencing the value of land in agricultural and residential use, β represents parameters to be estimated, and ε is a vector of random error terms representing unobserved characteristics and is assumed to be normally distributed. However, parcel characteristics that affect the market price may be spatially correlated. If some of these characteristics are not observable, then there may be spatial dependence across error terms. Many of the market transactions are in the same general region, thus they have similar unobservable characteristics. Given this possibility, the empirical problem becomes

$$\begin{aligned} P &= X\beta + \varepsilon \\ \varepsilon &= \rho W \varepsilon + \mu, \end{aligned} \tag{2a}$$

and can be estimated as (Whittle 1954; Cliff and Ord 1973),

$$P = X\beta + (I - \rho W)^{-1} \mu \tag{2b}$$

where W is a spatial weight matrix, ρ is a scalar parameter to be estimated, μ is a vector of random error terms assumed to have a mean of 0 and variance-covariance matrix $\sigma^2 I$, and ε is a vector of random error terms with mean 0 and with variance-covariance matrix $\sigma^2 (I - \rho W)^{-1} (I - \rho W')^{-1}$ (Kelejian and Robinson 1993; Bell and Bockstael 2000). The correlation between the errors of the observations is assumed to decrease as the distance between the observations increases. Thus the spatial-weight matrix is of a distance-decay format defined to be the inverse function of the distance between observations. A distance-decay matrix is different from the type of matrix often used when estimating regressions thought to have spatial correlation. Usually, spatial dependence is assumed to be 1 for adjacent observations such as those with common borders, and 0 for other observations (Anselin and Florax 1995).

   However, given the micro-level data used in this analysis, a distance-decay format is assumed to be more appropriate. A distance-decay format of the spatial-weight matrix assumes that those observations closest to a parcel are more highly correlated than those observations farther away. Defining d_{ij} as the distance between parcel i and parcel j, and c as the distance after which no spatial correlation is expected, the elements of W for the inverse-distance matrix equals $w_{ij} = \dfrac{1}{d_{ij}}$ if $d_{ij} < c$, and $w_{ij} = 0$ if $i = j$, or if $d_{ij} > c$ (Bell and Bockstael 2000). Using similar Maryland land values, Bell and Bockstael (2000) find that after 600 meters little to no spatial dependence is apparent. Given this result and the fact that as c becomes larger the matrix becomes less sparse, we set $c = 1,609.27$ feet (490 meters) or about one-third of a mile.

DATA

The Maryland Department of Planning (MDP) has geographically identified data from the Maryland Division of Tax and Assessment. This data contains transaction and parcel information, including the X and Y coordinates of the parcel that enables us to use a geographic information system (GIS) to extract additional parcel-level and geographic information. The spatially explicit dataset includes 709 parcels in Calvert County, 1,028 parcels in Carroll County, and 855 parcels in Howard County, which have arm's-length transactions that took place between July 1993 and June 1996. We include one-acre-or-larger parcels that have their parcel coordinates attached. Because the primary interest is the value of the land only, we subtract the assessed value of the residential structure from the market price if the parcel contains a house. The dependent variable is the natural log of the per-acre real-estate transaction price in 1996 dollars discounted using the Urban Annual Consumer Price Index (CPI). Thus the average per-acre price for the land was U.S. \$59,612 in Calvert County, U.S. \$61,208 in Howard County, and U.S. \$47,368 in Carroll County (Table 16.1).

These CPI averages are much higher than the average prices calculated using sales transactions for just farmland, which had an average sales price of U.S. \$8,998 (Nickerson and Lynch 2001). These private market county averages are also much greater than the average easement payments received in the same counties, which was U.S.\$2,403 in Calvert County, U.S. \$4,685 in Howard County and U.S. \$1,165 in Carroll County (Lynch and Lovell, 2002). For larger parcels with 20 or more acres, however, the per-acre average prices are lower but similar to the easement prices: U.S. \$5,203 in Calvert County, U.S. \$9,764 in Howard County, and U.S. \$5,620 in Carroll County.

When explaining the variation in per-acre transaction prices, we include characteristics that affect both the agricultural value and the developmental value of the land. Because the MDP has geo-coded the centroid of these land parcels, we are able to access other geographic data. Thus using Geographic Information System software (ARCINFO), parcel characteristics from digitized maps were added to the dataset. These include the percent of prime soil; the distance in miles to the nearest metropolitan area (Washington, D.C. or Baltimore), to the nearest town, and to the nearest road; the percent of area within a 100 square meter radius (approximately 1 acre) around the centroid that is water, wetlands, or beach (Chesapeake Bay or Patuxent River) in Calvert County; the

290 _Lynch and Lovell_

Table 16.1 Descriptive Statistics for Private Real Estate Transactions from 1993 to 1996 for Calvert, Howard, and Carroll Counties in Maryland

Variables	Calvert (N[a] = 709) Mean	Std. dev.[b]	Howard (N = 855) Mean	Std. dev.	Carroll (N = 1,028) Mean	Std. dev.
Price per acre	59,612[c]	35,950[c]	61,208[c]	37,807[c]	47,368[c]	27,800[c]
Miles to city	40.617	8.697	21.532	4.414	29.273	5.825
Miles to town	3.823	1.566	1.59	1.639	4.358	2.019
Miles to major road	0.572	0.475	0.473	0.436	0.723	0.656
Number of acres	2.231	5.355	2.966	5.323	4.059	11.772
Percent prime soils	51.10	0.373	76.01	0.285	56.50	0.401
Percent agricultural land use	8.50	0.193	13.92	0.231	21.50	0.288
Percent preserved for agriculture	1.20	0.083	2.91	0.13	6.30	0.213
Percent of water or beach	0.80	0.045				
Hooked to sewer	1.00	0.099	7.95	0.271	3.40	0.181
Future sewer planned	4.40	0.205	8.77	0.283	5.30	0.223
Size less than minimum rural zoning	93.50	0.246	45.61	0.498	89.20	0.311

[a]Number of parcels, [b]Standard deviation, [c]U.S. dollars.
Source: Authors' computations from Maryland Department of Planning Property View data (1993 to 1996) and from MDP Land-Use Maps.

percent of area within a 100 square meter radius around the centroid that is permanently preserved agricultural and forest land having an easement attached; the percent of area of the current land-use (pasture, row crops, vegetable crops, and forest); and if the parcel is connected to the county or to the municipal sewer system or if plans exist to connect the parcel to the sewer system in the future. Summary statistics for the data are presented in Table 16.1. The spatial variables are consistent with those used in previous analyses of parcel-level farmland values on the urban-rural fringe, including proxies for agricultural values and development values (Shi et al. 1997; Bell and Bockstael 2000; Nickerson and Lynch 2001; Lynch and Lovell 2002).

Proxies for development value and timing include the distance to the nearest employment center (either Washington, D.C., or Baltimore) and the distance to the nearest town (measured as a straight line). Land close to the city and thus to employment opportunity, or land near a local town is hypothesized to receive a higher per-acre sales price. However, the relationship between distance to these areas and the market value may not be linear (i.e., the effect of the city or town on the value could dissipate as the distance increases). Therefore we include both distances as a logged variable to allow for possible non-linearity. We also expect that parcels close to a road will receive a higher per-acre price. Surprisingly, we find that on average preserved parcels in Calvert County are closer to the city than are the nonpreserved ones in this analysis (i.e., 36 miles as opposed to 41 miles). Calvert County parcels next to a high percentage of waterfront property may have a higher market value. Therefore, the percent of water, wetland, or beach (the Chesapeake Bay or a major tributary) for Calvert County parcels is included as an explanatory variable in the analysis.

Larger parcels usually receive a lower per-acre price when sold in the land market. Thus we hypothesize that parcels with fewer acres will have a higher per-acre price. We also include a variable to indicate when the parcel size is less than the minimum rural-zoning acreage, which is 5 acres in Calvert County, 3 acres in Howard County, and 6 acres in Carroll County. This permits the rural area to have a different intercept. This binary variable indicating whether the parcel is less than minimum rural-zoning acreage introduces non-linearity into the acreage variable. We do not have a hypothesized sign for the estimated coefficient on this variable. The rural areas have had lower per-acre prices, but many individuals have been attracted to this type of location recently; therefore, we might find either a positive or negative effect from this variable.

The designers of easement programs typically express interest in preserving productive farms. Therefore, we proxy net agricultural returns by the size of the farm, the proportion of the parcel in agricultural uses (row crops, vegetables, and pasture), and the percentage of prime soils (e.g., Lynch and Lovell 2002). Data on seven soil characteristics (agricultural productivity, erosion susceptibility, permeability, depth to bedrock, depth to water table, stability, and slope) were extracted. Following the Maryland classification system, we define prime soils as agriculturally productive, permeable, with limited erosion potential, and with minimal slope (MDSP 1973). The desirable soil characteristics are aggregated into one variable—e.g., percentage of prime soils. A higher percentage of prime soil indicates higher productivity, and thus higher net agricultural returns, which could delay parcel conversion. Prime soils may increase the development value of the farm since it is often less costly to build on these soils. Therefore, parcels may receive higher returns in development due to this attribute and may have converted already.

Calvert County parcels we include in this analysis have 51 percent prime soils, Howard County parcels have 76 percent, and Carroll County parcels have 57 percent. These compare to Calvert County's preserved parcels with 43 percent prime soils, Howard County's with 82 percent and Carroll County's with 39 percent. In Calvert and Carroll counties, it appears that the land with the best soils was converted to housing the earliest. Row crops, vegetables, and pasture-land uses have lower conversion costs than do forested parcels, therefore, they may be considered more desirable. If a house had been constructed already, then the presence of land in a crop use could increase the land value if people desire to be near open farmland, or could decrease the value of the land if people perceive agricultural land to be smelly, noisy, dirty, and prone to attract insects. Similarly, land with a high percentage of agricultural use may not have been subdivided or improved for developed use, so we hypothesize that its value is lower.

In addition, land already enrolled in a land-preservation program and land that had sold or donated its development rights may have a lower per-acre price. While Nickerson and Lynch (2001) find little statistical evidence that easement restrictions lower the market price, include the percentage of land permanently preserved as a proxy for the absence of these rights. Permanently preserved open space is defined as farms and other land having easements prohibiting residential, commercial, and industrial development. These easements are either purchased by a county or state agricultural land-preservation program, or donated to a land trust or to the Maryland Environmental Trust.

Separate equations are run for each county because of the differences in the average returns landowners expect from the sale of their land, from county-level services, from permitted zoning densities, and from alternative opportunities such as preservation programs.

ESTIMATION AND RESULTS

We estimate a separate regression model for each of the counties using SpaceStat Version 1.9 (Anselin, 1998). Tests for spatial dependence using a spatial-weight matrix are also conducted. The spatial-weight matrix contains the inverse distance between parcel i and parcel j when the parcels are less than 1609.27 feet apart. The matrix is row standardized. We use the Robust Lagrange Multiplier test (LM-test) to determine spatial correlation (Anselin 1988a; Anselin and Bera 1997). When the LM-test is significant, we use an iterated Generalized Moments (GM) estimator to calculate the spatial error model. Due to the large sample size, the GM estimator provides statistically valid results (Bell and Bockstael, 2000).

In all three models (i.e., for Calvert, Howard, and Carroll counties) we find evidence of spatial dependence (*Robust* $LM_{(1)} = 20.67$; *Robust* $LM_{(1)} = 17.78$, *Robust* $LM_{(1)} = 7.19$). Therefore, iterated GM models are estimated for all three. While spatial correlation is identified as a problem in these models, qualitatively and quantitatively the estimated coefficients do not change dramatically between the corrected and uncorrected models (Tables 16.2, 16.3, and 16.4).

The significance of certain variables and the overall fit of the estimated model varies by county. For Calvert, the Buse R^2 is 0.68, for Howard it is 0.74, and for Carroll it is 0.55. The Buse R^2 is adapted to the error structure of the spatial-error model (Anselin 1988a). The correction for spatial correlation results in higher R^2 values for Carroll, Howard, and Calvert counties when compared to the R^2 values of the OLS models. Lynch and Lovell (2002) find a different pattern of overall fits using similar characteristics to explain easement values. For Carroll, spatial correlation is found and the Buse R^2 is 0.62; for Howard, no spatial correlation is found and the R^2 is 0.87; and for Calvert, no spatial correlation is found and the R^2 is 0.32.

In all three counties' regression models, the estimated coefficients on the distance-to-the-city variable suggests that the closer the parcel is to a city, the higher is its market price. A change in the distance has the biggest impact in Carroll County, where proximity of a parcel that is 1 percent closer to the city raises the price of the parcel 0.46 percent (Table 16.5). The easement value in Carroll County is almost 1 percent higher, as the parcel is 1 percent closer to Baltimore. The per-acre price increases 0.29 percent in Calvert County if the parcel is 1 percent closer to the city, compared to 7 percent for the easement

Table 16.2 Estimated Coefficients Explaining Private Market Real-Estate Transactions in Calvert County with Ordinary Lease Squares (OLS) and Corrected Spatial Model

Variables	OLS[a] Estimated Coefficient	Std. dev.[b]	Corrected Spatial Model Estimated Coefficient	Std. dev.
Constant	10.993	0.452	10.831	0.506
Log of miles to city	-0.334**	0.110	-0.285**	0.124
Log of miles to town	0.127**	0.052	0.121**	0.056
Miles to major road	-0.056	0.044	-0.059	0.049
Number of acres	-0.028***	0.004	-0.028***	0.004
Percent prime soils	-0.039	0.056	-0.034	0.060
Percent agricultural land use	0.085	0.108	0.06	0.112
Percent preserved for agriculture	0.160	0.208	0.068	0.201
Percent of water or beach	3.511***	0.476	3.333***	0.467
Hooked to sewer	0.343	0.219	0.333	0.214
Future sewer planned	-0.104	0.110	-0.113	0.114
Size less than minimum rural zoning	1.030***	0.092	1.015***	0.092
ρ			0.187	0.000
R^2	0.357			
R^2 (Buse)			0.684	
Spatial statistics for OLS	d.f.[c]	Value[d]		
Robust *LM* (error)	1	20.668***		

[a]Ordinary least-square regression, [b]Standard deviation, [c]Degrees of freedom, and [d]Value of the *t*-statistic.
**Denotes statistical significance at the 0.05 level of confidence.
***Denotes statistical significance at the 0.01 level of confidence.
Source: Authors' computations.

value for a similar change in distance. Of course the easement values are much lower. A 7 percent change in the easement value of U.S. $2,403 is U.S. $168, while a 0.29 percent change in the market price of U.S. $59,612 is U.S. $173. In Howard County, there is a 0.24 percent increase in the per-acre price (U.S. $147) if the parcel is 1 percent closer to a metropolitan area as compared to a 1.8 percent increase in the parcel's easement value (U.S. $84). The relationship between distance to the nearest city and the prices received in the 3 counties is illustrated in Figure 16.1. Interestingly, Calvert County has higher prices than Carroll County even for those parcels farther away from the city. Calvert County's nearest city is Washington, D.C., which may have more employment opportunities than Baltimore does, which is the closest city to Carroll County. Alternatively, Calvert County's location near the Chesapeake Bay may increase the desirability of its land. In addition, the road networks to Washington may result in a similar commute time.

The effect of distance to town varies by county. In Calvert County, the coefficient is positive and significant, suggesting that the farther the parcel is from a town, the higher its price. Distance to the city and distance to town in Calvert

Table 16.3 Estimated Coefficients Explaining Private Market Real-Estate Transactions in Howard County with Ordinary Least Squares (OLS) and Corrected Spatial Model

Variables	OLS[a]		Corrected Spatial Model	
	Estimated Coefficient	Std. dev.[b]	Estimated Coefficient	Std. Dev.
Constant	11.370	0.342	11.320	0.379
Log of miles to city	-0.251[***]	0.107	-0.240[**]	0.120
Log of miles to town	0.034	0.024	0.032	0.027
Miles to major road	-0.017	0.037	-0.012	0.038
Number of acres	-0.029[***]	0.003	-0.028[***]	0.003
Percent prime soils	0.086	0.058	0.075	0.060
Percent agricultural land use	-0.169[**]	0.073	-0.133[*]	0.074
Percent preserved for agriculture	-0.100	0.130	-0.068	0.135
Hooked to sewer	-0.317[***]	0.082	-0.297[***]	0.087
Future sewer planned	-0.239[***]	0.071	-0.251[***]	0.077
Size less than minimum rural zoning	0.727[***]	0.036	0.751[***]	0.037
ρ			0.175	0.000
R^2	0.464			
R^2 (Buse)			0.739	
Spatial statistics for OLS	d.f.[c]	Value[d]		
Robust LM (error)	1	17.780[***]		

[a] Ordinary least-square regression, [b]Standard deviation, [c]Degrees of freedom, and [d]Value of the t-statistic.
[*]Denotes statistical significance at the 0.10 level of confidence
[**]Denotes statistical significance at the 0.05 level of confidence.
[***]Denotes statistical significance at the 0.01 level of confidence.
Source: Authors' computations.

County are inversely related. In Carroll and Howard counties, the estimated coefficient for the distance-to-town variable is not significant. In the estimated county-level regressions of the easement value, the distance to town is not significant when explaining the easement-value. In a pooled model with all three counties, closeness to a town decreases the easement value.

In the Calvert County model, we find that having a high percentage of beach or water attached to the parcel increases the value of that parcel. One percent more beach or water increases the per-acre price 3.3 percent. While measured differently in the easement analysis, proximity to the Chesapeake Bay or to the Patuxent River also increases the easement payment. It appears that even preserved properties receive value for this attribute, which is also rewarded in the private market.

In all three models, parcels below the minimum-zoning acreage receive higher prices. In addition, the number of acres in a parcel affects the price of that parcel negatively. Larger parcels receive lower per-acre prices, which is consistent with other land-market studies. Thus there is either a rural land market (wherein minimum zoning is in force and where the land is valued lower), or there is a non-linear effect of parcel size on the price, or both.

Table 16.4 Estimated Coefficients Explaining Private Market Real-Estate Transactions in Carroll County with Ordinary Least Squares (OLS) and Corrected Spatial Model

Variables	OLS[a] Estimated Coefficient	Std. dev.[b]	Corrected Spatial Model Estimated Coefficient	Std. dev.
Constant	11.026	0.358	11.016	0.392
Log of miles to city	−0.461***	0.099	−0.459***	0.110
Log of miles to town	−0.022	0.032	−0.026	0.035
Miles to major road	−0.103***	0.027	−0.106***	0.030
Number of acres	−0.017***	0.002	−0.017***	0.002
Percent prime soils	0.109**	0.043	0.110**	0.045
Percent agricultural land use	−0.097	0.062	−0.090	0.063
Percent preserved For agriculture	−0.120	0.084	−0.123	0.087
Hooked to Sewer	0.026	0.099	0.013	0.105
Future sewer planned	0.129	0.084	0.122	0.089
Size less than Minimum zoning	1.321***	0.066	1.327***	0.066
ρ			0.120	0.000
R^2	0.486			
R^2 (Buse)			0.550	
Spatial statistics for OLS	d.f.[c]	Value[d]		
Robust LM (error)	1	7.187		

[a] Ordinary least-square regression, [b] Standard deviation, [c] Degrees of freedom, and [d] Value of the *t*-statistic.
**Denotes statistical significance at the 0.05 level of confidence.
***Denotes statistical significance at the 0.01 level of confidence.
Source: Authors' computations.

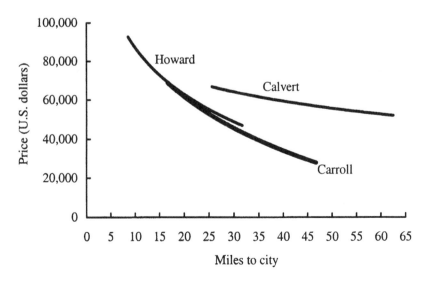

Figure 16.1 Market Price and Miles to Nearest City (by County)

In this chapter, we found that having a lot size of less than 5 acres increases the land value by U.S. $73,176 in Calvert County. Having a lot acreage that is fewer than 6 acres in size increases the land value by U.S. $120,225 in Carroll County. And having a lot less than 3 acres in size increases the land value by U.S. $112,389 in Howard County. In addition, as acreage increases by 1 percent, the land price falls 0.062 percent (U.S. $37) in Calvert County, whereas the easement value decreases 0.12 percent (U.S. $3). In Carroll County, the effect of acreage is similar, with a price drop of 0.068 percent (U.S. $32) in the real estate market and a 0.16 percent drop for the easement value (U.S. $1.9). The private market in Howard County also pays less per acre for larger parcels. Acreage does not affect the easement value of Howard County parcels, however, because Howard County uses a point system to determine the easement value. By so doing, Howard County may be overpaying this parcel attribute to attract larger farms to join the preservation program.

Table 16.5 Elasticities from Estimated Models by County

Continuous variables	Calvert	Howard	Carroll
Miles to city	–0.285	–0.240	–0.459
Miles to town	0.121	0.032	–0.026
Miles to major road	–0.034	-0.006	–0.077
Size in acres	–0.062	-0.083	–0.068
Percent prime soils	–0.034	0.075	0.110
Percent agricultural land use	0.06	-0.133	–0.09
Percent of water or beach	3.333		
Change in value for Binary = 0 to Binary = 1			
Hooked to sewer	$42,403	-$37,344	$1,846
Future sewer hook-up planned	$–11,547	-$32,193	$18,265
Size less than minimum zoning	$73,176	$112,389	$120,225

Source: Authors' computations.

We compare the relationship between the number of acres in a parcel of at least the minimum rural zoning size and the price received for that parcel in the real-estate market to the easement value of parcels in Howard and Calvert counties (Figure 16.2). In Howard County, parcels greater than 30 acres may receive higher per-acre payments in the agricultural-preservation programs than they do in the private market, unless the owner incurs the expense of subdividing and selling the parcel in smaller sections. Similarly, in Calvert County, parcels larger than 90 acres may find a developer who is willing to purchase the development rights in the TDR program for a higher price than the parcels would receive if they were sold in the rural land market.

While parcels with a higher percentage of prime soil command a significantly higher easement value estimate in the pooled model, in the county-level models only the estimated coefficients in Carroll county on prime soil are significant. Lynch and Lovell (2002) hypothesize that the soils within each county are not sufficiently different, and that this fact is reflected in the insignificant coefficients in the county-level regressions. In the Howard and Calvert real-estate market transactions, the soil quality has no impact. However, in Carroll County the per-acre price increases as the percent of prime soil increases. One percent more prime soil results in a market price that is 0.11 percent higher.

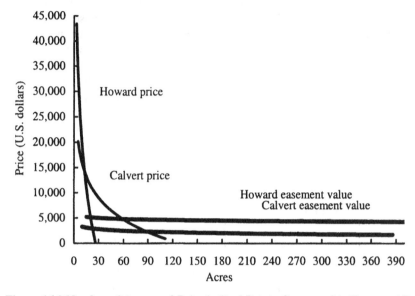

Figure 16.2 Number of Acres and Price in Real Estate Compared to Easement Payment (Howard and Calvert Counties)

Between 1993 and 1996, none of the counties' markets rewarded agricultural land use. Conversely, the percentage of agricultural land in a parcel decreased the market price in Howard county. This Howard county result is consistent with the finding that the percentage of agricultural land in a parcel decreases the easement value. Thus landowners who are planting crops have either a lower reservation price, or the preservation programs are not prioritizing enrollment of land being actively farmed. However, there was much less agricultural land use on the parcels included in this analysis. Only 9 percent on average of each Calvert parcel was in agricultural use from 1993 to 1996, Howard County had 14 percent, and Carroll County had 22 percent. Alternatively, for the parcels in the agricultural land-preservation program, Calvert County had on average 38 percent of the parcels, Howard County had 75 percent, and Carroll County had 85 percent.

We include a variable to proxy whether a parcel is enrolled in the agricultural land-preservation program, which we expected to decrease the parcel's value. This variable is not significant in any of the regressions. One interpretation, similar to the findings of Nickerson and Lynch (2001), is that the restrictions imposed by the easement provisions are not decreasing the price of preserved parcels. Alternatively, the variable might not be a good measure of these restrictions.

We also include several variables that we expected to affect the per-acre price in an urbanizing area: distance to major roads and having a sewer connection (or having a sewer connection planned). We find the distance between the parcel and the nearest major road does not affect the price in Calvert or Howard counties. In Carroll County, however, the closer the parcel is to a major road, the higher the price it receives. Thus when the parcel is 1 percent closer to the road, the market price is 0.077 percent higher. Surprisingly, being hooked up to a municipal or county sewer does not increase the land value in Calvert or Carroll counties. Nor does the anticipation of a sewer hook-up that is planned for the future increase the per-acre price. Even more surprising is that in Howard County, being hooked up to a sewer system or having a sewer planned for the land in the future actually decreases the value of the land. Being hooked up to a sewer system decreases the land value by U.S. $37,344, and having a sewer hookup planned for the future decreases the value by U.S. $32,193. Rural areas on a septic system must be considered more desirable, all else being equal.

CONCLUSIONS

In this analysis, we modeled the marginal contribution of different parcel characteristics to the market-based, per-acre land price in the Calvert, Howard, and Carroll counties of Maryland between 1993 and 1996. The results were compared to the marginal contribution of the same or similar characteristics to easement values paid by agricultural land-preservation programs investigated in an earlier paper (Lynch and Lovell 2002). Information on the similarities and differences between the factors affecting easement prices versus market prices can help formulate policy decisions to improve preservation programs. And it can help to ensure that the correct parcels are selected to achieve the stated goals.

Using spatially explicit data, hedonic models corrected for spatial correlation were estimated for three Maryland counties: Calvert, Howard and Carroll. The models explained 55 percent to 74 percent of the variation in the market price of parcels that were at least one acre in size. Spatial correlation was identified as a problem in each of the estimated models. However, except for the impact of agricultural land use on the price in Howard County, the estimated coefficients did not change dramatically—qualitatively or quantitatively—between the corrected and uncorrected models.

We found that strong similarities between the characteristics' effects on easement values and market prices do exist. These included the effects of distance to the city, size of the parcel, and proximity to water in Calvert County. For other characteristics, only one or two counties showed similarities between how prices in the market and values in the preservation programs were affected. For each of the three counties, both easement values and market-land prices were affected by the distance to employment centers, as measured to the nearest city (Washington, DC, or Baltimore). This distance was also one of the most important determinants of value and price, as measured by the elasticities. The magnitude of the distance effect in terms of percentage changes on easement values (i.e., the percentage decrease in easement value for a 1 percent increase in distance) was larger than for market prices in all three counties. The magnitude in terms of the actual dollar amount was remarkably similar. We found that distance to a metropolitan area is a significant determinant of easement value (Wichelns and Kline 1993; Plantinga and Miller 2001; Lynch and Lovell 2002). Program administrators appear to incorporate this market phenomenon into the values they are willing to pay for easements in PDR programs, which use either point systems or appraisal methods to determine easement values.

Another consistently strong result we found was that the per-acre market price declined as parcel size increased. The decline was higher in percentage terms and lower in dollar terms for easements than it was for market prices. Examination of the averages of the transactions was also illuminating. The average per-acre price for the land alone was U.S. $59,612 in Calvert County, U.S. $61,208 in Howard County, and U.S. $27,368 in Carroll County. Yet for parcels of 20 or more acres, per-acre prices were much lower: U.S. $5,203 in Calvert County, U.S. $9,764 in Howard County, and U.S. $5,620 in Carroll County. Per-acre easement payments also declined with larger sized parcels, except in Howard County.

As with distance to the city, administrators are following the market signal by paying less per acre for larger parcels. In Howard County, however, this appears not to be the case. A point system that assigns a higher weight to larger farms needs to be evaluated as to whether it is maximizing society's welfare. If larger farms were to contribute substantially more to goal achievement, such as maximizing total acres preserved or preserving productive farms, then the market solution of discounting the per-acre value due to large size may not be an optimal strategy. However, a program with limited resources could preserve more acres if it were to follow the private-market behavior of paying a lower per-acre price for larger parcels.

Both easement values and market prices were positively affected by their proximity to water in Calvert County. However, because most of the land preserved in Calvert County was through the TDR program, land with beach or water access is often not preserved due to its higher per-acre cost. Some people argue that the increase in impervious surfaces caused by increased development near a water body can damage significantly the water quality. Therefore, if one of society's goals were to protect water quality in addition to protecting open space, then a stronger mechanism may be needed to entice owners of farms near water to participate in the preservation program.

A higher level of prime soils does not receive a higher price in the real estate market except in Carroll County. This characteristic makes the land more desirable for both agricultural use and for residential or commercial use as it is easier to build on and easier to farm. Other parcel characteristics may have contributed more heavily than soil quality to determining the equilibrium market price, although soil quality played a role in the easement payment paid by preservation programs. If enrolling parcels with prime soils were to maximize society's goals, the preservation programs may be following the optimal strategy to reward this feature, even though only in Carroll County's private market does prime soil contribute to raising the market price.

Carroll County was also unique in that distance to a road had an effect on land price in the private market. Although some program administrators suggest that the distance to a road or to the road frontage should be included in the point system to determine the easement value, we found support for this suggestion only in Carroll County. Thus only in Carroll County should the programs be considering or consider increasing marginally the easement payment for proximity to a road.

What lessons can be learned from the above analysis? These results demonstrate that some parcel characteristics have a similar effect on both market prices and on easement values across all counties, but that the magnitude of the effects is not necessarily the same. As a consequence, program administrators who want to increase enrollment in preservation programs may need to adjust the payments to encourage program participation. Given the price that can be received for subdividing and improving land, especially in areas near employment centers, landowners will need to be adequately rewarded if they are to consider enrollment in the program.

The most pressing situation concerns those farms that are facing the greatest threat of development. If society determines that it wishes to preserve those acres because they are close to the cities and to urban populations and will thus provide more viewing and enjoyment possibilities, then some urgency is warranted. Program administrators are currently attempting to devise new methods and new payment schemes to attract these farmland owners to enroll. Easement programs may want to adjust their valuation of larger parcels in order to make them more comparable to the market valuation process.

As in this analysis, the use of hedonic models to determine the marginal value of certain land characteristics could provide some interesting information for easement-program administrators. While developed parcels may not be as comparable as unpreserved farms to preserved farms, the analysis sheds some light on how the real-estate market values certain characteristics. Future research using sales of undeveloped land may provide a more comparable analysis than one using both developed and undeveloped land, as in this study. On the other hand, landowners in these areas would consider the prices received by developed parcels as well as undeveloped parcels when evaluating the value of their land. In addition, while the data was unique and collected to do a microanalysis, refinement of certain variables such as preserved status would also be beneficial. Similarly, although spatial correlation was present in all three models, the analysis would have provided similar information and policy recommendations even if it had been estimated using OLS. Both qualitatively and quantitatively, the two sets of estimated coefficients were found to be similar in each county.

REFERENCES

AFT (American Farmland Trust). (2001a) "Status of Selected Local PACE Programs: Fact Sheet." Washington DC: American Farmland Trust.
_____. (2001b) "Status of State PACE Programs: Fact Sheet." Washington DC: American Farmland Trust.

_____. (2001c) "Transfer of Development Rights: Fact Sheet." Washington DC: American Farmland Trust.

Anselin, L. (1995) *SpaceStat Version 1.80 User's Guide.* Version 1.9 Update (1998). Internet Website: http://www.spacestat.com/docs/V180man.pdf.

_____. (1988a) *Spatial Econometrics: Methods and Models.* Dordrecht: Kluwer Academic Press.

_____. (1988b) "Model Validation in Spatial Econometrics: A Review and Evaluation of Alternative Approaches." *International Regional Science Review* 11: 279–316.

Anselin, L. and A. Bera. (1997) "Spatial dependence in linear regression models with an introduction to spatial econometrics." In *Handbook of Applied Economic Statistics*, A. Ullah and D. Giles,(eds). New York: Marcel Dekker.

Anselin, L. and R. Florax. (1995) *New Directions in Spatial Econometrics.* Berlin: Springer-Verlag.

Bay Journal. (1997) "Maryland Enacts Sweeping Growth Management Law." 7(3) May.

Beasly, S.D., W.G. Workman, and N.A. Williams. (1986) "Estimating Amenity Values of Urban Fringe Farmland: A Contingent Valuation Approach: Note." *Growth and Change* 17(Oct): 70–8.

Bell, K. and N. Bockstael. (2000) "Applying the Generalized-Moments Estimation Approach to Spatial Problems Involving Micro-Level Data." *The Review of Economics and Statistics* (February) 82(1): 72–82.

Bergstrom, J.C., B.L. Dillman, and J.R. Stoll. (1985) "Public Environmental Amenity Benefits of Private Land: The Case of Prime Agricultural Land." *Southern Journal of Agricultural Economics* 17(July): 139–50.

Bromley, D.W. and I. Hodge. (1990) "Private Property Rights and Presumptive Policy Entitlements: Reconsidering the Premises of Rural Policy." *European Review of Agricultural Economics* 17(2): 197–214.

Cliff, A. and J.K. Ord. (1973) *Spatial Autocorrelation.* London: Pion Publishing.

Drake, L. (1992) "The Non-Market Value of Swedish Agricultural Landscape." *European Review of Agricultural Economics* 19(3): 351–64.

Fischel, W.M. (1985) *The Economics of Zoning Laws: A Property Rights Approach to American Land Use Controls.* Baltimore: Johns Hopkins University Press.

Foster, J.H., J.M. Halstead, and T.H. Stevens. (1982) *Measuring the Non-Market Value of Agricultural Land: A Case Study.* Research Bulletin 672, Massachusetts Agricultural Experiment Station, University of Massachusetts, Amherst.

Gardner, B.D. (1977) "The Economics of Agricultural Land Preservation." *American Journal of Agricultural Economics* 59 (December): 1027–36.

Halstead, J.M. (1984) "Measuring the Non-Market Value of Massachusetts Agricultural Land: A Case Study." *Northeastern Journal of Agricultural Economics* 14 (April): 12–9.

Kelejian, H. and D. Robinson. (1993) "A Suggested Method of Estimation for Spatial Interdependent Models With Autocorrelated Errors, and an Application to a County Expenditure Model." *Papers in Regional Science* 72: 297–312.

Kline, J. and D. Wichelns. (1996) "Public Preferences Regarding the Goals of Farmland Preservation Programs." *Land Economics* 72(4): 538–49.

Land Trust Alliance. (2000) "More Than $7.3 Billion Committed To Open Space Protection." Internet website: http://www.lta.org/policy/referenda2000.html.

Lynch, L. and S.J. Lovell. (2002) "Hedonic Price Analysis of Easement Payments in Agricultural Land-Preservation Programs." Department of Agricultural and Resource Economics Working Paper #02-13, University of Maryland, College Park.

_____. (2001) "Factors Influencing Participation in Agricultural Land Preservation Programs." Department of Agricultural and Resource Economics Working Paper #01-05, University of Maryland, College Park.

Lynch, L. and W.N. Musser. (2001) "A Relative Efficiency Analysis of Farmland Preservation Programs." *Land Economics* 77(4): 577–594.

Maryland (1988) Michie's Annotated Code of the Public General Laws of Maryland, Agricultural Article, 2-501, subtitle 5. Lexislaw Publishing, Charlottesville, Virginia.

Maryland Department of Agriculture (DOA). (1998) Maryland Agricultural Land Preservation Foundation Report. Internet Website: www.mda.state.md.us/agland/preserva.htm

MDSP (Maryland Department of State Planning). (1973) *Natural Soil Groups Technical Report*. HUD Project Number: MD-P-1008-100. Maryland: HUD.

McConnell, K.E. (1989) "The Optimal Quantity of Land in Agriculture." *Northeastern Journal of Agricultural and Resource Economics* 18(October): 63–72.

Nickerson, C.J. and L. Lynch. (2001) "The Effect of Farmland Preservation Programs on Farmland Prices. " *American Journal of Agricultural Economics* 83(2): 341–351.

Plantinga, A. and D. Miller. (2001) "Agricultural Land Values and Future Development." *Land Economics* 77(1): 56–67.

Pruckner, G. J. (1995) "Agricultural Landscape Cultivation in Austria: An Application of the CVM." *European Review of Agricultural Economics* 22(2): 173–90.

Shi, Y.J., T.T. Phipps, and D. Colyer. (1997) "Agricultural Land Values Under Urbanizing Influences." *Land Economics* 73(February): 90–100.

Whittle, P. (1954) "On Stationary Processes in the Plane." *Biometrica* 41: 434–49.

Wichelns, D. and J.D. Kline. (1993) "The Impact of Parcel Characteristics on the Cost of Development Rights to Farmland." *Agricultural and Resource Economics Review* 22(2): 150–158.

Wolfram, G. (1981) "The Sale of Development Rights and Zoning in the Preservation of Open Space: Lindahl Equilibrium and a Case Study." *Land Economics* 57(3): 398–413.

ENDNOTES

[*] The opinions expressed here are those of the authors and not necessarily those of the U.S. Environmental Protection Agency.

Chapter 17

Efficiency, Equity, and Farmland Protection: An Economic Perspective

Lawrence W. Libby
The Ohio State University

INTRODUCTION

In this chapter, I consider how economics as a social science may help organize our thinking about farmland-protection policy in the United States. Every state and many localities have enacted policies to alter the pattern of development in the interest of keeping more land in farms (AFT 1997; Daniels and Bowers 1997). The policy instruments range from the regulation of the land market and the public purchase of the landowner's right to develop the land all the way to the organized wishful thinking about the future through land-use planning. All of these techniques alter the options and obligations of land buyers and sellers to achieve a land-use pattern more in line with social preferences. Many economists fret about these policy distortions of land markets as if there were markets without rules that would determine who can participate and for what. Other more enlightened economists avoid all the normative drivel about the sanctity of markets and the illegitimacy of farmland protection as a policy issue, preferring to consider the consequences of alternative means for achieving this public purpose.

I begin this chapter with a brief discussion of what economics is and what it is not as a discipline, with reference to farmland-protection policy. The two main analytical constructs of economics are efficiency and distribution. Both have strong normative underpinnings, the latter emerging as a particular distribution that is termed equity, and both are considered here. I conclude the chapter with some insights for future policy in this area.

ECONOMICS AS A LANGUAGE

Economics is a fairly useful discipline, if you do not take it too seriously. It can help anticipate the results of interactions (including competition) among people seeking their notion of improvement. Economists are notoriously disinterested in exactly why people do what they do, but economists observe that people collectively seem to act as if they are trying to be better off, and even as well off as possible. We do know that people are occasionally altruistic, even to the point of being cooperative with others. We also understand that human relationships matter in any particular instance. That is, I do feel good knowing that my neighbor is better off, so long as it is not always at my expense. Also, I would sell my car to my daughter at a lower price than I would to any of you. Now you might argue that such softness is really self-interest, since I expect reciprocity some day from both my neighbor and my daughter. In fact, even economists acknowledge that people do care about each other in any specific interaction.[i] But economists have observed that, over time and millions of land transactions, people tend to act as if they are interested in only themselves.

Economics is also a language of concepts that captures complex relationships among people in land transactions. Comparative advantage, for example, recognizes that resources are more productive in one use than they are in another and therefore there are gains from trade among individuals, regions, or nations. Opportunity cost is the assertion that every use of land has a price, which is the value of what might have been done on that parcel. Markets bring buyers and sellers together in an impressive fashion, but they achieve it with complex rules about who can negotiate about what. Markets are really collections of rules, including private property rights sanctioned and supported by governments to accomplish trade. Without government there would be no organized market.

This conceptual apparatus that is economics is helpful in understanding and predicting human interactions in land markets. In my view, it is much less helpful to say what should happen to farmland or to any other resource, or which particular distribution of land rights implicit in policy options is fair. People, including economists, have opinions about who should pay and who should gain from land-use rules, but the discipline itself can draw few conclusions on the matter.

EFFICIENCY AND FARMLAND POLICY

There are really three efficiency questions that may be applied to farmland use. Markets can help answer all three questions in the context of public policies that shape the rights and responsibilities of those who make the land transactions. These questions include how much farmland is enough; what is the best use for a particular parcel of farmland; and what is the best mix of farmland use overall. One might also ask the question "What is the most efficient way to protect farmland?"

Gardner (1977) concludes that if food production is the land-use purpose, there is no reason to assume that the market will not allocate sufficient land to that activity. Land prices reflect relative scarcity of land uses, and with a firm set of property rights, undistorted by zoning or other interventions, land will move to its most efficient use. Farmers will pay more for more productive land and will choose a variety of enterprises that are best for their particular location of farmland and for other aspects of the farmland market. Food scarcity will trigger higher prices, which will be capitalized in land values that will bring forth more food production. Substituting a bureaucratic decision process for a land market would be insufficient (and therefore unwise). But Gardner does acknowledge that there are some farmland services that have high exclusion costs for the landowner that cannot be withheld from those unable or unwilling to pay for them. Therefore, the market will allocate an inefficient amount of land to those joint products of food safety and various public amenity services. These other farmland services—including various ecological services (e.g., groundwater recharge and waste processing), habitat, the option value of having farmland available in case it is needed for future food production, and various landscape-amenity values—are difficult to impossible to sell because of high exclusion costs, but are nevertheless valued by consumers and are secured through the policy process. Libby (1997) argues that many people experience a certain sense of well being from knowing that their long-term food needs will be met. People will support efforts to maintain that sense of security. Thus, food security from farmland services is a pure public good in that enjoyment by one does not diminish the availability of its security to others and no one can be excluded from the use of food security as a public good.

The "how much" of farmland efficiency, then, requires that the high exclusion-cost services of farmland be part of the efficiency analysis, or too little land will be allocated to farming. And the best use for particular land parcels or acres must also consider these other services, or the resulting land-use pattern is inefficient. Highest and best use is a land-use concept that has both legal and economic meaning. It generally refers to the use that land will command the highest price between a willing buyer and seller. The concept has special status in both law and economics as the use of farmland that should prevail.[ii] The ability to bid more for a parcel of land than someone else is clearly a function of market rules concerning who has standing in that market and what the permitted uses for the land are. Thus, efficiency is a function of market rules—different rules imply a different definition of efficient land use for a particular parcel.[iii]

Similarly, market rules determine spatial patterns of land use. Patterns evolve, given the competition for farmland productivity and location. Thus, the greater the net monetary return, the higher is the price offered for a particular land parcel. Location in the form of distance to market also affects net returns. The intensity of land use tends to be greatest close to a market center, but its intensity declines as the distance into the farming area increases. The efficient land-use pattern, then, is reflected in the distribution of highest and best use over space.

Inclusion of the nonfood services of farmland secured through policy, however, alters the spatial pattern of land use to reflect what people really want from the land. Real efficiency of land use, then, requires a land market that includes joint products of farmed land such as countryside amenities or wildlife habitat. Much effort has gone into estimating the implicit willingness-to-pay for these amenity services as if efficiency were really the goal. One example of estimating willingness-to-pay is that of Bergstrom et al. (1985) and colleagues who ask residents of Greenville County, South Carolina what they would pay to stop the further development of specific farmland acres. Other examples are Halstead (1984) who poses various bid levels to the residents of Massachusetts to determine at what price development was more important than the price of protecting the farmland in question and Beasley et al. (1986) who uses an approach similar to that of Halstead for the state of Alaska. Costanza et al. (1997) go one step further to estimate the economic value of ecosystem services of farmland as the cost of providing those same services by artificial means. The socially efficient allocation of land to farming, then, would be that level at which no one could be better off without making someone else worse off. At the very least, those who gain from a specific land reallocation could pay off those who lose and still come out ahead (Varian 1996).

In fact, no one really cares about the efficiency of land use as such (except perhaps a few professional economists), instead they demand a mix of uses and a commitment of land to farming that reflects their interests. Land-use policy is concerned with securing those land services, and efficiency criteria are generally ineffectual when it comes to selecting among the many policy alternatives.

If the goals of farmland policy were clear and precise, we could compare alternative policy instruments for their cost to achieve a defined result. Perhaps zoning could protect 1,000 acres of prime farmland at a lower public cost than could the purchase of development rights (PDRs). Therefore, we could conclude that zoning would be more cost effective for protecting farmland, and, in some sense, it would be more efficient. The real question would be whether or not the same services could be acquired under zoning and PDR. Would zoned farmland, land restricted by easement, and land retained by bribing the farmer with use-value assessment produce the same services as farmland policy? I doubt it. Efficiency analysis is just not very helpful. The more meaningful question of policy choice relates to whose preferences are expanded and whose preferences are reduced under the various policy mixes.

EQUITY QUESTIONS

As noted above, equity implies that one distribution of the right to use land is the fairest. One is tempted to conclude that efficiency is fairest, and while economists studiously avoid making judgments about whose utility is more important, there is a strong presumption that efficient allocation is fair since no one could be made better off without someone else being worse off by any land-use or policy change. That logic aside, what questions on equity are there within farmland-protection policies?

Is Agricultural Zoning Fair?

The fairness of agricultural zoning is a matter of opinion. Many economists seem ready to accept that exclusive agricultural zoning is not fair, and that it could impose a wealth loss on farmers to provide the collective goods of farmed land for society as a whole (Gardner 1977). But in fact, the market value of farmland already reflects the impact of public spending for roads and for other infrastructure (Runge et al. 2000). Perhaps agricultural zoning just reclaims a portion of land value for the public in the interest of protecting certain nonexclusive land services while it protects the farmer's opportunity to farm. That, it seems, is a reasonable conclusion.

There is an important distinction here between large-lot rural zoning, which discourages residential development in farming areas, and truly exclusive agricultural zoning, which explicitly omits nonfarm residences and other such activities from the list of uses permitted in an agricultural district. Even at a 20-acre minimum lot size, there is real doubt that the large-lot approach achieves, or even could achieve, the stated public purpose of keeping land for farming. Many local agricultural-zoning ordinances impose few restrictions on residential or commercial activities within the agricultural zone and post minimum lot sizes of 3 acres or less. These districts essentially are holding areas for future development rather than they are areas of protection for farmland. As such, these ordinances may impose costs on some landowners who have land that has little public purpose. I would argue that such permissive restriction could constitute an unfair, even if short term, limitation on the landowner, with little contribution to the public good.

The major legal test of the fairness of agricultural zoning is the takings clause of the U.S. Constitution. Does the restriction so impinge on the owner's ability to earn income from the land that it constitutes a regulatory taking that requires just compensation under the Fifth Amendment of that U.S. Constitution? The relevant tests for legal acceptability of a zoning ordinance seem to be the Lucas test (Cordes 2002) of all economic viability and the somewhat less restrictive Penn-Central test of investment-backed expectations. While the U.S. Supreme Court has not applied these two filters for a specific takings case, lower courts have consistently held that exclusive agricultural zoning does not deprive the owner of all economic potential since farming is a viable activity. Conceivably, there could be circumstances under which that condition would not hold, but none have come before the courts as yet. The Penn-Central test is seldom applied, since farmland is bought with farming or zoning restrictions already in place that remove any reasonable expectations of a higher value use for that land, which might induce greater investment by the owner (Cordes 2002). A 2001 U.S. Supreme Court decision *Palazzolo v Rhode Island* in Rhode Island has the potential for affecting exclusive agricultural-zoning ordinances by holding that a land-use restriction in place at the time of purchase need not preclude a takings claim by the new owner. This case seems to strengthen the two-fold legality test of Lucas and Penn Central for all future takings cases. The decision goes on to insist, however, that existing land-use restrictions be factors in any future examination of investment-backed expectations. Thus, the Rhode Island case does not ignore the notice of existing zoning when considering investment behavior of the landowner, but it does go beyond previous lower-court precedents when protecting the owner's right to a takings claim. At the same time, this case strengthens the Lucas filter of economic viability by sustaining that test even when existing zoning would seem to preclude a higher value use.

Is Agricultural Zoning Legally Fair?

Cordes (2002: 9) argues "Even though [zoning] might have a significant economic effect on regulated landowners if done properly, it should not constitute a regulatory taking nor be viewed as inherently unfair." No landowner should expect to be totally free of the restriction to protect the rights of others in today's society. Thus, any policy change implies gainers and losers, and over time everyone should expect to come out ahead once in a while.

Several states have enacted property-rights protection statutes to alter the balance of fairness between landowners and community, which give landowners greater access for the legal redress of alleged unfairness of regulations that affect an owner's options. The Florida statute [Section 70.001 (2), 1995], for example, established a procedure through which landowners, whose property rights were inordinately burdened by government regulation, could seek compensation. A majority of Florida legislators and their supporters feel presumably that open-land zoning is unfair and that constitutional protection is too burdensome and uncertain for the landowner to pursue. The major effect of the law is to dissuade some local governments from trying to preserve farmland or trying to control urban growth (Libby 1996). The catch is that the law applies only to state and local laws passed after 11 May 1995 and excludes transportation regulations. If the landowner rejects what the circuit court feels is a reasonable compensation, the landowner pays the court cost. Also, in Texas a statute was established in which a 25 percent threshold on land-value reduction attributable to regulation was enacted and the burden of proving harm was to be borne by the landowner.

The most recent addition to state landowner-protection actions is the constitutional amendment passed by Oregon-state voters on 7 November 2000, which protects landowners from any loss of value that can be proven to be attributable to state or local government regulation. Washington-state voters defeated a similar measure in 1995. Wording of the ballot initiatives in both states had much to do with the different results—a yes vote in Oregon required that state and local governments pay the property owner if policy regulation reduced his or her property value; a no vote rejected that requirement. The Washington-state vote began as a law passed by state legislators that was designed to have the same sweeping effect as the Oregon amendment, but it failed. The public referendum in the state of Washington was an action to repeal that legislation, thus opponents had a role in framing the issue on the ballot. The language of the public referendum is that "The Washington State Legislature has passed a law that restricts land-use regulations and expands government's liability to pay for reduced property values of land or improvements thereon caused by certain regulations for public benefit. Should this law be approved or rejected?" (Kendall and Dorman 2001). The referendum passed effectively repealing the statute requiring government compensation. In these cases, language did make the difference between acceptance and rejection of the issue at hand.

Any conclusions we might draw about public views on the fairness of zoning and other regulations in Washington state and Oregon must be modified by the information given to the voters in the two states. Washington-state voters were asked to trade off public gain with private cost while Oregon voters simply responded to the suggestion that taxpayers should pay whenever public regulation reduced private-property value. The Oregon Supreme Court on 4 October 2002 under the single amendment requirement found that Measure 7 involved two changes to the Oregon constitution that were substantial and not closely related. Several environmental groups joined this action. A major consideration, though, was the likely public cost of defending against an endless stream of lawsuits. Virtually no public restriction of land use, no matter how reasoned or appropriate, could survive the zero-impact test. On 13 January 2003, House Bill 2137 was introduced to allow landowners to claim compensation if government were to enact land-use restriction that would lower the fair market value of the land by more than 10 percent (72nd Oregon Legislative Assembly).

What about the economic aspects of fairness? The usual assumption is that any policy-driven restriction on the options available to the landowner would mean a loss of wealth to the owner. In fact, zoning can expand opportunities in some cases by avoiding conflict with non-farm neighbors and by creating a positive environment for investment in a farm enterprise. U.S. Department of Agriculture (USDA) rural-zoning expert Erling Solberg (1967) presented the argument that "Farmers need first to realize that absence of zoning offers no protection ... it permits their neighborhoods to become dumping grounds for ... activities that are prohibited elsewhere." Further, lack of clarity of future land-use patterns can discourage on-farm investment. Why invest if the future is so uncertain? The faint possibility of a high-value sale for PDR can seem more attractive to some than can continued farming.

Farmers have expressed a willingness to share in the cost of guiding land-use regulation. A national survey of farmland owners indicated that while 13 percent of landowners preferred letting the market determine how to protect farmland from development, 58 percent of landowners preferred consistently implemented regulation (Esseks et al. 1998). More than 50 percent of Ohio Farm Bureau members of county advisory councils supported stronger county or township zoning (OFBF 1998). These farm people seem to feel that rural zoning is both effective and fair.

The land market may pick up some benefits of land restriction. Analysis of land values on farms zoned for agriculture in Baltimore County, Maryland reveal that the more restricted land actually sold for a higher price than did the less-restricted land. [Those parcels zoned to permit one residence per 5 acres sold for U.S. $6,282 per acre, while farmland zoned for one residence per 50 acres brought U.S. $7,097 an acre (Bowers 2001)]. Also, Henneberry and Barrows (1990) find a similar market effect in their analysis of farmland in Wisconsin by restricting future development.[iv] Zoned farmland offers a different mix of services than does land in an area without zoning, and reduced uncertainty of future land use may be worth something to buyers. Some of that price effect may be explained by the uneven implementation of local zoning. Thus if landowners feel that a rezoning or variance may be granted when future development is likely, they will be willing to pay a little more to capture that possibility for specific acres. No assumptions are made here regarding zoning implementation in particular places, but willingness-to-pay for specific land parcels is a function of expected future returns to those parcels within the prevailing institutional setting. At least some of that price effect reflects demand for land that has a relatively secure future in productive farming without the conflicts of expectation and of values that come with scattered rural development. The Henneberry and Barrows study (1990) reveals that zoned farmland farther from a city had a greater increase in farmland value than did smaller parcels of farmland that were nearer the city where the future in farming was already less certain. Thus there is no evidence in law or economics to suggest that agricultural zoning is inherently unfair.

Is Payment for Purchase of Development Rights Fair?

The general response about the fairness of payment for PDR rights of farmland for future food security from economists is "of course," since society is paying for the nonmarket services it gets from the farmland. "Preserving farmland for future food security is a public benefit, appropriately paid for by the public" (Tweeten 1998: 20). Economic-welfare criteria to support the central icon of efficiency include the possibility that gainers can compensate losers from a particular change and society can achieve the great normative goal of efficiency.

Since all PDR programs are voluntary, it would be difficult to conclude that PDR is unfair to the landowner. The landowner could sell development potential at a price that would reflect the difference between the land's farmland value and its full-market value, or he or she could choose not to sell his or her land. The landowner could decide whether recouping the current development value of farmland would be more attractive than waiting for the possibility that someone in the future could really capitalize that higher value. The problem with the latter option is that the farmer must change his way of life to cash in on the full development value of his farmland, but there would be no assurance that such development would really occur on his land anyway. In addition, the farmer must pay capital gains tax on the PDR payment, although the payment may come in installments over several years. The landowner may choose to donate some portion of the development value of his land by reducing his or her tax exposure up to 30 percent of the household income over his or her 6-year average income.

But, you might say, selling development rights lowers the total property value by that amount, which affects the owner's borrowing capacity, net worth, and other indicators of well-being. That, of course, would depend on whether future returns to development value are greater than the returns to the PDR payment that the farmer would receive. Further, there is evidence that sale of PDR does not always reduce property value, at least not by the full sale price. Nickerson and Lynch (2001) find that Maryland farmland on which development rights had been sold retained much of that value in the larger parcels that were farther from a city. Thus the PDR restrictions were not fully capitalized in land value. They suggest that some of that effect may come from the expectation by landowners that PDR is not really permanent and that some day the landowner may again be able to sell his land for development. Further, a protected parcel of good farmland with a residence may be especially attractive to a land baron from Washington, D.C. or from Baltimore who seeks a rural estate with little development around it. Such a person would likely pay the maximum for that opportunity.

Most agricultural conservation-easement programs have escape clauses. If there is clear hardship or the protected farmland is surrounded by development and becomes nonviable, the easement may be extinguished under rules that vary from state to state. As an example, Massachusetts requires specific legislation to reverse an easement sale. Also, Maryland and Pennsylvania require a 25-year waiting period. Further, New Jersey has no escape clause at all (Daniels and Bowers 1997). The possibility of extinguishment could increase land value slightly and could make easement programs somewhat more attractive to the farmer. Thus, if PDRs were unfair to farmers, the farmer would not realize it because there are more farmers willing to sell their development rights than there is money to buy them.

What about the income-distribution effect of PDR programs? Since purchase dollars must come from some tax source, conclusions about fairness would depend on the particular revenue instrument used and who would get the payment. PDR programs seeking to protect agriculture generally give priority to large parcels of productive land close to other protected land with the evidence that they have a good future in farming. We might predict that these are the farms producing some of the highest farm incomes and are the farms that can gain the most from the tax advantages available. Scenic quality is important, but PDR programs are generally sold as ways to help preserve an important agricultural industry. Smaller farms on less productive soil, particularly in areas with little development pressure, will not see much of the PDR money. PDR programs are not intended as assistance to low-income or to small farms, and it is unlikely that PDR programs would accomplish that purpose. Large livestock farms, on the other hand, often carry political baggage with them and would be less desirable for PDR payments than would farms that have only a few cows or horses, which contribute to the scenery.

Local programs drawing on a targeted increase in sales tax may have a negative income-distribution effect meaning that lower income people would bear a higher proportionate share of the cost of protecting farmland than would higher income people. Also, special increments on the local property tax may have a similar distributional impact. Pennsylvania has a surtax on cigarettes that is used as a source of funds for buying development rights that could cost the poor more than it could cost the rich. Also, Maine depends on interest on debt incurred on a state-issued credit card and Maryland raises funds through a special farmland-protection license plate. In addition, several states use transfer taxes on all real-estate transactions and some have special taxes that are imposed on the conversion of farmland. The distributional effects of programs that rely on the sale of state or municipal bonds are less clear and would depend on the tax structure of the governmental unit paying the interest on bonds.

What About Smart-Growth Laws?

What are the equity implications of urban-growth boundaries and other tools of suburban-growth management? A key purpose for these programs in Oregon, Maryland, Kentucky, and Tennessee is to protect farmland and other open space by controlling metropolitan expansion. A favorite theme for those opposing controls on the pattern and pace of development outside of urban areas is that such controls tend to be elitist, exclusionary, and anathema to affordable housing[v]. The argument is that attempts to concentrate development, to protect open space, and to prevent sprawl reduce the supply of developable land, which simultaneously causes price increases in the face of excess demand and raises the value of land close to open spaces to reflect that amenity (Brueckner 1997). Growth controls also reallocate the gains to land value with the effect that some landowners gain while others lose. The theory of growth controls is that they will reduce the externalities of unguided growth. If these external costs and benefits were actually part of the land price, an efficient land-use pattern would evolve without zoning or other regulation, leaving the consideration of whether or not the resulting distribution of impacts is fair.

Affordable housing is a target for policy in some areas that have strong growth controls. For example, New Jersey courts have more than a 30-year history of scrutinizing local zoning statutes for their impacts on affordable housing. The New Jersey Fair Housing Act of 1985 established procedures by which each municipality would accept an obligation for providing housing for low-income people. The Council on Affordable Housing certifies that housing guidelines, which are contained within a particular zoning ordinance, are adequate and providing a shield against possible litigation. The usual guideline for acceptable projects is that 20 percent of housing units within a municipality are affordable and these quotas are tradable across community boundaries. Thus, New Jersey decided that any local zoning that fails to meet housing needs is unfair (Rubin et al. 1990). Also, Montgomery County, Maryland has its Moderately Priced Dwelling Unit Ordinance, passed in 1973, that accomplishes the same thing. Under Maryland's program, builders must give first-purchase options to the Housing Commission of a county for a portion of the 15 percent of all new housing that is targeted for moderate-income households, which are those in the lowest third of income levels. These units become part of the county's inventory for low-income rentals (Rusk 1999: 183–200). By interspersing low-income housing throughout the community, there is no impact on value of unsupported units or poverty concentration in certain areas. These are extraordinary efforts taken by New Jersey and Maryland to create affordable housing and thereby to improve the fairness of urban growth controls.

Various articles in the popular media have referred to the effect of the Portland, Oregon urban-growth boundary on housing prices (e.g., Lacago 1999; Nelson et al. 2002). The general message is that while urban-growth limits have important aesthetic and developmental benefits for existing housing, there is a downside in new-housing cost unless there is specific policy to counter that tendency. The Flint (2001) analysis points to the utility of urban sprawl, since it provides more low-income housing than is available under many growth-management schemes. The study compares spread-out cities to cities in which growth is more tightly managed and concludes that the opportunities for low-income people, particularly minorities, were better with sprawl.

INSIGHTS FOR FUTURE FARMLAND POLICY

Conclusions about both efficiency and equity are central to judgments about the effectiveness of future state and local land-use policy. Efficiency implies that the benefits of farmland protection exceed the costs, to whomever they accrue. The idea is that somehow we are better off as a society with these programs in place than we are without them.

Equity is concerned with who will pay (or be inconvenienced) to provide any net benefits of farmland-protection programs. That is the politics of the issue. In principle, people tend to oppose those changes that cost them more than they gain. The rational homebuilder will oppose growth controls unless they are part of a long-range planning strategy that meets the housing needs of an area. The rational farmer will support those measures that enable him to make the choices necessary for him to remain viable, including the opportunity to sell to the highest bidder under the right circumstances. The rational land-use attorney will keep the pot boiling, and articulate forcefully on all sides of the issues to keep the opportunities for litigation alive and well. The rational economist will continue to seek truth, preferably measured in monetary units. I offer the following observations about the future of farmland policy:

Rationale for Farmland Protection is Strong

There are plenty of understandable reasons for having farmland-protection policies that fit the conventional economic paradigm as land-use services that are not reflected in land price. Services in this instance include an intergenerational sense of food security such that future consumers will have enough farmland on which to grow food. Services also include our interest in an open and attractive countryside, a more thoughtful pattern of urban development, and various ecological services of the farmed lands. The food-supply issue is largely a red herring in this discussion. Critics expend much energy trying to show that there is neither food nor cropland shortage (Staley 2000). That is true, but it is largely beside the point. Farmland-protection policy is on the local, state, and national agenda because of the many amenities and ecological services that farmland provides. All observers agree that food supply is not a problem for the foreseeable future.

Priority on Purchase of Development Rights will Continue, but with a Stronger Regulatory Component

PDR programs are popular and will continue to be so because they seem both efficient (by capturing the external benefits of farmland) and equitable (by forcing taxpayers to pay for the land-use patterns they prefer). But unless PDR programs are integrated with effective and flexible rural zoning, their effect in land use will be minimal. There is simply too much land to buy and prices are too high. Scattered islands of farmland under easement will accomplish little. In fact, these protected farms will attract suburban development that seeks a farmland view. Instead of using PDR to scatter farmland parcels and to consider farmland only as a pleasing extension to the surroundings of the wealthy, PDR priority should go to lands identified in a thoughtful planning process as having a future in farming, identified in meaningful blocks as agricultural-security areas or a similar designation, and identified as a protected area with true agricultural zoning. However, even though PDR without good zoning is better than nothing, it cannot achieve its full land-use potential when it is used to buy a way out of bad zoning (Bowers 2001). Hence farmland-policy techniques must be complements, not substitutes.

Exclusive Agricultural Zoning is Both Valid and Acceptable

It is easier and perhaps more fair to establish real agricultural zoning initially, rather than to later correct it with down-zoning practices. Successful down-zoning practices may require PDR transfer or mitigation to work politically. But, zoning is basically a reasonable expression of the public interest in the services generated by private lands. It can give assurance both to the farmer and to the community and need not entail a major wealth transfer from owner to community. The key to effective zoning is for agriculture consistent implementation once an ordinance is in place. The authority to zone is available just about everywhere, its application is well tested in the courts and with very few exceptions it has been determined to be reasonable and fair.

Move Toward Full Marginal-Cost Pricing of Development

There is ample evidence that new residential development outside of the municipality costs more to service than it generates in new property-tax revenue (AFT 1997: 7–8). The net public cost of new housing depends on the type and value of housing and on the services demanded by its residents. Even most economists seem comfortable with the notion that a fair development process will have new residences pay the additional cost they impose on the community. This could be accomplished through a system of developmental-impact fees that would be area-wide or statewide, which would avoid shifting housing developments to communities that have no fees. These development-impact fees would be tied to the measured costs of new development and would be important sources of revenue to finance development and internalize at least some of the external costs of developing rural areas (Brueckner 1997).

Learn from International Experience

Other nations of the developed world have done far more with growth-management and farmland-protection regulations than the United States has. In the first place, no nation other than the United States must deal with the takings issue that absorbs so much U.S. intellectual energy. No such presumption of overriding private rights exists in Europe. Further, home-rule and annexation authorities have no counterparts in any country other than in the United States, so unplanned farmland conversion is virtually unheard of elsewhere. Worldwide, farmland protection takes on many forms. Tough provincial-level controls exist throughout Canada. Also, there are strong limits on farmland conversion in Germany, Israel, Japan, and the United Kingdom. There is strong emphasis on public subsidy of farmer behavior that will assure provision of the amenity values in these other countries (OECD 1998). Thus Centner (1993) proposes to link the German concepts of protected agricultural areas with strong right-to-farm provisions in the gamut of U.S. policy options.

Canada, Europe, Isreal, Japan and the United Kingdom are reasonably civilized and successful nations of the world, with strong protection for civil rights and for the market system. They have reached accommodation with demands for food production and the amenity services of farmed land that can be instructive for the United States. They employ a mix of generous amenity subsidies and regulated conversion that is becoming more evident in emerging U.S. policy.

REFERENCES

AFT (American Farmland Trust). (1997) *Saving American Farmland: What Works.* Northampton: The American Farmland Trust.

Beasley, S., W. Workman, and N. Williams. (1986) "Estimating Amenity Values of Urban Fringe Farmland: A Contingent Valuation Approach," *Growth and Change* 17: 70–8

Bergstrom, J., B. Dillman, and J. Stoll. (1985) "Public Environmental Amenity Benefits of Private Land: The Case of Prime Agricultural Land." Southern Journal of Agricultural Economics 17(1): 139–49.

Bowers, D. (2001) "Oregon Property-Rights Measure Ruled Unconstitutional." *Farmland Preservation Report* 11: 5.

Brueckner, J. (1997) "Infrastructure Financing and Urban Development: The Economics of Impact Fees." *Journal of Public Economics* 66: 383–407.

Centner, T. (1993) "Circumscribing the Reduction of Open Space by Scattered Development: Incorporating a German Concept in American Right-to-Farm Laws." *The Journal of Land Use and Environmental Law* 8(2): 307–24.

Cordes, M. (2002) "Agricultural Zoning: Impacts and Future Directions," Presented at the Swank Program entitled *Protecting Farmland at the Fringe: Do Regulations Work?* Colombus: The Ohio State University. (September)

Costanza, R., R. d'Arge, R. deGroot, S. Farber, M. Grasso, B. Hannon, K. Linsburg, S. Noeem, R. O'Neill, J. Paruelo, R. Raskin, P. Sutton, and M. van den Belt. (1997) "The Value of the World's Ecosystem Services and Natural Capital." *Nature* 387(15): 253–60.

Daniels, T. and D. Bowers. (1997) *Holding Our Ground: Protecting America's Farms and Farmland.* Washington, DC: Island Press.

Esseks, D., S. Kraft, and L. McSpadden. (1998) "Owners' Attitudes toward Regulation of Agricultural Land." CAE Working Paper 98-3, DeKalb: Center for Agriculture in the Environment (May).

Flint, A. (2001) "Sprawl Lets Blacks Buy Homes, Study Says." *The Columbus Dispatch* 9: (D4).

Frank, R., T. Gilovich and D. Regan. (1993) "Does Studying Economics Inhibit Cooperation?" *Journal of Economic Perspectives* 7(2): 159–71.

Gardner, D. (1977) "The Economics of Agricultural Land Preservation." *American Journal of Agricultural Economics* 59(5): 1027–36

Halstead, J. (1984) "Measuring the Non-Market Value of Massachusetts' Agricultural Land: A Case Study." *Journal of Northeastern Agricultural Economics Council* 13: 226–40.

Henneberry, D. and R. Barrows. (1990) "Capitalization of Exclusive Agricultural Zoning into Farmland Prices." *Land Economics* 66: 249–58.

Kendall, D. and C. Dorman. 2001. "Taking Community Rights by Initiative: Lessons from Oregon, Arizona, and Washington." Washington, DC: Community Rights Counsel Mid-Year Seminar (May).

Lacayo, R. (1999) "The Brawl over Sprawl." *Time* March 22: 45–48.

Libby, L. (1997) "In Pursuit of the Commons: Toward a Farmland Protection Strategy for the Midwest." CAE-Working Paper 97-2. DeKalb: Center for Agriculture in the Environment.

_____. (1996) "Property Rights: the Public-Private Balance." (1996) In *Land Use Decision Making: Its Role in a Sustainable Future for Michigan* Conference Proceedings, S. Batie, K. Norgaard, and M. Wyckoff, eds., East Lansing: Michigan State University Extension.

NAHB (National Association of Home Builders). (2000) *Smart Growth*. Washington, DC: National Association of Home Builders. Internet Website: http://www.nahb.org/generic.aspx?genericContentID=3519

Nelson, A.C., R. Pendall, C.J. Dawkins, G.J. Knaap. (2002) "The Link Between Growth Management and Housing Affordability." Discussion paper for the Brookings Institute Center on Urban and Metropolitan Policy, Washington, DC.

Nickerson, C. and L. Lynch. (2001) "The Effect of Farmland Preservation Programs on Farmland Prices." *American Journal of Agricultural Economics* 83(2): 341–51.

OECD (Organization for Economic Cooperation and Development). (1998) *Adjustment in OECD Agriculture: Reforming Farmland Policies*. Paris: OECD.

OFBF (Ohio Farm Bureau Federation). (1998) "Farmland Preservation: Defining our Community." Report of Advisory Council Discussions. Columbus: Ohio Farm Bureau Federation (January).

Rubin, J., J. Seneca, and J. Stotsky. (1990) "Affordable Housing and Municipal Choice." *Land Economics* 66(3): 325–40.

Runge, S., T. Duclos, J. Adams, B. Goodwin, J. Martin, R. Squires, and A. Ingerson. (2000) "Public Sector Contributions to Private Land Value, Looking at the Ledger." In *Property and Values: Alternatives to Public and Private Ownership*, edited by C. Geisler and G. Daneker. Washington, DC: Island Press.

Rusk, D. (1999) *Inside Game, Outside Game: Winning Strategies for Saving Urban America*. Washington, DC: Brookings Institution Press.

Solberg, E. *The How and Why of Rural Zoning*. (1967) Agriculture Information Bulletin 196. Washington, DC: USDA/ERS.

Staley, S. (2000) "The Vanishing Farmland Myth and the Smart Growth Agenda." Reason Public Policy Institute, Policy Brief Number 12. (January)

Tweeten, L. (1998) *Competing for Scarce Land: Food Security and Farmland Preservation*. Department of Agricultural, Environmental, and Development Economics Occasional Paper ESO #2385. Columbus: The Ohio State University (August).

Varian, H. (1996) *Intermediate Microeconomics*. New York: W.W. Norton and Co.

ENDNOTES

[i]Empirically, however, economists are less cooperative than other people. Frank et al. (1993) compare economics graduate students and faculty with other people in their responses to various free-rider and prisoner's-dilemma games in which the tendency to cooperate is the key variable. They find that economists are demonstrably and consistently less cooperative. They conclude that training in economics tends to create, or at least reinforce, an unwillingness to cooperate.

[ii]Land taxes are levied based on the value of the highest and best use of the farmland parcel in question. There is considerable normative significance to the use that can pay the most for given acres as being the best use of that land and the use that should prevail. Devoting land to an activity of lower value than its highest and best can be costly for the owner.

[iii]It is also true that land price reflects various other policy interventions that are capitalized into land value. Public spending for highways, sewers, and water systems gives land added value. Federal-backed credit systems (e.g., the Federal National Mortgage Association, the Veteran's Administration, and even the tax deduction for mortgage interest) encourage the existence of new single-family homes. Most of these homes are at the periphery of urban areas (Rusk 1999).

[iv]That result in Wisconsin may be partially explained by the fact that use-value assessment in that state is tied to the existence of local agricultural zoning. Thus the higher land price may in part reflect capitalized property-tax savings.

[v]The National Association of Home Builders (NAHB) has devised its own definition of smart growth to emphasize freedom of location and of housing selection. "NAHB recognizes the right of every American to have a free choice in deciding where and in what kind of home to live... Despite concerns about growth, the American dream of owning a detached single-family home is alive and well, and people overwhelmingly reject higher density housing for both themselves and their communities" (NAHB 2000: 10 and 14).

Chapter 18

Urban Influence: Effects on U.S. Farmland Markets and Value

Charles Barnard, Keith Wiebe, and Vince Breneman[*]
United States Department of Agriculture, Economic Research Service

INTRODUCTION

Urban-related development at the edges of cities and in rural areas continues to contribute to a long-standing pattern of public concern and economic study related to loss of productive farmland and open space. Concern is manifest in an array of laws enacted to protect farmland from urban conversion (AFT 1997). Studies of the relationship between urbanization and agriculture include the examination of both the extent of urbanization (Fischel 1982; USDA/NRCS 2000; Barnard 2000; Heimlich and Anderson 2001) and its effects on agricultural land markets and prices (Chicoine 1981; Heimlich and Barnard 1992; Shi et al. 1997; Hardie et al. 2001; Plantinga et al. 2002).

Until about a century ago, the United States was largely an agrarian nation. By 2000, however, approximately 80 percent of the population resided in urban areas (USDOC/Census 2000). Census-defined urbanized areas include city centers and adjacent, densely settled territory at the core of metropolitan areas. These urbanized areas have grown in number from 106 in 1950 to 466 in 2000. Indeed, 836 of the 3,141 U.S. counties are now classified by the U.S. Bureau of the Census as Metropolitan Statistical Areas (Ghelfi and Parker 1997). The dramatic concentration of the U.S. population into urban areas, though, has been accompanied by an equally dramatic diffusion of population within those urban areas. The population density of city centers has been dispersed into the surrounding countryside (suburbs) causing this diffusion, and has lowered the population density of urbanized areas by nearly 50 percent—from 8.4 people per acre in 1950 to 4.47 people per acre in 2000 (U.S. HUD, 2000; USDOC/Census 2000). "For instance, central cities accounted for 57 percent of Metropolitan Statistical Area (MSA) residents and 70 percent of MSA jobs in 1950: by 1990, however, those percentages had changed to about 37 and 45" (Mieszkowski and Mills 1993: 135). During the 1990s, more than 80 percent of new homes were built in the suburbs (Hircshhorn 2000).

The forces that have driven urban growth during the twentieth century, and particularly since World War II, are quite well known and understood (Brueckner 2000). The ultimate driver has always been population growth, but the dynamics of household formation, consumer preferences for larger houses and lots, and increased wealth and income have combined to influence urban form and to increase the relative amount of land consumed for urban purposes (Heimlich and Anderson 2001). Recent trends indicate that urban land consumption is exceeding the change in population in relative terms. For instance, while the population of Philadelphia increased by 2.8 percent between 1970 and 1990, the city's developed area increased by 32 percent (OTA 1995). Between 1970 and 1990, the amount of total U.S. developed land in urbanized areas expanded by 74 percent, while population in urbanized areas grew by only 31 percent (GAO 2000).

Since many cities are located on what once was productive farmland, expansion of the cities has necessarily meant a loss of cropland and other types of farmland. The level terrain, well-drained soils, and availability of water that make cropland especially advantageous for agricultural production, also make the same land attractive for urban population expansion. This is particularly true in those areas of extended periods of warm weather (especially in the winter) that are exceptionally well suited for the production of the highly valued fruit and vegetable crops. Compared to wooded land, cropland and other types of farmland are relatively inexpensive to develop for commercial, industrial, or residential uses.

The same economic and demographic forces that drive urban expansion have led to an expanded, low-density urban fringe in which agricultural and urban-related activities are interspersed. During much of the twentieth century, the rural nature of the United States meant that there was limited interaction between rural and urban populations, and little overlap between rural and urban land uses. As the United States has become increasingly more urban, however, residential and commercial developments have spread farther from city centers in a form that increasingly intersperses urban activities with farmland activities in traditionally rural areas. The unplanned, relatively low-density growth is often typified by noncontiguous residential development interspersed with fragmented farmland, forestland, and idle land.

The process and pattern of urbanization has had three particularly noteworthy effects. First, conflicts over land use and the role of farming in the rural economy have intensified. Farmers at the fringe of urbanizing areas may have constraints placed on their farming activities when new residents object to odors, dust, or agricultural practices that are part of crop and livestock production (Associated Press 2002). In other cases, recurring air pollution, trespassing, vandalism, and loss of farm-support businesses may reduce the productivity of crop and livestock production.

Second, land assessments inevitably begin to rise above their value in agricultural production when land becomes appraised for its future use in nonfarm activities. Parcel characteristics that are unrelated to agricultural production, such as urban proximity and the potential for recreational use, become important determinants of farmland values[i] (Shi et al. 1997; Hardie et al. 2001; Plantinga et al. 2002).

Third, consumers and rural residents alike may feel that the process of urbanization (often characterized as sprawl) reduces the quantity and quality of a set of rural-amenity benefits that they may have previously taken for granted (Hellerstein et al. 2002, Libby and Irwin 2003). In recent years, there has been a renewal and intensification of interest in farmland preservation, including activity at the federal level. The urban and built-up area as defined by the National Resources Inventory (Heimlich and Anderson 2001: 10–11) is a relatively small percentage of the U.S. land area, so doubling or tripling the urban area would hardly affect aggregate U.S. agricultural output. Apparently, consumers are reacting to the loss of rural amenities that occurs as development proceeds.

Although the area classified as urban by the Census accounts for only 2.9 percent of the U.S. land base (Vesterby and Krupa 2001), the shadow of influence this urbanized area casts on the nation's agricultural land is much larger. The interspersion of urban activities with farm activities in traditionally rural areas creates a zone of urban influence that is much more widespread than the urbanized areas statistically defined by the Census. In this sense, urban influence refers to farmland that is subject to the spreading economic and social influence of urbanized areas.

This chapter first provides an overview of the agricultural land market in the rural areas subject to urban influence, and then documents the amount, intensity, and spatial location of urban influence on the Nation's farmland base. We focus on farmland, using national survey data to measure the acreage of farmland in the urban-influence zone for nine resource regions of the United States. Once an urban-influence zone is defined, the change in farmland values attributable to that urban influence can also be measured. We also examine change in size of the urban-influence zone over the 1990 to 2000 period. A final section of the chapter discusses policy implications related to the supply of rural amenities and the costs that urban influence imposes on traditional agricultural enterprises.

"URBAN INFLUENCE" IS NOT URBAN DEVELOPMENT

Most previous estimates of urbanization used surveys to measure the absolute amount of land that has been converted from rural or farmland use to urban or developed use, or measured the rate of that conversion (Fischel 1982; USDOC/Census 1999; USDA/NRCS 2000; USDOC/Census 2000; Vesterby and Krupa 2001). These surveys provide information on the aggregate amount of land converted or the rate of conversion, but do not identify or measure the remaining farmland that is economically subject to urban influence and hence, has a higher probability of conversion. Essentially, the concepts of "urbanized" and "developed" areas measure the footprint of development.

In contrast, "urban influenced," is an economic concept, relating to the zone in which urban-related activities have changed the value of farmland and other open land, changed the resource allocation decisions of farm operators and landowners, and where farmland has an elevated probability of conversion to urban-related uses.

Empirical Measurement of Urban Development

Though measures of urban development provide only a look backward—i.e., a look at urbanization that has already occurred—such measures do provide a perspective for comparing the magnitude of the area that is urban influenced. The Bureau of the Census and USDA's National Resources Inventory (NRI) (USDA/NRCS 2000) measured trends in the amount of urban and developed land in the United States, each using different procedures, data, and concepts/definitions. Census defines total urban area as comprising all continuously built-up areas with a population of 50,000 or more (one or more central places and the adjacent, densely-settled surrounding area), plus all places of 2,500 or more inhabitants outside of such urban areas (UAs). Total Census urban area more than doubled over the last 40 years from 25.5 million acres in 1960 to 55.9 million acres in 1990 (Table 18.1). USDA's NRI defines "urban and built-up areas" to include those Census areas classified as urban (by definition), plus some developed plots outside the Census urban areas. NRI "urban and built-up area" was estimated as 76.5 million acres in 1997. The NRI "developed" land statistic includes "urban and built-up land" and land devoted to rural transportation. Rural transportation land includes highways, roads, railroads and rights-of-way outside of urban and built-up areas. "Developed" area was estimated at 98.3 million acres in 1997. The trends since 1960 in all of these statistics are provided in Table 18.1.

Table 18.1 Trends in U.S. Urban Development, 1960 to 2000

Year	U.S. Bureau of the Census urban area	USDA/NRI urban and built-up areas	USDA/NRI developed area
		Million acres	
1960	26		
1970	34		
1980	47		
1982		52	73
1987		58	80
1990	56		
1992	57	65	87
1997[a]	62	76	98
2000[b]	65	79	107

[a]Census urban for 1997 estimated.
[b]All data for 2000 estimated.
Source: Heimlich and Anderson (2001).

In summary, previous studies estimate the proportion of U.S. land that is "developed." The range of estimates depends partly upon the definition of "developed" or "urbanized," whether the basis of comparison is private land, farmland, or total land area, and whether the comparison includes Alaska and Federal land (Vesterby and Krupa 2001). Heimlich and Anderson (2001: 10–11) provide a more complete review of the geographical concepts and terminology associated with definitions of urban, developed, metropolitan, and rural land. In 1990, "total urban area" measured by Census made up less than 3 percent of U.S. land area (excluding Alaska). NRI's "developed area" made up nearly 5 percent of U.S. land area in 1992. Heimlich and Anderson (2001: 11) maintain that the NRI estimates of "large urban and built-up areas" are higher than the Census "urban area" estimates for nearly all States.

Defining Urban Influence

When urban development spreads to rural areas, the demand for farmland for nonagricultural uses increases. Urban fringe farmland becomes valuable for future commercial, industrial, and residential uses. Characteristics unrelated to agricultural production become important determinants of its value. For most parcels in urban-influenced areas, crop and livestock production generate relatively less in net returns per acre than do nonagricultural uses. Consequently, the price of urban-influenced farmland inevitably rises above the price at which it is economical for sustained use in agricultural production.[ii] In regions where farmland is in great demand for conversion to urban use, relatively large proportions of the market value of farmland is attributable to nonfarm demand (Barnard 2000).

Although the rise in farmland values is probably the easiest effect of urban influence to observe, urban influence also affects the economic costs and returns associated with fringe farm operations in a number of more subtle ways (Nehring and Barnard 2003). In some cases, the rural/urban conflicts that arise from the increased interspersion of urban and agricultural activities, plus the economics of rising property taxes, give some farmers incentive to sell farmland for nonfarm development. In other cases, adaptation of the type and intensity of agricultural production offers offsetting advantages that allow land near urban areas to remain longer in agricultural production. For instance, Gardner (1994), Heimlich and Barnard (1992), and Vesterby and Krupa (1993) discuss increased production of high-value fruit and vegetable crops as the character of farming changes with the advance of urbanization.

AN INDEX OF URBAN INFLUENCE

Identification of farmland that is urban influenced was empirically implemented in this study by analyzing 1990 population data from the U.S. Bureau of the Census within a Geographic Information System (GIS) framework. In the initial step we created an index of urban influence, which is essentially a measure of urban proximity derived from Census block population data using GIS-based statistical smoothing techniques for the total U.S. land area. In geographers' terminology, the index is derived from a gravity model of urban development, which provides measures of accessibility to population concentrations (Shi et al. 1997).[iii] Recently, indices constructed on the basis of the gravity concept were used in econometric analyses of factors (including urbanization) affecting farm-land values (Barnard et al. 1997; Shi et al. 1997; Hardie et al. 2001). Song (1996) evaluates variations of the gravity index along with alternative accessibility measures. A gravity index accounts for both population size and distance of the parcel from that population. The index, which we label population-accessibility index, increases as population increases and/or as distance from the parcel to population decreases. Our construction of the population-accessibility index is calculated on the basis of population within a 50-mile radius of each parcel. The study by Song (1996) indicates that distance, rather than distance squared, most effectively captures changes in population accessibility. For that reason, we choose to weight population by distance.

In order to classify parcels as urban influenced (UI) or non-UI based on this continuous index, we establish thresholds that distinguish between rural reference or background levels of the index and levels that indicate potential UI. This accounts for a background level of population density (accessibility) that would likely exist in the absence of urban influence. The background level includes population that supports an active commercial farming industry, including employees of input and output industries that support production agriculture as well as other population associated with the rural-community infrastructure. That background level can be expected to vary regionally due to differences in the productivity of farmland. Consequently, we establish thresholds for each of the nine U.S. Department of Agriculture, Economic Research Service, Farm Resource Regions (USDA/ERS/FRR) (Heimlich 2000). These U.S. farm-resource regions include the Basin and Range, Eastern Uplands, Fruitful Rim, Heartland, Mississippi Portal, Northern Crescent, Northern Great Plains, Prairie Gateway, and Southern Seaboard (Figure 18.1).

In order to establish the regional threshold for the background levels of each index, we examine levels of the population-accessibility index in areas that clearly had not been subject to urban influence. Cromartie (2001) and Cromartie and Swanson (1996) identify Census tracts[iv] that are totally rural, which are based on 1990 commuting data and U.S. Census Bureau geographic definitions. The term totally rural means that the tract does not contain any part of a town of 2,500 or more residents and the primary commuting pattern was to sites within the tract. Assuming that the most rural census tracts defined by Cromartie (2001) and Cromartie and Swanson (1996) provide a suitable reference or background level for the index, any parcel whose index exceeds the background level of these rural tracts is considered urban influenced. Thresholds for individual FRR regions were established at the 95th percentile of the distribution of index numbers for 5-kilometer grid cells in the set of totally rural tracts in the region containing the land parcel. Land was classified as urban influenced if the population-accessibility index exceeded the associated regional threshold.

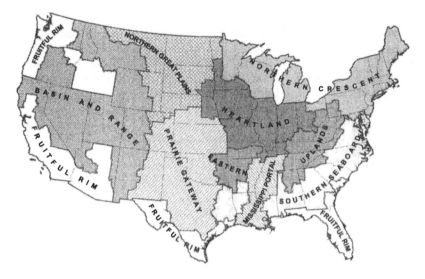

Figure 18.1 USDA/ERS Farm-Resource Regions

MEASURING THE EXTENT OF URBAN INFLUENCE EMPIRICALLY

The U.S. Department of Agriculture, National Agricultural Statistics Service (USDA/NASS) collects annually information on farmland acreage and values via the national June Agricultural Surveys (JAS).[v] Data from JAS, pooled for 1994 to 1996, consist of more than 75,000 observations geo-referenced to the approximate parcel location (latitude and longitude). Each parcel is classified as urban influenced or rural, and is based on the population-accessibility index described above. The latitude and longitude information associated with each JAS sample point enables the GIS software to link each sample point to its corresponding population-accessibility index described above. Within each USDA/ERS/FRR, the JAS sample points classified as urban influenced are sorted by their associated population-accessibility index number and the distribution is split into three categories (low, medium, high), each containing equal numbers of JAS sample points.[vi]

The statistically designed properties of JAS are then used to estimate the number of farmland acres subject to each category of urban influence, using the rural or UI classification scheme described above. The analysis is conducted for each of the nine FRRs, and for each of the low, medium, and high levels of urban influence mentioned above. We estimate, based on data from 1994 to 1996, that approximately 143 million acres of U.S. farmland are urban influenced, which represents 17 percent of the U.S. farmland acres estimated from the JAS survey data.

Of course, a national perspective obscures wide regional variation in urban influence, both across regions and within UI zones (Figure 18.2 and Table 18.2). In the Northern Great Plains region, only 8 percent of acreage is classified as urban influenced. In the Heartland region, about 23 percent of acreage is urban influenced, which affects nearly 3 times the acreage of that in the Northern Great Plains region. Parts of the Northern Crescent region contain some of the most UI regions in the United States, with urban influence affecting about 34 percent of farmland. In the remaining regions, the percentage of farmland acres that are urban influenced ranges from 11.9 percent to 19 percent.

Degree of Urban Influence, 1990

Urban Influence
   Rural
   Low
   Medium
   High or Developed

Source: ERS analysis of 1990 Census population date, by block group.

Figure 18.2 1990 Degree of Urban Influence

URBANIZATION AND FARMLAND VALUES

The urban-fringe land market is an extremely complex mechanism for allocating a scarce resource (i.e., location near a center of economic activity) between competing rural and urban uses. Each site is allocated to an economic activity that yields the highest economic rent over the long run. The economic rent generated at a site by a given rural or urban use is a function of the bundle of sites and of the accessibility characteristics embodied in the parcel. Since land is a durable factor of production, the rent generated by the combination of parcel characteristics is subsequently capitalized into the value of the land resource. If the most profitable expected use of the land is nonagricultural, then the current market value of that land will tend to be higher than the agricultural-use value (Phipps 1984).

Table 18.2 Extent of Urban Influence by U.S. Department of Agriculture, Economic Resource Services, Farm Resource Region, 1994 to 1996

(JAS[a] Code) and Region		Regional acreage total	Total rural land	Total UI[b] land	Relative UI		
					Low UI land	Medium UI land	High UI land
(1) Heartland	Million acres	140.3	108.5	31.8	11.6	10.5	9.7
	Region total	100%	77%	23%	8%	8%	7%
(2) Northern Crescent	Million acres	50.2	33	17.2	6.1	6.0	5.2
	Region total	100%	66%	34%	12%	12%	10%
(3) Northern Great Plains	Million acres	144.9	133	11.9	4.6	3.4	3.9
	Region total	100%	92%	8%	3%	2%	3%
(4) Prairie Gateway	Million acres	220.2	185.2	34.9	13.1	11.7	10.1
	Region total	100%	84%	16%	6%	5%	5%
(5) Eastern Upland	Million acres	50.5	41.9	9.4	3.2	3.3	2.8
	Region total	100%	81%	19%	6%	7%	6%
(6) Southern Seaboard	Million acres	50.4	40.8	9.6	4.1	3.4	2.1
	Region total	100%	81%	19%	8%	7%	4%
(7) Fruitful Rim	Million acres	87.7	70.8	16.9	6.7	5.3	4.9
	Region total	100%	81%	19%	8%	6%	6%
(8) Basin and Range	Million acres	65.3	57.9	7.4	2.4	4.0	1.1
	Region total	100%	89%	11%	4%	6%	2%
(9) Mississippi Portal	Million acres	29.7	26.1	3.7	1.3	1.3	1.1
	Region total	100%	88%	12%	4%	4%	4%
48 U.S. States	Million acres	839.1	696.3	142.8	53.0	48.8	41.0
	Region total	100%	83%	17%	6%	6%	5%

[a]June Agricultural Survey code.
[b]Urban influenced.
Source: Authors' computations from USDA/NASS/FRR (1994 to 1996).

Hardie et al., (2001: 121) state that:

> ...population growth increases the distance for the new housing at the rural-urban fringe from the urban center. Faced with higher costs of commuting this greater distance, housing consumers become willing to pay more for a site that is closer to the urban center. Because the value of agricultural land depends on future rents, this marginal increase in location rent also increases the market price of agricultural land.

Shi et al. (1997: 93), building on the work of Coughlin and Keene (1981) state that:

> Persons wishing to acquire farmland for nonagricultural uses (or persons wishing to acquire farmland for both farming and nonfarming purposes) generally must pay a premium to bid the land away from its agricultural use or agricultural owners. The sale of land at prices above those that had prevailed in an

area will tend to increase the value of all land since prices
convey information and owners will therefore raise their expectations. Thus, even if only relatively small amounts of land
are sold for nonagricultural uses or to nonagricultural purchasers, land values in the affected area will tend to rise.

Numerous studies have demonstrated the negative relationship between urban-fringe land values and distance from a central business district (CBD), in
both theoretical and empirical contexts (Muth 1961; Alonso 1964; Schmid 1968;
Downing 1970; Clonts 1970; Hashak 1975; Chicoine 1981; Reynolds and Tower
1978; Hashak and Sadr 1979; Shonkwiler and Reynolds 1986; Folland and
Hough 1991). These studies often use variables that serve as proxies for urban
proximity (distance from the CBD, distance from other nearby urban areas,
highway frontage, or distance from a highway), in conjunction with other land
characteristics to model land prices in UI areas. Many of these studies have empirically demonstrated that nonfarm factors, particularly urban proximity, are
important when explaining farmland values in urban-fringe areas where there is
potential for conversion to suburban residential or other nonfarm use (Table 18.2).

Shi et al. (1997) survey a number of the econometric models, types of data,
and estimation techniques that have been utilized. Hardie et al. (2001) extend
urban-growth models pioneered by Muth (1961), Alonso (1964), Mills (1972), to
consider the joint determination of farmland values and housing prices within a
simultaneous equation model. Both of these studies find that an index of urban
proximity similar in concept to the population-accessibility index described
above was strongly significant statistically. In both cases, the results provide
empirical support for the efficacy of using the population-accessibility index in
analyses of the effect of urban influence on farmland prices.

SURVEY ESTIMATES

In general, the studies mentioned above use a model-based approach to estimate
an elasticity of farmland prices with respect to a change in the level of urban
proximity. A decided advantage of the techniques applied is the capability of
estimating the effect of urbanization, all else being equal. A disadvantage is that
this marginal effect is usually estimated for selected states or regions. For instance, the Shi et al. (1997) model is estimated for West Virginia, while the Hardie et al. (2001) model is estimated for a mid-Atlantic region. Similarly, the dependent variable in such models is often the county-average farm real-estate
value from the U.S. Census of Agriculture, (2001) which also includes the value
of farm buildings and operator dwellings (Plantinga et al. 2002). Such models
are not easily generalized for farmland in other regions.

In this chapter, we take a design-based approach, using data from the national June Agricultural Surveys pooled for 1994 to 1996 to analyze parcel-level,
land-value data with respect to a rural category and to three levels of urban influence—low, medium, and high. The national scope of the JAS survey, combined with available geo-coding and a GIS system, allows us to estimate the

farmland-value effects of urbanization consistently across the United States, and to display those results in a spatially explicit format. A limitation of this design-based approach is that other factors, such as income, housing prices, and agricultural productivity, are not available from the survey.

We use the statistical-design properties of JAS, in conjunction with the same rural/UI classification scheme described above, to estimate the average per-acre market value in the rural category and in each of the low, medium and high UI categories, for the 9 FRRs. Because the average market value in the rural category for each region is assumed to approximate the agricultural value in each of the three UI categories for that same region, we estimate the per-acre effect of urban influence for each region as the difference between the average market values in the rural category and the average market values in each of the UI categories. We summarize the results at both the regional and U.S. levels.

For the United States, the average value of farmland parcels classified as rural (or non-urban influenced) is U.S. $640 per acre (Figure 18.3). The average value is nearly three times higher for farmland parcels classified as urban influenced, and is U.S. $1,880 per acre. Combining those two categories to get an average for all U.S. farmland yields a per-acre average value of U.S. $850.

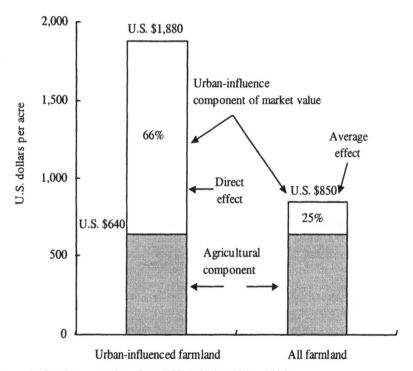

Source: Authors' computations from USDA/ERS (1994 to 1996).

Figure 18.3 Urban-Influenced (UI) and Rural Components of Average U.S. 1994 to 1996 Farmland Values.

We estimate the effect of urbanization on farmland values by assuming that the value of agricultural parcels not subject to urban influence represents the value of land for purely agricultural use. The effect of urbanization is estimated by finding the difference between the farmland market value and its agricultural value. This is also illustrated in Figure 18.3 in which the effect of urbanization on the value of agricultural land is viewed from an average perspective and a direct perspective.

First, urban influence accounts for 25 percent of the market value of all U.S. farmland (U.S. $210 of the U.S. $850 per acre average). We label this the average effect. Second, by assuming that the rural or agricultural value of those parcels classified as urban influenced is the same as the agricultural value of those parcels classified as rural, we determine (as a residual) the effect of urban influence on the value of just those parcels that are subject to its influence. The results indicate that for parcels within the UI zone, urban influence constitutes 66 percent of market value (U.S. $1,240 of the U.S. $1,880 per-acre average). We refer to this as the direct effect of urbanization on farmland value.

For the United States, the average value of parcels subject to urban influence is nearly three times higher than for parcels not subject to urban influence. On a regional basis, the percentage of the per-acre average value of all farmland accounted for by urban influence (the average effect) ranges from 45 percent in The Northern Crescent region to 9 percent in the Northern Great Plains region and 10 percent in the Mississippi Portal region (Table 18.3).

The direct effect of urbanization—the increase in market value per acre above agricultural value for UI parcels—also varies by region, but is considerably higher than the average effect. The effect also gets larger as the UI category moves from a low to medium category and from a medium to high category. For instance, the direct effect for the low-UI category in the Northern Crescent region is 45 percent (Table 18.3). The direct effect of urbanization increases to 68 percent and 82 percent, respectively for the medium-UI and high-UI categories. In the Heartland region, this direct effect increases from 26 percent for the low-UI category to 62 percent for the high-UI category. The effects in the Northern Great Plains region are even larger.

The change in direct effect is most dramatic for the change from the medium-UI to the high-UI category. The average increase in the direct effect between the low-UI and medium-UI category is 38 percent.[vii] For the difference between the medium-UI and high-UI categories, the equivalent increase is 55 percent. Based on our estimates, U.S. farmland has a total market value of U.S. $959 billion (Table 18.4). It appears that U.S. $360 billion of that farmland (38 percent of total market value) is urban influenced. Of that amount, U.S. $237 billion (Table 18.5) is attributable to urban influence, which is 66 percent of the U.S. $360 billion (Table 18.4) of that farmland that is urban influenced. The U.S. $237 billion that urbanization has added to agricultural or rural value accounts for 25 percent of the overall market value of all U.S. farmland.

The Fruitful Rim is the region in which urban influence has added the largest absolute amount to the agricultural value of farmland, contributing U.S. $69 billion of a total market value of U.S. $182 billion (Tables 18.4 and 18.5). Urban influence has added U.S. $45 billion in the Northern Crescent and U.S. $38 billion in the

Table 18.3 Per-Acre Value of Farmland and Indicators of Urban Influence by USDA/ERS Farm Resource Region

(JAS code) and region		Regional average (A)	Rural Average (B)	Urban Influence (C)	Relative UI Low UI category	Med. UI category	High UI category
(1) Heartland	Farmland value[a]	1,870	1,598	2,801	2,150	2,247	4,183
	Average effect[b]		15%				
	Direct effect[c]			43%	26%	29%	62%
(2) Northern Crescent	Farmland value	2,009	1,109	3,737	2,018	3,502	6,038
	Average effect		45%				
	Direct effect			70%	45%	68%	82%
(3) Northern Great Plains	Farmland value	319	291	630	454	572	887
	Average effect		9%				
	Direct effect			54%	36%	49%	67%
(4) Prairie Gateway	Farmland value	574	500	965	667	1,067	1,232
	Average effect		13%				
	Direct effect			48%	25%	53%	59%
(5) Eastern Upland	Farmland value	1,446	1,130	2,832	2,268	2,606	3,739
	Average effect		22%				
	Direct effect			60%	50%	57%	70%
(6) Southern Seaboard	Farmland value	1,506	1,222	2,712	2,560	2,414	3,473
	Average effect		19%				
	Direct effect			55%	52%	49%	56%
(7) Fruitful Rim	Farmland value	2,070	1,288	5,344	3,211	5,159	8,409
	Average effect		38%				
	Direct effect			76%	60%	75%	85%
(8) Basin and Range	Farmland value	737	587	1,912	1,744	1,171	4,919
	Average effect		20%				
	Direct effect			69%	66%	50%	88%
(9) Miss. Portal	Farmland value	1,324	1,194	2,243	1,515	1,594	3,773
	Average effect		10%				
	Direct effect			47%	21%	25%	68%
48 U.S. States	Farmland value	1,143	861	2,519	1,762	2,255	3,811
	Average effect		25%				
	Direct effect			66%	51%	62%	77%

[a]U.S. dollars per acre.
[b]Average effect = percent of regional average value due to urban influence (Col. A – Col. B/Col. A).
[c]Direct effect = percent of value of UI parcels due to urban influence (Col. C – Col. B / Col. C).
Source: Authors' computations based on analysis of USDA/ERS (1994 to 1996).

Table 18.4 Inflation-Adjusted Total Market Value of Farmland for ERS Farm Resource Regions (By Degree of Urban Influence)[a,b]

(JAS code) and region	Total	Rural	Urban-influenced	Urban-influenced category		
				Low	Medium	High
			Billion dollars			
(1) Heartland	262.5	173.4	89.1	25.0	23.7	40.5
(2) Northern Crescent	100.9	36.6	64.3	12.3	20.9	31.1
(3) Northern Great Plains	46.2	38.7	7.5	2.1	1.9	3.5
(4) Prairie Gateway	126.4	92.7	33.7	8.7	12.5	12.5
(5) Eastern Uplands	73.0	46.4	26.5	7.3	8.7	10.6
(6) Southern Seaboard	75.8	49.9	26.0	10.4	8.2	7.4
(7) Fruitful Rim	181.5	91.1	90.3	21.4	27.4	41.6
(8) Basin and Range	48.2	33.9	14.2	4.1	4.6	5.4
(9) Mississippi Portal	39.4	31.1	8.3	1.9	2.0	4.3
All 48 States	959.0	599.2	359.8	93.3	110.1	156.4

[a]Based on analysis of USDA (1994 to 1996) June Agricultural Surveys (JAS)
[b]1994 to 1996 average regional values from JAS were adjusted to January 1, 2001 using the average percentage change in USDA/NASS/ERS State farmland value series for those states containing parts of each region.

Table 18.5 Inflation-Adjusted Total Market Value Due to Urban Influence, by Urban Influence Category[a]

(JAS code) and region	Urban influenced				Due to government payments
	Low	Medium	High	Total	
	Billions of dollars				
(1) Heartland	6	7	25	38	40
(2) Northern Crescent	6	14	25	45	2
(3) Northern Great Plains	1	1	2	4	3
(4) Prairie Gateway	2	7	7	16	9
(5) Eastern Uplands	4	5	7	16	1
(6) Southern Seaboard	5	4	5	14	2
(7) Fruitful Rim	13	21	35	69	2
(8) Basin and Range	3	2	5	10	<1
(9) Mississippi Portal	0	1	3	4	3
48 States	48	68	121	237	62

[a]1994 to 1996 average regional values from JAS were adjusted to January 1, 2001 using the average percentage change in NASS/ERS state farmland value series for those States containing parts of each region.

Heartland region. As a proportion of farmland value, the urbanization effect is larger in some other regions, but those regions have smaller acreages of farmland subject to urban influence and/or relatively lower cropland value. In comparison, Barnard et al. (2001) estimate that federal-commodity program payments in 2001 added nearly U.S. \$40 billion to the total value of farmland in the Heartland (Table 18.5). The combined impact of these two effects is far-and-away the largest in the Heartland, where together these two effects have contributed nearly U.S. \$80 billion to total farmland value.

CHANGES IN URBAN INFLUENCED AREA, 1990 TO 2000, USING GIS SOFTWARE AND NRI DATA

We use data from the 1997 NRI survey (conducted by the USDA/NRCS) to estimate the change in UI acres during the 1990/2000 period. Using GIS techniques, NRI survey points or locations (latitude/longitude) are intersected with the estimated 1990 population-accessibility index surface created to classify the JAS points (as described above). The classified NRI points are then used with their associated 1997 NRI parcel to estimate 1997 NRI-defined farmland acres in each of the three urban categories.[viii] The same NRI points were then intersected with a new population-accessibility index surface estimated from 2000 population data. The reclassified points are then used to estimate 1997 NRI-defined farmland acres in each of the three urban categories. This reclassification permits the estimation of acres that change the UI status for 1990/2000. In a sense, the GIS-generated surfaces based on 1990 and 2000 population were overlaid to identify changes in UI status. Using this GIS procedure, we estimate the farmland acres that change the UI status, and create map images of those changes. The results (Figure 18.4 and Table 18.6) provide estimates of the changes in UI acres that are attributable to the measured change in population accessibility generated by population growth over the 10-year period from the 1990 census to the 2000 census.

Using the procedure described in the previous paragraph, we estimate that an additional 27 million rural acres became newly urban-influenced land between 1990 and 2000. An additional 22 million acres moved from the low-UI to the medium-UI category, and approximately 12 million acres moved into the highest category of urban influence. The 27 million acres that became newly urban influenced over the 1990 to 2000 period is an average rate of roughly 2.7 million acres per year. This compares with a rate of 2.2 million acres per year that USDA/NRCS estimates from NRI data as the change in developed acres from 1992 to 1997. One implication of this comparison is that urban influence appears to be spreading at a slightly faster rate than is the footprint of development as measured from NRI and other sources. (The NRI category of developed land varies from that used by some other data collection entities. For instance, the intent of the NRI is to identify which lands have been permanently removed from the rural land base.) Also, the results imply that on average for 1990 to 2000 an additional 3.4 million acres per year have moved into a higher-intensity UI category.

Figure 18.4 Change in U.S. Urban Influence, 1990 to 2000

Table 18.6 Change in Urban-Influenced Farmland Area, 1990 to 2000 (By Category)

(JAS code) and region	Change from rural to urban influenced	Change from low to medium	Change from medium to high
	Acres		
(1) Heartland	3,317,100	2,813,200	1,884,600
(2) Northern Crescent	1,749,700	1,617,900	1,052,200
(3) Northern Gt. Plains	773,700	761,100	238,100
(4) Prairie Gateway	4,906,500	4,128,800	2,789,400
(5) Eastern Uplands	2,705,100	1,719,200	769,800
(6) Southern Seaboard	2,492,200	1,951,500	1,156,900
(7) Fruitful Rim	4,678,800	4,207,500	2,957,700
(8) Basin and Range	5,505,000	3,995,400	938,600
(9) Mississippi Portal	1,359,800	608,200	298,400
Total: 48 States	27,487,900	21,802,800	12,085,700

Source: Authors' computations based on 1997 USDA/NRCS (2000) and population-accessibility codes.

On a regional basis, the largest acreage of newly urban-influenced land (5.5 million acres) from 1990 to 2000 was located in the Basin and Range region (Table 18.6). More than 4 million acres became newly urban-influenced land in both the Prairie Gateway and Fruitful Rim regions. Both of the latter regions also experienced the largest acreage moving from a medium-UI to a high-UI category, and from a low-UI to a medium-UI category. Changes in acreage from the rural to UI category and from lower to higher levels of urban influence are visualized for NRI acres in Figure 18.4.

POLICY CONSIDERATIONS

Increased farmland values, especially those generated by demand for nonfarm uses, are double-edged swords for American farmers who compete in a global economy. For many individual farm operators, farmland-value increases are favorable. Farm real-estate value underlies the financial stability of those farm businesses whose investment portfolios derive a large proportion of their value from farm real estate. In addition, farm real estate is often the principal source of collateral for farm loans, which enables many farm operators to finance the purchase of additional farmland and equipment or to finance current operating expenses. In addition, many farm operators consider farmland as a retirement instrument, funded by accrued capital gains realized by sale of the land.

Population growth and other factors are expected to increase the level of urban influence and, ultimately, the conversion of additional agricultural lands to urban uses. The U.S. Bureau of the Census predicts a U.S. population increase of nearly 50 percent by 2050, from 273 million in 1999 to 404 million by 2050 (GAO, 2000). Several other factors that contribute to lower density development of urban areas are also contributing to increased urban influence on agricultural lands and to the interspersion of agricultural and urban activities. Household size has been declining, requiring more houses for the same number of people. Average lot size has increased since the 1970s, with a large and increasing proportion of average lot-size growth accounted for by proliferation of isolated, large-lot housing development (1 acre or more), much of which has occurred in rural areas beyond the urban fringe (Heimlich and Anderson 2001).[ix] An increasing fraction of land area used for single-family housing has been for the largest lot sizes (Hirschhorn, 2000).

The above factors contribute to reduced development density, which reinforces a trend toward a more spatially dispersed economy that has been set in motion by the advent of the automobile. Consistent with this phenomenon, density gradients for U.S. urban areas have flattened (Mieszkowski and Mills 1993; Gordon and Richardson 1997; Kim 2002). The flattened gradients, in turn, have extended the spatial area subject to urban development and to urban influence farther and farther from CBDs. According to Dyckman (1976), the effective radius of U.S. cities in 1890 was approximately 2 miles, limited largely by walking distance. By the 1970s, public transit and urban freeway systems had moved the city edge 20 to 24 miles beyond the CBD. During the 1980s, the existence of cities near the outer edge of the 1970s suburban ring had the effect of increasing the urban radius by an additional 50 miles. This effect has reinforced the American consumer's increasing demand for larger residential spaces (including the housing site) and amenities associated with open space. Both of these characteristics are more widely and cheaply available in suburban, exurban, and rural locations.

Though very little is known about the eventual effect on urban-settlement patterns of recent microelectronic advances in communication technologies, OTA (1995) argues that a wholesale dispersal of commercial and residential activity to rural areas will not occur. Webber (1993), on the other hand, notes that as the barriers to communication and commerce continue to be diminished, urban form may evolve beyond the current polycentric pattern to one of generalized dispersion. One implication is that access to city centers and other nodes of economic activity may no longer be as important (commanding less rent or land-price premiums), and that residential and business location may be determined by other factors, including the type and supply of amenities.

These trends in density, which are indirectly reshaping American cities, raise questions about whether or not increased interspersion of agricultural and urban-related activities portend significant impacts on production agriculture. Given the potential prevalence of this type of development, is there a particularly large but hidden cost to production agriculture that arises from the interspersion of agricultural and urban activities in a large, growing, and permanent fringe? Though population growth has positive implications for some types of agriculture (Berry 1978; Heimlich and Barnard 1992; Lopez et al. 1988; Larson et al. 2001), a common hypothesis expressed in both the popular and professional literature is that urban interspersion in rural areas reduces the competitive position of those farms that continue to produce traditional products. Traditional agricultural enterprises are likely to incur much of the costs, while reaping few of the benefits. For instance, urban proximity is unlikely to bring increases in market prices for traditional field crops and livestock, especially in comparison to vegetable and specialty crops that can be marketed directly to consumers. On the other hand, firms producing field crops and livestock are likely to bear many of the costs associated with constraints on agricultural practices and the disappearance of input suppliers and output markets. Farmers, in particular, often feel that federal commodity-program payments enable continued farming in UI areas, thus slowing the rate of farmland conversion (Marbella, 2002).

Nehring and Barnard (2003) use stochastic and nonparametric estimation techniques to examine the impact of urban influence on the costs of producing corn and soybeans, which are two traditional field crops in the U.S. Heartland region. (The measure of urban influence used in the Nehring and Barnard study is identical to that discussed and used in earlier parts of this chapter.) They find that rural farms in the Heartland region are more technically efficient than are the more UI farms. The findings confirm that traditional farms in urban areas are facing challenging economic adjustments, particularly with respect to labor, off-farm employment, and livestock enterprises.

CONCLUDING REMARKS

There appears to be widespread agreement among economists that even with the continuation of the rapid development that has occurred over the 1980s and 1990s, the actual conversion of rural land to urban-related uses will not endanger an adequate cropland base to meet foreseeable food demand. Thus, the primary concerns surrounding the development patterns arising from innovation in the information and communications technologies of the new economy are three-fold. First, do these development patterns imply significant loss of important historical, cultural, and natural resources as increasing amounts of farmland, forest land, wetlands, and other forms of wildlife habitats and open space are converted to urban-related uses in the proximity of major metropolitan areas? Second, do these development patterns portend a particularly large loss of forest-land and scenic amenities due to tele-work and recreational-type second homes in high amenity areas? A considerable amount of development is taking place that is not associated with population concentrations. That development may be for second home or recreational use, and as a consequence, may never reach urban densities, yet may contribute significantly to the (perceived) loss or degra-dation of rural amenities (Hellerstein et al 2002). Activities associated with this type of development may also be imposing costs on nonresident consumers of the public goods associated with rural landscapes. And finally, the question arises as to whether the interspersion of agricultural and urban activities will become so widespread and particularly concentrated in productive agricultural areas, as to increase costs of production to levels that command policy attention.

REFERENCES

AFT (American Farmland Trust). (1997) *Saving American Farmland: What Works?* Northhampton: American Farmland Trust.
Alonso, W. (1964) *Location and Land Use.* Cambridge: Harvard University Press.
Associated Press. (2002) "Weekly Farm: EPA seeks to crack down on spray drift from farms." Internet Website: http://usda.newsedge-web.com/NewsEdge/FullStory/ (8 April).
Barnard, C.H. (2000) "Urbanization Affects a Large Share of Farmland." *Rural Conditions and Trends* 10(2): 57–63.
Barnard, C.H., R. Nehring, J. Ryan, and R. Collender. (2001) "Higher Cropland Value from Farm Program Payments: Who Gains?" *Agricultural Outlook,* USDA/ERS/AGO-286. Washington D.C.: USDA/ERS.
Barnard, C.H., G. Wittaker, D. Westenbarger, and M. Ahearn. (1997) "Evidence of Capitalization of Direct Government Payments into U.S. Cropland Values." *American Journal of Agricultural Economics* 79(5): 1642–50.
Berry, D. (1978) "Effects of Urbanization on Agricultural Activities." *Growth and Change* 9(3): 2–8.
Brueckner, J.K. (2000) "Urban Sprawl: Diagnosis and Remedies." *International Regional Science Review* 23(2): 160–71.
Chicoine, D.L. (1981) "Farmland Values at the Urban Fringe: An Analysis of Sales Prices." *Land Economics* 57(Aug.): 353–62.

Clonts, H.A., Jr. (1970) "Influence of Urbanization on Land Values at the Urban Periphery." *Land Economics* 46(November): 489–97.

Coughlin, R.E. and J.C. Keene. (1981) *The Protection of Farmland: Report to the National Agricultural Land Commission*. Washington, DC: U.S. Government Printing Office.

Cromartie, J. (2001) "Data: Rural-Urban Commuting Area Code." Washington, DC: USDA/ERS. Internet Website: http://www.ers.usda.gov/emphases/rural/data/ruca/ (26 July).

Cromartie, J.B. and L.L. Swanson. (1996) "Census Tracts More Precisely Define Rural Populations and Areas." *Rural Development Perspectives* 11(3): 31–9.

Downing, P.B. (1970) "Estimating Residential Land Values by Multivariate Analysis." In *The Assessment of Land Value*, D. Holland, ed. Madison: University of Wisconsin Press. (101–24)

Dyckman, J.W. (1976) "Speculation on Future Urban Form." Working Paper, Center for Metropolitan Planning and Research. Baltimore: Johns Hopkins University.

Fischel, W.A. (1982) "The Urbanization of Agricultural Land: A Review of the National Agricultural Lands Study." *Land Economics* 58(2): 236–59.

Folland, S.T. and R.R. Hough. (1991) "Nuclear Power Plants and the Value of Agricultural Land." *Land Economics* 67(February): 30–6.

GAO (U.S. General Accounting Office). (2000) *Community Development: Local Growth Issues-Federal Opportunities and Challenges*. GAO/RCED-00-178, GAO (September)

Gardner, B.L. (1994) "Commercial Agriculture in Metropolitan Areas: Economics and Regulatory Issues." *Agricultural and Resource Economics Review* 25(April): 100–9.

Ghelfi, L.M. and T.S. Parker. (1997) "A County-Level Measure of Urban Influence," *Rural Development Perspectives, Economic Research Service*, USDA, Washington, D.C. Vol.12, No. 2.

Gordon, P. and H.W. Richardson. (1997) "Alternative Views of Sprawl: Are Compact Cities a Desirable Planning Goal?" *Journal of the American Planning Association* 63(1): 95–105.

Hardie, I.W., T.A. Narayan, and B.L. Gardner. (2001) "The Joint Influence of Agricultural and Non-farm Factors on Real Estate Values: An Application to the Mid-Atlantic Region." *American Journal of Agricultural Economics* 83(1): 120–132.

Hashak, L.J. (1975) "The Urban Demand for Urban-Rural Fringe Land." *Land Economics* 51(May): 112–23.

Hashak, L.J. and K. Sadr. (1979) "A Spatial Model of Land Market Behavior." *American Journal of Agricultural Economics* 61(November): 697–701.

Heimlich, R.E. (2000) "Farm Resource Regions." Economic Research Service, Washington, D.C. ERSAIB760. Internet Website: www.ers.usda.gov/publications/aib760/aib-760.pdf

Heimlich, R.E. and C.H. Barnard. (1992) "Agricultural Adaptation to Urbanization." *Northeastern Journal of Agricultural and Resource Economics* 21(April): 50–60.

Heimlich, R.E. and W.D. Anderson. (2001) *Development at the Urban Fringe and Beyond: Impacts on Agriculture and Rural Land*,USDA/ERS/AER-803. Washington, DC.: USDA/ERS.

Hellerstein, D., C. Nickerson, J. Cooper, P. Feather, D. Gadsby, D. Mullarkey, A. Tegene, and C. Barnard. (2002) *Farmland Protection: the Role of Public Preferences for Rural Amenities*. Agricultural Economics Report No. 815 (October). Washington DC: USDA/ERS.

Hirschhorn, J.S. (2000) *Growing Pains: Quality of Life in the New Economy*. Washington, DC: National Governor's Association.

Kim, S. (2002) "The Reconstruction of the American Urban Landscape in the Twentieth Century." NBER Working Paper 8857. Cambridge: National Bureau of Economic Research.

Larson, J.M., J.L. Findeis, and S.M. Smith. (2001) "Agricultural Adaptation to Urbanization in Southeastern Pennsylvania." *Agricultural and Resource Economics Review* 30(1): 32–43.

Libby, L.W. and E. G Irwin. (2003) See chapter 19 this volume.

Lopez, R.A., A.O. Adelaja, and M.S. Andrews. (1988) "The Effects of Suburbanization on Agriculture." *American Journal of Agricultural Economics* 70(2): 346–58.

Marbella, J. (2002) "Reaping the Harvest of Federal Funds." *The Sun.* Baltimore: The Baltimore Sun Company. 21 April(1A).

Mieszkowski, P. and E. S. Mills. (1993) "The Causes of Metropolitan Suburbanization." *Journal of Economic Perspectives* 7(3): 135–147.

Mills, E.S. (1972) *Studies in the Structure of the Metropolitan Economy.* Baltimore: Johns Hopkins University Press.

Moss, C.B., G. Livanis, V. Breneman, and R. Nehring. (2001) "Productivity *v* Urban Sprawl: Spatial Variations in Land Values." American Agricultural Economics Association (AAEA) Symposium, Chicago, Illinois.

Muth, R.F. (1996) *Cities and Housing.* Chicago: University of Chicago Press.

_____. (1961) "Economic Change and Rural-Urban Land Conversion." *Econometrica* 29: 1–23.

Nehring, R. and C. Barnard. (draft 2003) "Does Urban Influence Increase Costs of Production in the Corn Belt? A Frontier Approach." Draft journal submission. Washington DC: USDA/ERS.

OTA (U.S. Congress, Office of Technology Assessment). (1995) *The Technological Reshaping of Metropolitan America.* Washington, DC: OTA.

Phipps, T.T. (1984) "Land Prices and Farm-Based Returns." *American Journal of Agricultural Economics* 66(November): 422–29.

Plantinga, A., R. Lubowski, and R. Stavins. (2002) "The Effects of Potential Land Development on Agricultural Land Prices." Faculty Research Working-Paper Series RWP02-012. Kennedy School of Government, Harvard University, Cambridge, MA.

Reynolds, John E. and D.L. Tower. 1978. "Factors Affecting Rural Land Values in an Urbanizing Area." *Review of Regional Studies* 8(Winter): 23–34.

Schmid, A.A. (1968) *Converting Land from Rural to Urban Use.* Baltimore: The Johns Hopkins Press.

Shi, U.J., T.T. Phipps, and D. Colyer. (1997) "Agricultural Land Values under Urbanizing Influences." *Land Economics* 73(1): 90–100.

Shonkwiler, J.S. and J.E. Reynolds. (1986) "A Note on the Use of Hedonic Price Models in the Analysis of Land Prices at the Urban Fringe." *Land Economics* 62(February): 58–61.

Song, S. (1996) "Some Tests of Alternative Accessibility Measures: A Population Density Approach." *Land Economics* 72(4): 474–82.

USDA/ERS. (U.S Department of Agriculture, Economic Research Service). (1994 to 1996) Agricultural Surveys. U.S. Department of Agriculture, Washington, D.C.

USDA/NASS (U.S. Department of Agriculture, National Agricultural Statistics Service) (2001) Census of Agriculture 1997. Internet Website: www.nass.usda.gov/Census

USDA/NRCS (U.S. Department of Agriculture, Natural Resources and Conservation Service) and Iowa State University Statistical Laboratory. (2000) *Summary Report, 1997 National Resources Inventory (revised, December 2000).* Internet Website: http://www.nhq.nrcs.usda.gov/NRI/

USDOC/Census (U.S. Dept. of Commerce, Bureau of the Census). (2000) *Geographic Areas Reference Manual*. Internet Website: www.census.gov/geo/www.garm.html.

_____. (1971) *1970 Census of Population. Volume 1: Characteristics of the Population, United States Summary*. PC70-1-A1.

_____. (1993) *1990 Census of Population and Housing. Volume 2: Population and Housing Unit Counts, United States Summary*. 1990 CPH-2-1.

USDOC/Census (U.S. Department of Commerce, Bureau of the Census) and U.S. Department of Housing and Urban Development, Office of Policy Development and Research. (1999) *American Housing Survey for the United States, 1997 Current Housing Reports*. H150/97 (September).

USHUD (U.S. Department of Housing and Urban Development). (2000) *The State of the Cities 2000: Mega-Forces Shaping the Future of the Nation's Cities*. Internet Website: http://www.huduser.org.publication/polleg/soc2000_rpt.html.

Vesterby, M. and K.S. Krupa. (2002) "Rural Residential Land Use: Tracking its Growth." Agricultural Outlook, AO293, August. Internet Website: www.ers.usda.gov/publications/agoutlook/aug2002/ao293e.pdf

Vesterby, M. and K.S. Krupa. (2001) Major Uses of Land in the United States, 1997. USDA/ERS/SB-973. Washington, DC: USDA/ERS.

_____. (1993) "Effects of Urban Land Conversion on Agriculture." In *Urbanization and Development Effects on the Use of Natural Resources: Proceedings of a Regional Workshop*, Thunberg, E.M. and J.E. Reynolds, eds. Mississippi State, MS.: Southern Rural Development Center. DRDC No. 169.

Webber, M.M. (1993) "The Marriage of Autos and Transit: How to Make Transit Popular Again." Presented to the Fourth International Research Conference, Center for Transportation Studies, University of Minnesota, St. Paul.

ENDNOTES

* The opinions expressed here are those of the authors and not necessarily those of the U.S. Department of Agriculture.

[i] Soil productivity and slope, both of which may be related to agricultural productivity, may also influence the value of land for urban development. Relatively flat, well-drained cropland, for instance, may be less expensive to develop for residential purposes than steep, rocky hillsides that may more commonly have been used for pasture.

[ii] Moss et al (2001) discuss the effect that proximity to urban markets may have upon the agricultural value of farmland that is urban influenced.

[iii] The gravity model is based conceptually on the Newtonian physical principle that "the gravitational attraction of physical objects is directly proportional to the product of their masses, but declines with increases in the distance between them" (Shi et al. 1997). Enhanced computing capabilities coupled with GIS advances have made application of the concept to countrywide data feasible.

[iv] Census tracts are small, relatively permanent statistical subdivisions of a county. Census tracts usually have between 2,500 and 8,000 persons and, when first delineated, were designed to be homogeneous with respect to population characteristics, economic status, and living conditions. The spatial size of census tracts varies widely depending on the density of settlement.

[v] JAS is based on a probability-area frame with a 1995 sample of approximately 14,603 segments or parcels of land that average approximately 1 square mile. Segments are

selected to represent approximately 1 percent of the total land area in the coterminous 48 States, with 20 percent of the segments replaced each year. Enumerators conducting the area survey contact all farmers having operations within the sampled segments and collect (among other data) opinions of farmland values for their operations.

[vi] Alternatively, the distribution could have been divided to achieve an equal number of acres in each category.

[vii] For two regions, the change was (surprisingly) negative. We attribute this to the simplified assumption used to estimate agricultural value for urban-influenced parcels. Predictions of agricultural value from a hedonic model might yield better results.

[viii] NRI farmland acres were defined as cultivated and uncultivated cropland, rangeland, pastureland, CRP, farmsteads, and other farmland.

[ix] Nearly 80 percent of the acreage used for recently constructed housing is land outside urban areas or in non-metropolitan areas: 94 percent of that land has a lot size of 1 acre or more and 57 percent of that land has a lot size greater than 10 acres (Heimlich and Anderson 2001: 2).

Chapter 19

Rural Amenities and Farmland Values

Lawrence W. Libby and Elena G. Irwin
The Ohio State University

INTRODUCTION

Farmland value reflects the economic returns from the various products and services available from a parcel of land. Buyers and sellers agree on a price that captures expected income from the relevant outputs. Among the farmland services are certain amenities, some of which the landowner may withhold from non-payers and others for which exclusion costs are very high.

In this chapter, we consider the economic importance of both categories of amenity services of actively farmed land—those that are private goods, captured by the owner and priced for sale to consumers, and those that are public goods, nonexclusive and/or non-rival in consumption and whose values are not reflected in market price. Included in the former category are hunting rights, on-farm recreation, farmers markets, and certain landscape features that may make one parcel of land more valuable than another. In the latter category are such ecosystem services as groundwater recharge, nutrient cycling, reduced flood risk, and the overall cultural or heritage value of farms in a rural setting. Both categories of land service are important and will affect land-use patterns, but only private goods are directly exchanged in a market. Public-good services of privately owned land are secured through policy.

While the landowner has the right to determine the mix of products and services flowing from a particular parcel of land, these output choices are made within an opportunity set shaped by various laws and institutions that establish relationships among people in a modern society. Those who do not own farmland but value certain ecosystems or other amenity services will express those preferences through the policy process. Policy changes will alter the opportunity sets of farmland owners. Examples of such policy changes are the Endangered

Species Act of 1974 that requires the farmer to avoid actions that could destroy habitat, the Environmental Quality Incentives Program in the 1996 Farm Bill that pays farmers to avoid actions that may pollute the waterways, and agricultural conservation easement programs that purchase the owner's right to develop the land. Each of these institutions affects land use by adjusting the choice sets of owners.

Policies, rules, and incentives that influence the owner's land-use options will likely be reflected in land prices. If a farmer sells his or her development rights, the market price of that land will generally decline accordingly. If agricultural zoning limits the list of acceptable uses of farmland, its market price will generally decline. In some situations, however, the price effects of these policies may be modest or nonexistent if land bidders expect that the policies will not be enforced permanently. In other situations, development restrictions in an area may make the land more attractive to a buyer primarily looking for amenity value. In the latter part of this chapter, we examine the evidence of the impact of amenity-securing institutions on the market value of farmland. We conclude the chapter with some general observations regarding the efficacy of these programs based on the empirical evidence that exists in the literature.

CONCEPTUAL FRAMEWORK

In the simplest model of land valuation, land is treated solely as a productive asset used to produce marketable goods and services (Randall and Castle 1985). Because land is a fixed factor of production, land rents $R(t)$ are generated by any economic surplus associated with the production of market goods

$$R(t) = \left[\left(p_z^* Z(t) - p_x^* X(t) \right) \right] / L,$$

where Z is a vector of market goods, X is a vector of inputs (excluding land), L is land, and p_z and p_x are the output and input price vectors, respectively. Rents are capitalized into the value of the land, so that the price of land is determined by the present discounted value of rents over an infinitely long time horizon associated with the land parcel

$$V_0 = \int_0^\infty e^{-rt} R(t) \, dt. \tag{1}$$

Following Ricardo's (1821) basic insight, negative rents from the land are determined by differences in land quality and positive rents accrue to scarce, high-quality land. Land quality is broadly defined as parcel-level characteristics that determine the productive capacity of the land parcel to generate marketable goods and services. The most obvious of these agricultural-use characteristics is soil fertility, but other land characteristics may matter as well when other land uses are possible. If the land parcel is in a potentially urban area, land characteristics that influence the potential returns or the costs of urban development will generate rents (e.g., impervious soil and steep slopes will add to the cost of development, so positive rents will accrue only to level land parcels that have soils that permit effective drainage). In addition, rents will accrue to land based on its relative location. Land located close to a central market will generate positive rents because of transportation-cost savings (von Thuenen 1842). Other location characteristics, including externalities from neighboring land uses and access to highways, may matter as well. Denoting on-site characteristics as $H(t)$ and location characteristics as $D(t)$, we determine rents that can be generally specified as a function of parcel-level and regional-level variables

$$R(t) = R\big[Z(t), G(t), p_z p_x\big],$$

where $Z(t) = Z\big[H(t), D(t), X(t), L\big]$, and $G(t)$ include regional factors that may influence rents (e.g., regional population, preferences, and government policies).

Because rents accrue according to the actual and potential land uses of a parcel, $R(t)$ can be decomposed into actual and potential rents that accrue to the land in its existing and alternative uses. For simplicity, assume that land rents can be from either an agricultural use $R_A(t)$ or an urban use $R_U(t)$. In rural areas located far away from urban activities, land rents will be determined primarily by $R_A(t)$, whereas in urbanizing areas, it is likely that $R_A(t)$ will be relatively small compared to $R_U(t)$. The primary characteristics influencing $R_A(t)$ are soil fertility and other physical features of the land that determine agricultural productivity. However, to the extent that secondary goods and services can be produced from agricultural land, other features of the land may matter as well. For example, the opportunity to rent land out for hunting may increase the rents from pasture or wooded areas. Locations nearer to urban areas may increase the returns from agricultural land because, for example, proximity to urban populations increases the potential rents from agri-tourism activities on the land.[i] Using the subscript A to denote variables that generate $R_A(t)$ and the subscript U to denote variables that generate $R_U(t)$, we can further specify $R(t)$ as

$$R(t) = \sum_k R_k\big[Z_k(t), G_k(t), p_{zk}, p_{xk}\big], \qquad (2)$$

where

$$Z_k(t) = Z_k\big[H_k(t), D_k(t), X_k(t), L\big]$$

and $k = A, U$. When facing a decision regarding the optimal use of land, a private landowner will consider the expected rents from the land over time and will choose the optimal land use or optimal sequence of land use that will maximize this discounted stream of returns over time.

So far we have ignored externalities and public goods associated with land that may generate a range of external benefits and costs. For example, as is discussed in more detail below, agricultural land generates nonmarket benefits, including open-space amenities that provide scenic views and reduce the overall level of development and congestion in an area. While these benefits do not accrue to the private landowner, they do generate benefits for others and therefore they influence the socially optimal allocation of land across different uses. For this reason it is important to consider these additional sources of value and the extent to which they augment (or diminish, as in the case of external costs and public bads) the rents that accrue to private landowners in a particular land use. Because we are concerned primarily with rural amenities and farmland values in this chapter, we focus on the social benefits that are associated with farmland.

The social benefits associated with farmland can be categorized as local distance-based benefits that decay as distance increases from the farmland and as distance-independent benefits that accrue to the community or region as a whole. The former are largely externalities, such as open-space amenities, that provide the greatest benefit to those who are immediately adjacent to the parcel and whose benefits lessen as distance from the farmland increases. Because they are somewhat excludable, in the sense that not everyone can locate immediately adjacent to farmland, these benefits are a form of an impure public good. The latter, on the other hand, are benefits that are independent of distance from the farmland and accrue equally to members of the community or perhaps region and therefore more closely resemble a pure public good. Examples include provision of wildlife habitat areas, biodiversity, and groundwater recharge.

These social benefits can be incorporated into the asset-based definition of rents in Equation (2) to define the social returns of farmland $R_A^S(t)$ as returns that include both private rents from productive activities that accrue to the landowner and social benefits from the land being in an agricultural use that accrue to society

$$R_A^S(t) = R_A^S \left[Z_A(t), G_A(t), E_A(t), p_{zA} p_{xA} \right] \tag{3}$$

where $E_A(t)$ is a vector of distance-decaying externalities and distance-independent public goods that generate the social benefits from farmland. As discussed below, depending on the type of social benefit, revealed preference methods may or may not be able to identify the contribution of $E_A(t)$ to $R_A^S(t)$. For example, hedonic methods are commonly used to estimate the distance-based externalities associated with farmland, most notably its value as an open-space amenity. However, stated preference methods are necessary for identifying nonuse values, including option and existence values, associated with the social benefits of farmland (e.g. the value of preserving wildlife habitat).

AMENITIES AS PRIVATE GOODS

Some farmers in urbanizing areas have learned that the most valuable outputs of their land may be the amenity services. Customers will pay for the privilege of picking their own corn or apples. They will pay for the right to hunt deer or other game. They will choose the farm setting as a tourism destination, if the owner provides those opportunities.

Farmers Markets and Consumer Supported Agriculture

People will pay for the opportunity to buy farm-fresh produce direct from the farmer. The owner is thus selling the consumer an additional farmland service—direct personal contact with the production site or process. There were fewer than 100 farmers' markets in 1980. The Agricultural Marketing Service of the U.S. Department of Agriculture (USDA) began keeping track of the numbers in 1994, and by 2000 the number of farmers' markets had increased to nearly 3,000, providing the sole marketing outlet for about 19,000 farmers throughout the United States (CAST 2002). Consumers know that they are paying more for the produce than they would in a supermarket, but feel that the sweet corn, blueberries, garden vegetables, and other products are somehow fresher and better flavored. In the Stephenson and Lev (1998) study, 80 percent of consumers shopping at farmers' markets indicated willingness to pay a premium over supermarket prices, 44 percent expressed the desire to support local farmers as another reason for paying higher prices at the farmers market, and 30 percent of respondents referred to the buying experience as one service they were acquiring from farmers. Consumers using farmers' markets will also buy other products (jams, baked goods, baskets and other crafts) that they associate with the farm, whether or not the farm family actually prepared the product. Urban residents view the trip to the farmers' market as recreation, an alternative to a day at the library, shopping mall, or other venue. Participants in the women, infants, and children and food-stamp programs of the USDA may spend their coupons at the farmers' markets. The income from these coupons, which totals more than U.S. $100 million a year, is then added to the revenue of these markets. Farmers markets typically support food-bank and gleaning programs running in various states.

Consumer-supported agriculture (CSA) is a means by which consumers may purchase a share of produce from a farm and share in the risk of production variability at the same time. The farm becomes theirs at some level and they may work on the farm to improve chances for a good crop. The shared ownership and chance to work in the fields for their own output are services that consumers also purchase from the farmer (Henderson and van En 1999).

Most CSAs employ organic or other ecologically friendly production systems. Consumers who insist on organically grown produce are buying a sense of personal security from the farmer. They exhibit a certain willingness to pay for the marginal increment of security from produce raised without pesticides. Organic certification and labeling enable the farmer to deliver a specific service for a price. Stahlbush Island Farms in Western Oregon, for example, will establish a contract with consumers assuring them that produce from that farm is grown with green technologies, including organic methods in some cases (CAST 2002). The buyer is paying for produce, personal security, and a farming system that is ecologically sound. These farmland services, bought and paid for, become part of the income flow to that farm that is capitalized in land value.

The Value of Agri-tainment

People will pay for the opportunity to visit a farm and experience a little of the farming life. Visitors to Ohio's predominately Amish counties will pay for the chance to stay at a farm, help with farm chores, enjoy family-style Amish meals and generally experience the Amish farm life that feels like traditional farming was before the advent of huge combines and inorganic fertilizers. The farm is an alternative to other forms of weekend recreation and the purchased services have little if anything to do with the farm crop itself.

Graf Growers, located partially inside the city limits of Akron, Ohio, is an example of a farm that sells all sorts of recreation and educational experiences that focus on the farm operation. There are seminars on flower arranging, bee keeping, tree grafting, and care of bulbs. There are special farm events for bus-loads of local school children (hayrides, corn roasts, and other special events). The farm specializes in sweet corn and apples. The Graf Grower label on these products assures their quality and commands a price increment. Craig Graf knows that his competition is not other farms but parks and other recreation sites. Graf Growers is an intensive use of farmland, generating income from a variety of services connected to the farm, competing favorably with development that is not tied to farm production.

Vermont dairy farmers will alter the milking schedule to permit visitors "to join Bob for the floorshow with the girls down in the barn" (Russell 1997). The Economic Development Office in Loudon County, Virginia arranges self-directed tours of Virginia fruit farms and vineyards, and arranges Fall-Color tours with stops at selected farms. The right to hunt on farmland is valued by many, both as individuals and in large hunting clubs. Farmers may grant permission to hunt on their land for a price. One Illinois farmer earns U.S. $600 a day for access to prime deer habitat along the Mississippi River. A Kansas pheasant ranch attracts more than 800 bird hunters a year (Miller 1999).

These are separable, rival-in-consumption farmland amenity services that are priced and are part of farm income flow capitalized in land value. They become parts of adaptive strategies for farmers at the rural-urban fringe and represent additional income sources that maintain farm viability in times of low product prices.

Composting and Other Ecosystem Services

Some farmers can market the nutrient-cycling capacity of land. Farmers in several states contract with nearby municipalities to receive wastewater for secondary treatment as irrigation water. Farmland is the essential component of wastewater treatment for hundreds of communities and subdivision developments that employ package systems. Price-Barnes Organics in Central Ohio composts lawn waste from nearby communities, converts recycled newsprint into litter for the hog operation, uses certain food-processing by-products as an energy supplement for dairy heifers, and in other ways employs the nutrient-cycling capacity of farmland to earn income and good will among nonfarm neighbors. The composted mixture of farm animal manure and community-generated lawn and garden waste is then packaged and sold to nurseries and to garden stores.

AMENITIES AS PUBLIC GOODS

In addition to the range of private amenity-based goods, farmland generates a variety of externalities and public goods. Because of the nature of these goods and services, the benefits are not capitalized into the private returns from the farmland and therefore drive a wedge between the private and socially optimal amount and distribution of farmland within a region. Should the net external benefits of farmland be positive implies that $R_A^S(t) > R_A(t)$, and therefore that farmland will be under-provided in the absence of policies that attempt to correct this market failure. Quantifying the magnitude and nature of these external benefits is important for designing policies aimed at providing the socially optimal amount of farmland.

As discussed above, the social benefits associated with farmland may be distance-based benefits that are essentially local and decay with increasing distance from the farmland or are distance-independent benefits that accrue to the community or region as a whole. The former are largely open-space amenities that provide the greatest external benefit to those who are immediately adjacent to the parcel. Because they are excludable, in the sense that not everyone can locate immediately adjacent to the farmland, these benefits are capitalized into the property values of neighboring land parcels and therefore it is possible to use revealed preference techniques (namely hedonics) to estimate the direction and magnitude of these distance-based amenities. On the other hand, any nonuse benefits that are associated with farmland, including those that are independent of distance and that accrue more or less equally to members of the community (e.g. existence value of farmland), can be estimated only via stated preference methods, namely contingent-valuation or contingent-choice (or conjoint) analysis. In what follows, we briefly review each of these methodologies and the empirical evidence with respect to the nonmarket amenity value of farmland.

Hedonic Pricing Method

The hedonic pricing method is a revealed-preference technique that relies on the fact that the spatial immobility of land results in the capitalization of location amenities, such as open-space spillovers from nearby farmland, into the sales price of residential homes. Therefore, surrounding land uses create amenity (and dis-amenity) effects that are included in the bundle of characteristics that describe a residential location and create amenity effects that are traded off with other features when households make residential-location choices. Hedonic pricing models offer a means to estimate the marginal implicit prices of characteristics associated with a differentiated market good, such as housing, based on the interactions of many buyers and sellers in the market. The hedonic-pricing function posits price as a function of the quantities the attribute of a good. The marginal implicit price of any of the attributes is found by differentiating the hedonic-price function with respect to that attribute, which, when evaluated at an individual's optimal choice, represents the individual's marginal willingness-to-pay for the attribute. As such, it provides a means for uncovering the marginal value associated with distance-dependent externalities that are generated by nearby farmland.

Using some of the notation that was introduced earlier, we specify the hedonic residential pricing model as $P_i = P(H_i, D_i, F_i, Q_i \cdot \beta, \lambda, \delta, \theta)$, where P_i is the residential sales price of the ith property, H_i is a vector of on-site characteristics associated with the land, D_i is a vector of locational variables, F_i is a scalar that measures the proportion of farmland within the local neighborhood of parcel i, Q_i is a vector of structural attributes associated with the house, and β, λ, δ, and θ are the respective parameter vectors that we estimate.

Several econometric issues arise when estimating hedonic models, including functional form, extent of the housing market, multicollinearity, and spatial-error autocorrelation. As discussed by Irwin (2002) and Irwin and Bockstael (2001), an issue that is specific to the estimation of open-space externalities using hedonic pricing models is the identification of these effects given the potential endogeneity and spatial-error autocorrelation of the neighborhood open-space variables. These problems arise if the neighboring open space, such as farmland, can be converted to residential use at any point in the future. This capability for conversion implies that farmland is part of the regional market for residential land and is subject to the same economic forces that determine a location's residential value.

Such a situation would imply that variables measuring the influence of this developable open space on neighboring residential property values will be endogenous in a hedonic-pricing model, thus identification problems arise that bias the open-space coefficients. The problem is further complicated by the presence of spatial-error autocorrelation, which is likely to be correlated with the endogenous measures of open space, and therefore introduces a second source of bias into the estimation. While the issues of functional form, extent of the market, and multicollinearity are well recognized in the hedonic literature, the empirical challenges that arise from spatial-error autocorrelation and the potential endogeneity of neighboring open-space land have not been addressed by most empirical studies.

Finally, Riddel (2001) notes that the amenity value of proximity to open space may accrue over time due to housing-market inefficiencies that lead to time lags in the full capitalization of environmental amenities into housing prices. In this case, cross-sectional data may be insufficient for revealing the true marginal value of a neighboring environmental amenity.

Several studies have employed the hedonic-pricing method to estimate the marginal contribution of neighboring farmland to residential property values.[ii] These studies use the results from the first-stage estimation to calculate the implicit price associated with an acre of neighboring farmland. For example, Ready et al. (1997) use a hedonic-pricing model of housing expenditures and hourly wages to estimate the amenity effects associated with proximity to horse farms using county-level data from a national sample of households within metropolitan areas. They use the results to estimate the average marginal willingness-to-pay of Kentucky residents to prevent the loss of a horse farm, which they calculate to be an annual payment of U.S. $0.43.

Johnston et al. (2001) use data on market transactions of houses from Suffolk County, New York to estimate a hedonic-pricing model of residential property values in which proximity to farmland is included as a location characteristic. Contrary to expectations, they find that residential property values are negatively influenced by proximity to farmland. Specifically, the results indicate that a parcel of land that is adjacent to farmland has on average a 13.3-percent lower per-acre value than does a similar parcel that is not adjacent to farmland. However, because problems of spatial autocorrelation and potential endogeneity issues associated with neighboring farmland are not addressed, these results may not reflect the true marginal effects of neighboring farmland. Irwin (2002) uses a hedonic-pricing model of residential property values to estimate the influence of surrounding cropland, pastureland, and preserved agricultural land on the residential values of neighboring homes in a Central Maryland region. Problems of endogeneity of surrounding open-space measures and spatial-error autocorrelation are dealt with by using an instrumental-variable approach and a spatial-sampling routine that omits nearest-neighbor properties. Results show that conversion of one acre of either surrounding cropland or pastureland to low-density residential development decreases the value of the mean residential property by U.S. $1,530, whereas conversion to a commercial or industrial use reduces the mean residential property value by U.S. $4,450. These estimates can be interpreted as an individual's marginal willingness-to-pay to avoid the loss of an additional acre of either surrounding cropland or pastureland and to demonstrate that this value depends on the alternative land-use state.

Results also indicate a premium associated with protected agricultural land. If an acre of surrounding land is preserved as a permanent agricultural easement, the value of neighboring residential property will be estimated to increase by U.S. $4,523. Extending this result to a hypothetical 10-acre parcel of farmland that is located in the center of a low-density residential development, the predicted benefits of preserving the land as an agricultural easement are found to range from U.S. $10,403 to U.S. $52,014 per acre of preserved open space, depending on the density of the neighboring residential development. These estimates can be interpreted as the total marginal willingness-to-pay of residents located sufficiently close to the preserved farmland to benefit from its amenity effects.

Stated Preference Methods

In contrast to the hedonic-pricing method, which relies on indirect methods and observed behavior to estimate the value of nonmarket goods or services, stated-preference methods rely on individuals' responses regarding their anticipated behavior in a hypothetical market situation. Two primary methods have been used to identify the nonmarket value of farmland—contingent valuation and conjoint analysis.

The contingent valuation method (CVM) relies on the consumer's direct evaluation of how a marginal change in the good or service of interest will influence his or her utility level in terms of the amount that the consumer would be willing to pay to obtain (or retain) a higher level of the good or service in question. This approach is dependent on the consumer having detailed information regarding the market situation that describes the good or service of interest [e.g., its current and alternative quantities, qualities, institutional arrangements that affect the provision of the good, geographical location, expected duration, and, most importantly, the stream of expected nonmarket benefits (and costs) that are generated by the good]. In the case of farmland, the CVM may entail describing the location, amount, spatial distribution, and type of farmland; the ownership of the farmland and the factors that influence the decision of the landowner to keep the land in farming; and the external benefits and costs that are conveyed by the farmland to the community. Given a full description of the good or service in question, the respondent is asked whether he or she would be willing to pay varying amounts of money to maintain the provision of the good in question. These responses are then combined with a variety of socioeconomic data on the individual respondents to estimate a willingness-to-pay function that describes the demand for the good in question and yields a total marginal willingness-to-pay measure that represents an estimate of the marginal benefits derived from the good. The primary advantage of this approach over revealed-preference methods is that nonuse values can be estimated. However, this approach suffers from a number of limitations, including concerns over whether potential strategic behavior on the part of the survey respondent, incomplete information, the payment vehicle, and the hypothetical nature of the question may introduce estimation biases.

Several studies have used the CVM to estimate the nonmarket benefits of farmland.[iii] These studies have elicited willingness-to-pay estimates either by asking whether the respondent would be willing to pay a specified amount to avoid the conversion of neighboring farmland to an urban use or by asking whether the respondent would vote positively or negatively to set aside a targeted amount of farmland. For example, Beasley et al. (1986) use photographs of different development-density scenarios and a bidding-game technique to elicit the respondent's maximum annual willingness-to-pay to preserve farmland as a means to prevent development of surrounding land. They find that willingness-to-pay for preserving farmland increases from U.S. $76 per household annually to U.S. $144 per household annually when the alternative to farmland is high- as opposed to moderate-density development.

Bergstrom et al. (1985) present individuals with a series of photographs that depict prime agricultural land and alternative states of development. They then solicit willingness-to-pay to prevent conversion of the prime agricultural land to development. They estimate that the mean annual willingness-to-pay of households ranges from U.S. $5.70 for 18,000 acres of preserved agricultural land to U.S. $8.94 for the preservation of 72,000 acres of agricultural land. As summarized by Heimlich and Anderson (2001), the range of values estimated by the CVM studies is in part due to variations in the study areas. The Beasley et al. (1986) study is in an agricultural region of South-Central Alaska, where agricultural land is relatively scarce, whereas the Bergstrom et al. (1985) study is in the

Piedmont region of South Carolina, in which agricultural land is probably more plentiful. Rosenberger and Walsh (1997) use a CVM study to estimate households' total annual willingness-to-pay to protect (or to avoid the loss of) valley ranchland for different regions in Colorado. They find that willingness-to-pay is the highest for ranchland in the valley that is under the most urbanization pressure. Thus average annual household willingness-to-pay to protect 25 percent, 50 percent, 75 percent, and 100 percent of the existing ranchland in the valley is U.S. $72, U.S. $102, U.S. $118, and U.S. $121, respectively.

Other CVM studies that have been employed to value the nonmarket benefits of farmland include those by Krieger (1999), Halstead (1984), and Bowker and Didychuk (1994). As reviewed by Heimlich and Anderson (2001), these studies are similar in terms of the type of questions posed to the respondents and range in estimated mean annual willingness-to-pay per 1,000 acres of farmland from U.S. $1.08 to U.S. $49.80. In contrast to most valuation studies, which posit development as the alternative state, Drake (1992) estimates households' willingness-to-pay in Sweden for keeping land in an agricultural use rather than letting it be converted into forest. He estimates the average annual individual willingness to pay to preserve the land in agriculture to be 541 SEK/person and finds that this value varies on a per-hectare basis depending on the location of the land and the type of agricultural land use.

Lastly, Ready et al. (1997) conduct a CVM analysis to compare these results from the findings of their hedonic analysis. For the case of Kentucky horse farms, they find that median willingness-to-pay per household per year varies from U.S. $0.49 to prevent the loss of one horse farm to U.S. $681 to prevent the loss of 75 percent of the horse farms within their county. This finding is consistent with the estimated marginal willingness-to-pay that was generated by the hedonic analysis.

An alternative method for eliciting stated preferences for social benefits associated with nonmarket goods is contingent choice or conjoint analysis, which relies on indirect methods in a hypothetical situation to elicit an estimate of marginal benefits. This method focuses on stated preferences between defined alternatives to estimate marginal values associated with a heterogeneous good (e.g., housing). These estimates can in turn be used to calculate willingness-to-pay measures and to predict demand for particular attributes. This method presents survey respondents with two or more alternatives, each of which is described by a set of attributes. For example, in the case of housing, this list may include house size, lot size, distance to work, the amount of surrounding farmland, the quality of local public services, and price. Respondents are asked to either rank the alternatives or simply to choose the alternative they like best. Because respondents are also given the price of each alternative, they can be viewed as making choices subject to a budget constraint and analysis can be carried out within a utility-theoretic framework. The resulting estimates can be used to calculate an individual's marginal willingness-to-pay for any of the attributes that were used to describe the hypothetical good and, summed over the entire population, this approach provides a means for estimating the total marginal benefits associated with the attributes of interest. While this method is open to some of the same criticisms as the CVM (e.g., it is derived from hypothetical and not observed behavior), it is based on an indirect method of eliciting preferences and therefore may not suffer from as many of the limitations that many believe hamper CVM.

A few studies have used contingent choice to investigate the nonmarket value associated with farmland as a neighborhood amenity. Johnston et al. (2001) estimate a contingent-choice model (in addition to the hedonic model discussed above) to estimate mean willingness-to-pay measures for the preservation of several different types of natural lands, including farmland. Contrary to the hedonic estimates, the per-household annual value of preserving an acre of farmland is estimated at U.S. \$0.143. As emphasized by Johnston et al. (2001), this difference reflects the different values that are captured by contingent-choice versus hedonic methodologies: the hedonic approach offers only a partial estimate of value, whereas the contingent-choice approach is able to capture both use and nonuse values. Roe et al. (2002) use conjoint data from a survey that collected responses to hypothetical housing choices to estimate the value of permanently preserved agricultural land around a home. The findings indicate that the conversion of 10 percent of all existing agricultural land within one mile of the house (about 188 acres) into permanent cropland has a value ranging from 1 percent to 4 percent of the value of the house being considered. For the median respondent and the median-valued house, this estimate is found to correspond to an annual mortgage payment increase of U.S. \$277, which translates into an annualized per-acre value of U.S. \$1.47 per household or, capitalized into the value of a home, a U.S. \$3,607 increase in price of the median house.

In summary, results from the different studies that have attempted to identify the amenity values associated with farmland vary with the type of methodology used, the accuracy of the data (e.g. parcel level versus county level), the alternative land use that is considered, and the location of the study area. Nonetheless, the general conclusion based on these results is that the amenities associated with farmland are positive, which indicates that because these values are not reflected in the market price of agricultural land, farmland is an underprovided good. For this reason, a variety of policies have been put forward to secure the amenities associated with farmland. We review these policies and review the empirical evidence regarding their efficacy in the following sections.

POLICIES FOR SECURING FARMLAND AMENITIES

Various regulatory, incentive, and acquisition devices are employed by governments at all levels to retain the many services of farmland, including amenities. By influencing the terms of trade in land markets, these institutions will be at least partially reflected in land value.

Regulations

Land-use regulations adjust the rights and responsibilities of landowners regarding public health, safety, morals, and general welfare to varying degrees in different places. Agricultural zoning and urban-growth boundaries are the most commonly used regulations to affect patterns and timing of land-use change in the interest of protecting the open-land amenities. When land-use options are constrained by regulation, there is presumably a wealth transfer from owner to the general public, measured as the increment of land value associated with the present value of expected future returns to that land in a foregone higher-value use. The equity question involves a judgment as to whether the owner's sacrifice is more than offset by gains to others. In the Kaldor-Hicks (Freeman 1993) sense, if those who gain from land regulation could compensate the owner and still come out ahead, the change in rules is a welfare gain, whether or not the compensation actually occurs.

Agricultural zoning was first used in Pennsylvania and California in the early 1970s; by 2002 there were 21 states with agricultural-protection zoning (AFT 1997). Most are inclusive agricultural zones permitting residential, commercial, and sometimes even industrial land uses in agricultural districts, relying on a minimum lot size of twenty acres or more to discourage but not to prohibit nonfarm development. A few states have exclusive agricultural districts (e.g., Pennsylvania, Maryland, and Oregon) that permit only those land uses consistent with active farming. The effect of a zoning ordinance on the market value of land would depend on how the presence of that land-use restriction influences willingness to pay in the land market. The buyer has to consider how limiting the rules really are and whether or not local authorities easily and frequently change the restrictions.

The other equity question when it comes to agricultural zoning is whether or not such restriction of private property rights may constitute a regulatory taking as defined under the Fifth Amendment to the U.S. Constitution. Evidence suggests that since farming is generally an economically viable use of land, agricultural zoning does not constitute a regulatory taking requiring compensation (Cordes 1999 and 2002).

Incentives

State-enacted property-tax reductions for active farms attempt to provide a management incentive that will keep the farmer farming longer than if full market value were the basis for farmland taxes. Since virtually all active farmland in all 50 states is eligible for use-value assessment or for a state income-tax credit if property tax exceeds a threshold level of household income, it is doubtful that these tax programs affect land value substantially. There are various forms of tax recapture among the states to discourage conversion and to reimburse other taxpayers for past incentives paid. In most cases, it is a 3-year to 5-year rollback of the difference between farm and full-market value. California requires the farmer to dedicate land to farming for a running 10-year period in return for lower taxes. Other states have a capital-gains-tax provision to enable the community to recapture a portion of the capital gain that comes when farmland is developed. The community generates much of that capital gain through investment in roads, sewer, water, and other infrastructure, and takes some of its investment back when the farmland is developed (AFT 1997).

Whatever the recapture provisions, tax incentives only retain farmland amenities so long as the value of the incentive, plus any penalties for change, is greater than the return to an alternative land use. Except for the California case of restrictive easements in return for lower taxes, full discretion remains with the landowner. That is the nature of incentives.

Purchase

Agricultural-easement or development-rights purchase programs are in place in 20 states. Farmers may sell the right to develop their land at a price reflecting the difference between the farm value and the full-market value. These programs are the beneficiary-pays strategy of farmland protection that generally constitutes permanent protection of the amenity services that flow from that land. Those who stand to gain from the amenity pay the farmer for these amenities, rather than having the farmer provide them purely in the public interest.

Once the development right has been transferred to a state or local government or to a qualified land trust, future land-use options are constrained, presumably affecting the market value of the land. As always, opportunities are in the eye of the beholder. Perhaps a buyer will place higher value on protected land because of those restrictions. Assurance that the land cannot be developed may be worth something to the buyer—not the usual discounted value of future income from the land but the value the buyer places on long term, often multigenerational amenity attributes. A very wealthy buyer may be looking for a suitable estate with plenty of space around the manor house. The land might be available at nearly farmland value if an easement is in place, although the buyer is interested only in the amenity services and not in the productive capacity of the land (USDA/ERS 2002).

In other cases, a buyer may be willing to pay more for use-restricted land on the assumption that nothing is really forever and that the restrictions can be changed in the future. All states except New Jersey have escape clauses built into their agricultural-easement programs that permit governments to release the restriction under extreme circumstances. Maryland and Pennsylvania require a 25-year waiting period before any such escape can even be requested. In Massachusetts, specific legislation to release an easement is required. Most programs allow a house or two for family members, with up to 10 such lots in Maryland. There is virtually no experience with escape from permanent easements, leaving judgment on the matter up to the buyer. But it is doubtful that the price of protected land will reflect the full value of the easement sold to the state or local government, at least not for long.

AMENITY-SECURING POLICIES AND FARMLAND VALUE: THE EVIDENCE

The effects of land-use restrictions on the market value of farmland are not as straightforward as one might expect. As noted above, buyers will consider various indicators of future earning potential when making a purchase offer, and policy-driven use restrictions are only part of the picture. Similarly, the seller's reservation price is affected by expectations of future earning as conditioned by many economic and social factors. As discussed above, home buyers value proximity to protected open land and will pay more for those homes (Nelson 1985; Pollakowski and Wachter 1990). Recent work by McGranahan (1999) shows that people do move to places that have significant amenities, and being close to open land has an impact on buyers' willingness to pay. As an example, there is a 25-percent difference between parcels just inside and just outside the urban-growth boundary (UGB) around Portland, Oregon (Nelson 1998). At least part of that difference can be attributed to the amenity value of nearby open space. Also, property values adjacent to the designated greenbelt around Boulder, Colorado were found to be 32-percent higher than those less than three-quarters of a mile away (Correll et al. 1978). Thus amenity-securing restrictions have value to those who can find unprotected land nearby to build on. The scarcity value of developable land must also be a factor in these cases.

Nonvoluntary development restrictions through zoning will likely cause a wealth transfer to unrestricted land. Growth restrictions accounted for more than two-thirds of the value difference of farmland in development as compared to nondevelopment zones in a suburb of Minneapolis (Gleeson 1979). Vaillancourt and Monty (1985) find a 15-percent to 30-percent reduction in farmland value as a result of exclusive agricultural zoning near Montreal in Quebec, Canada.

Farmland value is affected by the anticipation of regulations designed to protect amenity values. Beaton (1991) examines open-land price trends near the Pinelands area of New Jersey before, during, and after the enactment of growth restrictions as part of the Pinelands Protection Act of 1979. Prior to the enactment of the Protection Act, land value within the area later designated for preservation rose at a slower rate than it did for other land. During the several years of debate on this land-use control measure, land prices within the study area rose more rapidly than they did outside the Pinelands area. The market was picking up the possibility of future restrictions on land-use options. As the 1981 implementation date approached, land value in the area to be protected rose to a level 228 percent higher than similar land in a control area. After implementation of the Pinelands Protection Act of 1979, the value of land within the preservation area fell back to previous levels, tempered somewhat by a waiver provision to correct for any major hardship suffered by landowners who were caught by the restrictions.

Nelson (1986) finds little speculative value capitalized into rural land prices close to the UGB around Salem, Oregon within five years of the Pinelands Protection Act of 1979. In fact, farmland close to the boundary was less valued due to various spillovers from urban activities across the border. Rural land values increased with distance from the UGB. On the other hand, the market value of land on the urban side of the boundary increased due to its proximity to the amenities of open land. He finds that urban land within 1,000 feet of the boundary was worth U.S. $1,200 an acre more in 1979 than land was that was closer to the urban center.

Buyer and seller expectations about the real effect of land-use restrictions are further revealed with the UGB in Oregon. Knaap (1985) finds major differences in land-value impacts of the growth controls among different areas around Portland, depending on degree of enforcement. The purpose of the limitation on nonfarm development in the exclusive farm use (EFU) districts outside of Portland was to avoid direct displacement of farms and patterns of land use that seemed to signal the end of future active farming. Allowances were made for the construction of new homes related in some way to the farm. Also, allowances were made if the proposed new homes were built on less-productive land or if the construction would not destabilize the existing land-use pattern (Liberty 1999) A minimum parcel size of 80 acres was enforced in most of these agricultural zones. Again there were qualifiers (e.g., a county could petition to have a smaller minimum parcel size). The list of uses permitted within an EFU has gradually expanded over the years.

The many opportunities for making exceptions to the intent of urban-growth boundaries have planted the seed of potential higher-value use in the minds of farmers and developers alike. For example, 1,300 new farm dwellings were approved in several counties outside the Portland UGB while the actual number of farms and people declaring farming as their primary occupation declined in those counties during the same five-year period (Liberty 1999). Thus it appears that much of the land outside the UGB is available for nonfarm residential development, perhaps with a little extra effort by the developer. Further, the UGB is up for revision every few years, with much negotiation over how much additional urban land is needed to meet commercial and residential needs. Certainly the willing buyer and willing seller of Oregon farmland would be influenced by those possibilities.

The analysis by Nickerson and Lynch (2001) bears further evidence that buyers and sellers are willing to project more possibilities for protected land than may be suggested by the use-restricting instrument. Their study of development-rights purchase and transfer programs in three Maryland counties indicates that even with permanent liens against any future development there is "little statistical evidence that voluntary permanent preservation programs significantly decrease the price of Maryland farmland" (2001: 350). They offer several possible explanations for this result, including that land-market participants may not expect the restrictions on development to be truly permanent and that land buyers may buy up preserved farm parcels as hobby farms, thus bidding up the price of a restricted land parcel. More data are needed on sales of farmland carrying permanent restrictions against development to enable researchers to separate the component of land value attributed to buyer confidence that no land-use restriction is forever. In addition, challenges to such laws in court will help to clarify the permanency of such restrictions.

In Maryland's case, the development-rights purchase program has a proviso that allows withdrawal from the program after twenty-five years. Because the program has been in place less than twenty-five years, this proviso has not yet been challenged, and landowners do not know with certainty the stringency with which the restrictions will be enforced. Exclusive agricultural zoning is far less permanent than a development-right purchase program. Henneberry and Barrows (1990) find that the value of farmland near Beloit and Janesville, Wisconsin was actually enhanced by restrictive zoning. Land in larger parcels farther from these urban centers show a larger value increment from zoning than do smaller parcels closer to the cities, which is exactly the opposite from what we might expect based on location alone. The authors conclude that farmers are willing to pay more per acre for large parcels with development restrictions to avoid some of the pressures of urban incursion into farm areas. Further, the Wisconsin use-value assessment law requires that land must first be zoned exclusively for agriculture. Thus perhaps some of that price increment is really capitalized property-tax reduction.

In Baltimore County, Maryland, heavily restricted land sold for more than did land with less restriction. Farmland permitting one residence for every 5 acres sold for nearly U.S. $1,000 per acre less than did land permitting one residence per 50 acres (Bowers 2002). Observers in that state say that a country manor factor is at work in Baltimore County. Wealthy urbanites from Baltimore and Washington will pay more for a well-zoned rural estate than they will for one that is less restricted. They are paying for isolation rather than for development potential, and are confident that a segment of the market will always have such buyers.

Farmland enrolled in use-value assessment programs throughout the country generates land taxes based on the productive potential of that land as calculated by a net land-rent formula in each state. Farmland value calculated as such is typically only a fraction of the price a farmer would pay another farmer for the land. Purchase of development-rights programs, in the twenty or so states that have implemented such programs, pays the farmer the difference between the appraised full-market value and the appraised farm value of the land, which will nearly always exceed the value for use-value taxation. Thus, general use-value assessment is not an important factor in farmland value. According to Anderson and Bunch (1989), tax relief granted to land enrolled in the Michigan Farmland Protection Program does affect land value, and it accounts for more than 8 percent of the average price. This increment of land value captures at least a portion of benefits, amounting to an 80-percent to 90-percent reduction in taxes for the eligible farmer.

The Conservation Reserve Program (CRP) enacted as part of the 1985 Farm Bill, leases erodible land from farmers to capture an amenity or to reduce run off. Shoemaker (1989) determines that the CRP contract was capitalized into land value, but primarily as a result of the bidding procedure that paid farmers more for marginal land than farmers would have received in rents from farming the land.

CONCLUSIONS

Farmland value is a reflection of the returns to the various services that flow from that land. Among the attributes that contribute value to farmland are certain amenity characteristics. Some of the amenity services of farmland are exclusive and rival, and therefore can be captured by the owner and priced for purchase by consumers. Farmers in urbanizing areas find these farm-based amenities to be important income sources, enabling them to adjust to local economic realities. Various policies and programs help farmers to consider and to undertake these new enterprises. Other amenities are public goods that are nonexclusive and/or nonrival in consumption, and therefore are not reflected in the market price of farmland. These external benefits create a divergence between private and social returns to farmland and, as a result, a myriad of policies have attempted to correct for this market failure by targeting farmland amenities.

The U.S. experience with amenity-securing policies is extensive. Evidence of how those policies affect land-use patterns and value comes from case studies, statistical analyses, and general conclusions about changing patterns of land use. We draw the following conclusions from our brief review of this broad literature:

- Policy interventions by state and local governments to protect the amenity services of farmland do have a modest effect on land-use patterns, but much evidence suggests that this is true perhaps only in the short term. The regulations, incentives, and spending programs are well intended, but many times have been found to be unable to withstand intense and prolonged pressure for change. Competing land uses press the policy boundaries, and the value of farmland amenity services is implicitly compared with returns to an alternative use. Home rule, the annexation power, state property-rights protection statutes, local referendum authority, and the takings issue in general create a climate within which amenity protections are fragile at best. Other developed nations have far less trouble protecting rural land; home rule and takings are unique American institutions (Alterman 1997).
- Because U.S. amenity policies are often short lived and are readily adjusted by local governments, they appear to have little effect on farmland value. Buyer expectations of future returns include the likelihood of development that would displace the amenity.
- In some cases, amenities are the most important services of a land parcel, reflecting scarcity of open land and location relative to nonfarm population. The amenity is actually more valuable in the market than the possibility of development. Under those circumstances, policies designed to protect amenities will actually increase or at least not diminish the land value.

While we believe these conclusions apply broadly to many of the policy studies considered here, our conclusions are limited in several ways. First, we have ignored the external costs associated with farmland. A variety of negative externalities may be associated with farmland (e.g., odors, noises, and congestion on roads due to slow-moving farm vehicles). A comprehensive approach to optimal policy design would take these considerations into account.

Secondly, our discussion has virtually ignored the issue of how farmland should be spatially distributed relative to people in order to maximize social benefits. Studies that have estimated the benefits of farmland tend to represent location in relatively simplistic ways (e.g., distance to nearest farmland or the proportion of land within a given neighborhood that is farmland). While this gives some indication as to how individuals value the relative location of farmland, such findings do not provide any real insights into how farmland should be distributed at a regional level relative to the spatial distribution of population and other features of the landscape. For example, if proximity to farmland is valuable, then to what extent should plots of farmland be scattered throughout residential development to minimize any one person's distance to a plot of agricultural land? Or should farmland be preserved in contiguous tracts of land that enable economies of scale and preservation of a wider range of amenity services? Considerations of how policymakers should interpret individual-level values of farmland in terms of what this implies for optimal regional patterns of farmland preservation are complex, and the current body of literature does not provide sufficient insights into how to translate individual-level values into optimal patterns of farmland at a community level.

Despite these limitations, the empirical evidence on farmland amenity values and on the policies that have been used to secure these amenities provides insights for policymakers. They provide a range of evidence as to the positive amenity values associated with farmland, and therefore provide the rationale for programs that seek to secure these amenities.

REFERENCES

AFT(American Farmland Trust). (1997) *Saving American Farmland: What Works.* Northampton: The American Farmland Trust.

Alterman, R. (1997) "The Challenge of Farmland Preservation: Lessons From a Six-Nation Comparison." *American Planning Association Journal* (Spring): 220–43

Anderson, J. and H. Bunch. (1989) "Agricultural Property Tax Relief: Tax Credits, Tax Rates, and Land Values." *Land Economics* 65:1 (February): 13–22.

Beasley, S.D., W.G. Workman, and N.A. Williams. (1986) "Estimating Amenity Values of Urban Fringe Farmland: A Contingent Valuation Approach." *Growth and Change* 17: 70–78.

Beaton, W. (1991) "The Impact of Regional Land Use Controls on Property Values: The Case of the New Jersey Pinelands." *Land Economics* 67: 172–94.

Bergstrom, J., B. Dillman, and J. Stoll. (1985) "Public Environmental Amenity Benefits of Private Land: The Case of Prime Agricultural Land." *Southern Journal of Agricultural Economics* 17(1): 139–49.

Bonnieux, F. and P. Le Goffe. (1997) "Valuing the Benefits of Landscape Restoration: a Case Study of the Cotentin in Lower-Normandy, France," *Journal of Environmental Management* 50: 321–33.

Bowers, D. (2002) "Achieving Sensible Agricultural Zoning to Protect PDR Investment." *Protecting Farmland at the Fringe: Do Regulations Work?* Columbus: The Swank Program in Rural-Urban Policy, Ohio State University, Columbus, OH.

Bowker, J.M. and D.D. Didychuk. (1994) "Estimation of Nonmarket Benefits of Agricultural Land Retention in Eastern Canada." *Agricultural and Resource Economics Review* 23(2): 218–25.

Breffle, W.S., E.R. Morey, and T.S. Lodder. (1998) "Using Contingent Valuation to Estimate a Neighborhood's Willingness to Pay to Preserve Undeveloped Urban Land." *Urban Studies* 35(4): 715–27.

Cheshire, P. and S. Sheppard. (1995) "On the Price of Land and the Value of Amenities." *Economica* 62: 247–67.

Cordes, M. (2002) "Agricultural Zoning: Impacts and Future Directions." *Protecting Farmland at the Fringe: Do Regulations Work?* Columbus: The Swank Program in Rural-Urban Policy, Ohio State University, Columbus, OH.

____. (1999) "Takings, Fairness, and Farmland Preservation." *Ohio State Law Journal* 60: 1033–84.

Correll, M., J. Lillydahl and L. Singell. (1978) "The Effects of Greenbelts on Residential Property Values: Some Findings on the Political Economy of Open Space." *Land Economics* 54:2 (May): 207–17.

CAST (Council on Agricultural Science and Technology). (2002) "Urban and Agricultural Communities: Opportunity for Common Ground." Council on Agricultural Science and Technology, Task Force Report No. 138. Ames: CAST.

Drake, L. (1992) "The Non-Market Value of the Swedish Agricultural Landscape." *European Review of Agricultural Economics* 19: 351–64.

Freeman, A. (1993) *The Measurement of Environmental and Resource Values.* Washington, DC: Resources for the Future.

Garrod, G. and K. Willis. (1992) "The Environmental Economic Impact of Woodland: A Two-Stage Hedonic Price Model of the Amenity Value of Forestry in Britain." *Applied Economics* 24: 715–28.

Geoghegan, J. (2002) "The Value of Open Spaces in Residential Land Use." *Land Use Policy.* 19(1): 91-98.

Geoghegan, J., L.A. Wainger, and N.W. Bockstael. (1997) "Spatial Landscape Indices in a Hedonic Framework: An Ecological Economics Analysis Using GIS." *Ecological Economics* 23(3): 251–264.

Gleeson, M. (1979) "Effects of an Urban Growth Management System on Land Values." *Land Economics* 55(3): 350–65.

Halstead, J. (1984) "Measuring the Nonmarket Value of Massachusetts Agricultural Land: A Case Study." *Journal of Northeastern Agricultural Economic Council* 13: 226–47.

Heimlich, R.E. and W.D. Anderson. (2001) "Development at the Urban Fringe and Beyond: Impacts on Agricultural and Rural Land." Economic Research Service, U.S. Department of Agriculture AER No. 803. Washington, DC: USDA/ERS.

Henderson, E. and R. van En. (1999) "Sharing the Harvest: A Guide to Community-Supported Agriculture." White River Junction: Chelsea Green Publishing Company.

Henneberry, D. and R. Barrows. (1990) "Capitalization of Exclusive Agricultural Zoning into Farmland Prices." *Land Economics* 66(3): 249–58.

Irwin, E.G. (2002) "The Effects of Open Space on Residential Property Values." *Land Economics.* 78(4): 465–80.

Irwin, E.G. and N.E. Bockstael. (2001) "The Problem of Identifying Land Use Spillovers: Measuring the Effects of Open Space on Residential Property Values." *American Journal of Agricultural Economics* 83(3): 698–707.

Johnston, R., J.J. Opaluch, T.A. Grigalunas, and M.J. Mazzotta. (2001) "Estimating Amenity Benefits of Coastal Farmland." *Growth and Change* 32 (Summer): 305–25.

Knaap, G. (1985) "The Price Effects of Urban Growth Boundaries in Metropolitan Portland, Oregon." *Land Economics* 61(1): 26–35.

Krieger, D. (1999) "Saving Open Spaces: Public Support for Farmland Protection." Working Paper Series 99-1, Center for Agriculture in the Environment, April.

Liberty, R. (1999) "Oregon's Farmland Protection Program." *The Performance of State Programs for Farmland Retention: Proceedings of a National Conference.* Columbus: The Swank Program in Rural-Urban Policy, Ohio State University, Columbus, OH.

Mansfield, C., S. Pattanayak, W. McDow, and P. Halpin. (2002) "Shades of Green: Measuring the Value of Urban Forests in the Housing Market." Paper presented at the Association of Environmental and Resource Economists Session (AERE), 2002 Allied Social Sciences Association meetings, Atlanta, GA (January).

McGranahan, D. (1999) *Natural Amenities Drive Rural Population Change.* Agricultural Economics Report No. 781. Washington, DC: USDA/ERS.

Miller, D. (1999) "Farming the Wild Side." *Progressive Farmer* (August): 20–2.

Nelson, A. (1986) "Using Land Markets to Evaluate Urban Containment Programs." *American Planning Association Journal* (Spring): 156–71.

_____. (1985) "A Unifying View of Greenbelt Influences on Regional Land Values and Implications for Regional Planning Policy." *Growth and Change* (April): 43–8.

_____. (1998) "An Empirical Note on How Regional Urban Containment Policy Influences an Interaction Between Greenbelt and Exurban Land Markets," *American Planning Journal* (Spring), pp. 170-184.

Nelson, G. and J. Geoghegan (2002). "Deforestation and Land Use Change: Sparse Data Environments." *Agricultural Economics* 27(3): 201–16.

Nickerson, C. and L. Lynch. (2001) "The Effect of Farmland Preservation Programs on Farmland Prices." *American Journal of Agricultural Economics* 83(2): 341–51.

Pollakowski, H. and S. Wachter. (1990) "The Effects of Land Use Constraints on Housing Prices." *Land Economics* 66(2): 315–24.

Randall, A. and E. Castle. (1985) "Land Resources and Land Markets." In *Handbook of Natural Resource and Energy Economics* Vol. II, A.V. Kneese and J.L. Sweeney, eds. Amsterdam: Elsevier Science Publishers.

Ready, R.C., M.C. Berger, and G.C. Blomquist. (1997) "Measuring Amenity Benefits from Farmland: Hedonic Pricing *vs* Contingent Valuation." *Growth and Change* 28 (Fall): 438–58.

Ricardo, D. (1821) *Principles of Political Economy and Taxation.* London: John Murray.

Riddel, M. (2001) "Hedonic Prices for Environmental Goods." *Land Economics* 77(4): 494–512.

Roe, B., E.G. Irwin, and H. Morrow-Jones. (2002) "The Effects of Farmland, Farmland Preservation and Other Neighborhood Amenities on Proximate Housing Values: Results of a Conjoint Analysis of Housing Choice." Working Paper, Department of Agricultural, Environmental, and Development Economics, Ohio State University.

Rosenberger, R. and R. Walsh. (1997) "Non-Market Value of Western Valley Ranchland Using Contingent Valuation." *Journal of Agricultural and Resource Economics* 22(2): 296–309.

Russell, S. (1997) "Can Your Farm Compete with Disney World?" *New England Country Folks* 13(January): 14–6.

Shoemaker, R. (1989) "Agricultural Land Values and Rents Under the Conservation Reserve Program." *Land Economics* 65(2): 131–37.

Stephenson, G. and L. Lev. (1998) "Common Support for Local Agriculture in Two Contrasting Oregon Cities." Unpublished paper, presented at Annual Meeting of Rural Sociological Association, Portland, OR (August).

Tyrävinen, L. and A. Miettinen. (2000) "Property Prices and Urban Forest Amenities." *Journal of Environmental Economics and Management* 39: 205–23.

USDA/ERS (U.S. Department of Agriculture, Economic Research Service). (2002) "Farmland Protection: The Role of Public Preferences for Rural Amenities" (forthcoming) Rural Amenities Team, USDA.

Vaillancourt, F. and L. Monty. (1985) "The Effect of Agricultural Zoning on Land Prices, Quebec, 1975–1981." *Land Economics* 61(1): 36–42.

von Thuenen, J.H. (1842) "Der isolierte Staat." In *Sammlung Socialwissenschaftlicher Meister, Vol. XIII.* Hamburg: Muenster.

Willis, K.G. and G.D. Garrod. (1993) "Valuing Landscape: a Contingent Valuation Method." *Journal of Environmental Management* 37: 1–22.

Willis, K.G., G.D. Garrod, and C.M. Saunders. (1995) "Benefits of Environmentally Sensitive Area Policy in England: A Contingent Valuation Assessment." *Journal of Environmental Management* 44: 105–25.

ENDNOTES

[i] We discuss these and other market-based amenity services of farmland in the following section.

[ii] The following summary is focused on studies that have isolated the effects of farmland relative to other types of open space on residential property values. Papers that have used the hedonic-pricing method to study the externality effects of other types of neighboring open space or just open space in general include Cheshire and Sheppard (1995) (private versus public open space), Garrod and Willis (1992) (woodlands), Nelson and Geoghegan (2002) (permanent versus developable open space), Geoghegan, Wainger, and Bockstael (1997) (open space), Mansfield, et al. (2002) (urban forests), Riddel (2001) (open space), and Tyrävinen and Miettinen (2000) (forests).

[iii] Again, we focus on studies that have considered farmland explicitly versus other types of open space. Studies that have used the CVM to value the benefits from other types of open space include Willis et al. (1995) (environmentally sensitive areas), Bonnieux and Le Goffe (1997) (landscape restoration), Breffle et al. (1998) (undeveloped land), and Willis and Garrod (1993) (natural landscape).

Section VI:
Regional and International
Dimensions

Chapter 20

Micro-Markets for Farmland: The Case of Florida and California

John E. Reynolds and Warren Johnston
University of Florida
University of California - Davis

INTRODUCTION

Many farm real-estate markets are described as local markets in which active participants are predominantly from the surrounding area or neighborhood. This is generally true for agricultural areas where highest and best use is limited to a small portfolio of agricultural uses typical to the area (Raup 2003) and where there are insignificant or no near-term transitional urban or rural/urban uses (Barnard et al. 2003). The agricultural economy of neither Florida nor California operates in such splendid isolation from intermingled bustling development and population pressures. There, farmlands often have a larger set of agricultural-production opportunities. In addition, there is often competition for agricultural resources (i.e., land and water) from surrounding and expanding municipal, industrial, and environmental uses.[i]

The average per-acre statewide value of farm real estate for Florida's 44,000 farms or for California's 87,500 farms is a singular statistic of dubious value. It is akin to imagining whether or not Merriwether Lewis and William Clark, prior to their departure on their North American expedition 200 years ago, would have found it of value to know the average depth of the Missouri River from its headwaters in the Rocky Mountains to its junction with the Mississippi.

In the sections that follow, we describe characteristics of Florida and California land markets where highest and best use is determined by diversity in agricultural production possibilities and in resource competition arising from development and population pressures.

FLORIDA AGRICULTURAL LAND MARKETS

Florida Agriculture

Florida boasts a diverse agricultural economy (Figure 20.1). It has tropical fruits and vegetables produced in the Southeast Florida region, sugarcane produced in the South Florida region, citrus produced in the Central and South Florida regions and row crops (e.g., cotton, peanuts, corn, and soybeans) produced in the Northwest and Northeast Florida regions. Vegetables are grown in a number of areas across the state and the more traditional animal agriculture (e.g., beef and dairy) is produced throughout the state. In 2000, for example, Florida agriculture ranked ninth in the nation in cash receipts with sales of U.S. $6.95 billion. (About 80 percent of Florida cash receipts are derived from crops and 20 percent are from livestock and livestock products.) Also in 2000, Florida ranked first in citrus and sugarcane production, second in greenhouse and nursery products, tomatoes, and strawberries and fourth in aquaculture and honey (USDA/NASS 2002).

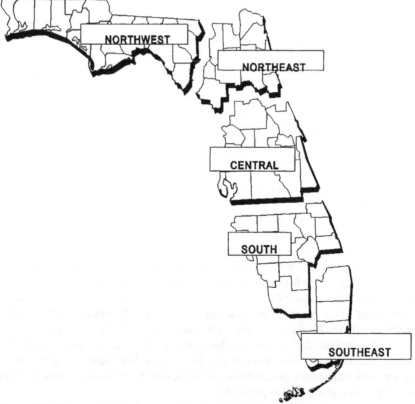

Figure 20.1 Geographic Regions used in the Florida Land-Value Survey

Data on agricultural land use and agricultural production often do not include data on land devoted to timber production. There are 14.6 million acres
classified as timberland in Florida, which is more than 40 percent of the total
Florida land area (Brown 1996). About 1.25 million acres of these 14.6 million
acres are devoted to national forests. In addition, large areas of North Florida are
taken up with timberland that is privately owned. Thus, land in Florida that is
utilized for agriculture and natural resources ranges from land used for timber
and pasture to land that is used for the highly intensive citrus, vegetable, and
ornamental crops. These diverse land uses result in many micro-markets for agricultural land in Florida.

Florida Farm Real-Estate Markets

U.S. Department of Agriculture (USDA) data on the aggregate farm real-estate
market in Florida indicates that the average value for agricultural land in Florida
was U.S. $2,800 per acre in 2002. This average increased from U.S. $2,085 per
acre in 1990—a compound rate of 2.5 percent per year (USDA/NASS 1998).
However, the aggregate average for the value of agricultural land in Florida can
be misleading because the factors that affect this value within Florida's many
different land types may impact differently the various micro-markets throughout the state. The value of agricultural land in Florida, therefore, varies by the
type of agricultural land use within these different Florida markets.

The aggregate per-acre average value of agricultural land in Florida is determined from a number of smaller micro-markets within the state. The demand
and supply of land for agricultural use varies considerably among these various
micro-markets. The amount of land available for agricultural use is very limited
along the Southeast coast of Florida where only 4 percent and 6.8 percent of the
land area is in farms in Broward and Dade counties, respectively. The median
size farm in 1997 in these two counties was 5 acres (USDA/NASS 1999b) and
the population density then was greater than 1,000 people per square mile. Although Dade County is highly urbanized, there is also a highly productive agricultural sector that produced more than U.S. $416 million of agricultural prod-
ucts in 1997 (UDSA/NASS 1999b). Contrast Dade County that had less than 7
percent of its land area in 1997 allocated to agricultural uses to many counties in
North Florida that were devoted almost entirely to agricultural and forestry uses.
For example, the population density in Liberty County in the Northwest region is
only 8 people per square mile and there is very little nonagricultural demand for
land. As a consequence of the varying agricultural and nonagricultural demands
for land and the availability of land for different uses, the value of agricultural
land also varies substantially.

The Florida Land Value Survey (Reynolds 1995) that is conducted annually
by the Food and Resource Economics Department at the University of Florida
provides estimates of the value of different types of agricultural land for geographic regions of the state that can be used to illustrate this variation. The state
is divided into five major regions for the survey—Northwest, Northeast, Central,
South and Southeast—based on agricultural production and on the impact of
urbanization (Figure 20.1). Survey respondents include rural appraisers, farm

lenders, real-estate brokers, farm managers, land investors, county extension agents, the Farm Services Agency and Natural Resource and Conservation Service personnel, county property appraisers and others who develop and maintain information about rural land values in their areas.

The results of the 2002 Florida Land Value Survey indicated that the value of agricultural land used for agriculture, which is likely to continue in that use at least in the short term, varied from an average of U.S. $1,165 per acre for unimproved pasture in the Northwest region to an average of U.S. $5,687 per acre for mature orange groves in the South region (Reynolds 2000). Also, agricultural land that was in a transition phase in 2002 (i.e., land being converted or is likely to be converted to nonagricultural uses such as sites for homes, subdivisions and/or commercial uses) averaged from U.S. $3,234 per acre for land located more than 5 miles from a major town in nonmetropolitan counties in the Northwest region to an average of U.S. $45,083 per acre for land located within 5 miles of town in metropolitan counties in the Southeast region. These data represent per-acre averages for the different types of uses and regions of the state. The individual estimates provided by respondents to the survey varied considerably from the aggregate averages presented for each region.

The 2002 survey results also indicated that the value of cropland and pastureland increased in all regions of the state when compared to the 2001 survey results. The value of cropland increased from 8 percent to 13 percent depending on whether or not it was irrigated and depending on the region in which the land was located. The value of pastureland increased 7 percent to 15 percent depending on the region and depending on whether it was improved or unimproved pastureland. However, the value of citrus land decreased substantially from the previous year, due primarily to the lower prices received by its growers. The per-acre value of orange groves declined 11 percent while the per-acre value of grapefruit groves declined 15 percent from 2001 to 2002. In summary, agricultural land values increased 7 percent to 15 percent for cropland and pastureland, but decreased 11 percent to 15 percent for citrus land. The aggregate estimate for the per-acre average value of agricultural land in Florida increased 7.7 percent from 2001 to 2002 (USDA/NASS 2002). This aggregate estimate reveals neither the large increases for some types of agricultural land nor the substantial decreases in the value of other types (e.g., citrus land).

The value of agricultural land has increased substantially over time as the agricultural use of land has intensified and as the population of Florida has mushroomed. In 1950, the average aggregate per-acre value of agricultural land in the United States was U.S. $65 and was higher than the U.S. $57 per-acre aggregate average for Florida's agricultural land. By 2002, the average value of an acre of agricultural land in Florida had increased to U.S. $2,800 per acre, 2.3 times greater than the average value for the United States (Figure 20.2). In both Florida and the United States, the average value of agricultural land increased steadily during the 1960s and 1970s and then increased rapidly during the 1980s.

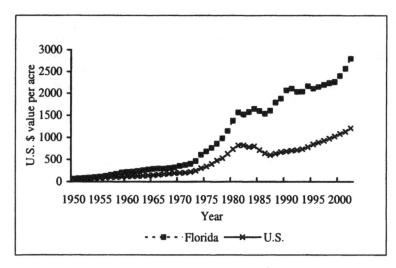

Figure 20.2 The 1950 to 2002 Value of Farmland in Florida and in the United States

During the 1950 to 1980 period, the average value of agricultural land in Florida increased at a compound annual rate of 11.2 percent per year whereas the aggregate U.S. compound annual rate grew at only 8.4 percent per year. The rapidly growing population of Florida during this period created a strong demand for the conversion of land to urban uses: For each person added to the population base, land in urban areas increased by 0.54 acres (Reynolds 1999). The population of Florida increased 6.98 million between 1950 and 1980, so approximately 3.7 million acres of rural land was needed for urban expansion during this period. The strong competition between urban and agricultural uses in and around urbanizing areas during this period increased land values in many micro-markets of the state much more than in others.

During the 1982 to 1987 period of the "farm crisis," the aggregate average value of agricultural land in the United States declined at a compound annual rate of 6.2 percent and declined at an even greater rate in certain regions. The average value of agricultural land in Florida was higher in 1987 than it was in 1982. However, agricultural land values in the North Florida region actually followed a declining trend similar to other parts of Southeast United States, then the North Florida region entered a period in which agricultural land values increased slowly. Agricultural land values in Central and South Florida also increased during the 1980s.

As a result of severe freezes in 1983, 1985, and 1989, many acres of citrus were frozen and the on-tree prices for citrus increased substantially. These high prices made citrus a very profitable enterprise and the value of producing groves increased rapidly. With the freezes destroying many groves in Central Florida and prices increasing, a massive replanting of citrus took place in South Florida. This relocation of citrus to South and Southeast Florida also created a strong demand for land suitable for citrus in these areas. As a result, the value of cropland and pastureland increased as did the value of existing citrus groves during this period. By 1990, with many of the new groves producing fruit, Florida's production of citrus reached the pre-freeze levels of production and on-tree prices began to fall. The per-acre average value of citrus groves in the South Florida region peaked in 1990 at U.S. $13,351 for orange groves and U.S. $12,169 per acre for grapefruit groves (Figure 20.3). During the 1990s, citrus land values declined as the low prices received by growers made the returns to groves decline substantially. By 2002, the average value of orange groves in the South Florida region had declined to U.S. $5,687 per acre and was only 42.6 percent of the average value in 1990—an annual decline of 6.9 percent in value. The per-acre aggregate average value of grapefruit groves in South Florida dropped 9.5 percent per year to U.S. $3,658 in 2002, which was 30 percent of its 1990 value.

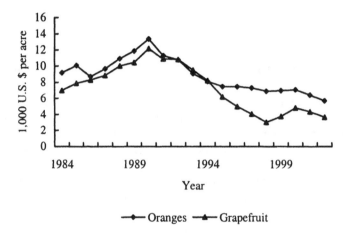

Figure 20.3 Value of Orange and Grapefruit Groves in South Florida, 1984 to 2002

Although the annual value of citrus groves in the Central and South Florida regions were declining during the 1990s, the value of cropland and pastureland was increasing from 3 percent to 5 percent in the North Florida region during this period, depending on the type of cropland or pastureland and depending on the region in which the land was located. The value of cropland and pastureland in the Central and South Florida regions declined from 1990 to 1998 as the demand for cropland and pastureland for conversion to citrus land dissipated when citrus land values declined. From 1998 to 2002, the average value of cropland and pastureland in the Central and South Florida regions increased 6 percent to 7 percent per year. Since 1990, the trends in the average value of agricultural land have varied considerably for the different uses of land.

The regional average value of agricultural land in the annual Florida Land Value Survey includes a range of survey responses for the different uses of land. In 1990, the per-acre value of agricultural land varied from an average of U.S. $13,351 for orange groves in the South Florida region to an average of less than U.S. $700 per acre for unimproved pasture in the Northwest Florida region (Reynolds, 1995).

The Florida Land Value Survey is a survey in which questionnaires are mailed to participants to ask them to provide *estimates* of the per-acre average market value for different types of land in their county. Therefore, the average values reported for a region vary widely. In 1990, the Florida Land Value Survey reported per-acre averages for the value of orange groves in the South Florida region as high as U.S. $20,000 and also reported the average value for unimproved pasture in the Northwest Florida region as low as U.S. $300 per acre. By 2002, the per-acre average value of orange groves in South Florida had declined to less than U.S. $5,700 per acre, while the average value of unimproved pastureland in the Northwest Florida region had increased to U.S. $1,165 per acre. From 1990 to 2002, the per-acre value of grapefruit groves in the South Florida region declined from U.S. $12,169 in 1990 to U.S. $3,658 in 2002, which was a decline of 9.5 percent per year. Also, the per-acre value of orange groves declined 7.1 percent per year in the Northwest Florida region while unimproved pasture there increased 4.3 percent per year during this same period.

The average per-acre value of agricultural land in Florida is such a diverse composite of the interaction of many agricultural and nonagricultural factors affecting the demand for agricultural land in many micro-markets that the aggregate value (state average) does not indicate what is happening in many of the state's micro-markets. The value of some type of agricultural land may be increasing in one micro-market while the value of other types may be decreasing in that or other markets. Thus the use of per-acre aggregate data to analyze the changes in the value of Florida's agricultural land is not able to detect and measure changes in micro-market data.

Micro-Market Agricultural Land-Value Data

The type of land-value analysis may also differ with the type of land-value data collected. The types of analyses conducted with aggregate data are different than analyses using parcel data based on land sales (micro-market data). Studies that use aggregate data can be classified into two groups: time-series data and county-level data (Reynolds 1986). The time-series data studies normally involve an analysis of the change in the average state or national value of agricultural land over a given time period. With this type of data, the principal factors or variables affecting land values generally involve aggregate measures and include some of the following variables: 1) income or some proxy to reflect income, 2) government payments, 3) capital gains, 4) rate of return or interest rate, 5) average size of farm, 6) number of farms, 7) land in cropland or other intensive use, 8) taxes, 9) debt, and other aggregate measures.

The studies that use county-level data have been primarily cross-sectional studies using census data and the principal explanatory variables have included: 1) some type of income or productivity variable, 2) percent of land in a specific land use class, 3) average size of farm, 4) population pressure, 5) distance or location variable, and 6) tax rate. There have also been studies that have combined time-series and cross-sectional characteristics by pooling census data for multiple census years. As an example, Shi et al. (1997) pooled census data from 1950 to 1992 that allowed them to use time-series variables (expected returns to land and capital gains) in their analysis to estimate expected net real returns to land.

Micro-market analyses of rural land values have used primary data collected on individual tracts of land that sold in those markets. The principal variables used in these analyses have been primarily the physical characteristics of those tracts of land and the economic and institutional variables that may affect land values in that micro-market. These variables have included physical characteristics such as size of tract, land use, productivity measures (type of soil or soil rating), and wetlands. Economic and institutional variables have included location, value of buildings and/or improvements, real-estate taxes or tax rates, crop allotments or quotas, and zoning. Those studies that focused on micro-market analyses in urbanizing areas have included other variables including neighborhood influences, aesthetics, road frontage and other variables that are specific to those micro-markets.

Reynolds and Regalado (2002) estimate the effect of the physical and locational characteristics of wetlands on the sales price of rural land in a four-county study area in Central Florida. Location, parcel size, capital improvements, proportion of land in intensive uses, and land area in wetlands explained more than 80 percent of the variation in the sales price of the 212 parcels in the study. This study classified wetlands into wetland systems and classes to test the effect of various types of wetlands on rural land prices. When all of the wetlands were aggregated into a total wetlands variable, a significant negative impact on rural lands was indicated. However, when wetlands were further subdivided into different systems and classes of wetlands, the impact of the physical and locational characteristics of wetlands ranged from small positive effects to significant negative effects.

Other studies in Florida examine the effect of physical and locational factors on the value of rural land in urbanizing areas. Analyses of sales of rural land in the Orlando area (Reynolds and Tower 1978) and in the Bradenton-Sarasota area (Shonkwiler and Reynolds 1986) indicate that the nonagricultural demand for rural land in urbanizing areas is an important factor that affects the per-acre value of rural land prices near urban centers. Locational factors are also important determinants of rural land prices as are distance to the metropolitan area and distance to the major transportation arteries. The effect of the distance variables decreases at a decreasing rate as the distance from the urbanizing area increases.

The Florida Land Value Survey provides annual estimates of the per-acre value of different types of agricultural land that is aggregated to the five regions of the state, that is, the estimates are averages for multi-county regions. Figure 20.3 illustrates how the value of orange groves increased rapidly during the 1980s and then declined since 1990. An analysis of the South Florida region indicates that from 1984 to 2002 the value of orange groves is a function of the on-tree prices that growers received for their crops:

Value of Grove = -264.03 + 1834.22 Weighted On-Tree Price for Oranges   (1)

In this equation, the weighted on-tree price is a four-year weighted average and is calculated by weighting the most recent price by 4 and each preceding year after that by one less and dividing the sum by 10. The four-year weighted average of past prices explains 95 percent of the variation in the value of orange groves during this period of increasing and then declining grove values. Without data from micro-markets these types of analyses are not possible.

CALIFORNIA AGRICULTURAL LAND MARKETS

California Agriculture

California, the nation's #1 agricultural producing state, had an agricultural gross farm income of U.S. $27.2 billion in 2000. With a farm income exceeding the combined farm income of Texas (#2) and Iowa (#3), California alone accounted for 12 percent of U.S. gross farm income.

California agriculture is a diverse, demand driven agricultural system that has by necessity been forced to respond to distant domestic and export markets from its outset (McCalla and Johnston 2003). It is unique, not only in its aggregate productivity, but also in the variety of crops it produces, its high yields from large capital investments in intensive-production practices, and its firm and industry structure which, combined with favorable climate, rich soils, and irrigation, is the envy of the world. With increasing population and competitive pressures, it operates in a dynamic, environment and faces numerous challenges not common to other production areas (Johnston and Carter 2000; Siebert 1997).

The state's physical dimensions (100 million acres) alone would make it a large aggregate producer. Its north-south length (828 miles from its Oregon coastal border to the southeastern point) and its topography (highest and lowest elevations in the coterminous United States) combine to yield climatic choices from Mediterranean to Alpine and about 1,200 different named soil series. Natural precipitation is abundant in the north but lacking in the south. The most productive California agricultural lands are the alluvial areas of its coastal valleys (also most attractive to the population inflow of the Twentieth Century) and the southern Central Valley. This mal-distribution of water demands and supplies had its Twentieth Century solution in water-transfer projects benefiting agriculture and growing municipal and industrial demands in southern California.

Production is predominantly an expanding list of intensive, higher-valued commodities reflecting an ability to produce for changing consumer demands. Whereas U.S. agricultural output has about equal 50-50 shares of crops and livestock, the proportions of California are 75 percent crops to 25 percent livestock and livestock products. Roughly 80 percent of marketings in 2000 were of intensive crops and products—28 percent fruits and nuts, 26 percent vegetables, 15 percent dairy products, and 11 percent nursery and greenhouse products. The remainder is extensive field crops and other livestock and livestock products. [ii]

The 1954 Census of Agriculture (USDA/NASS 1999a) identified 139 crops grown in significant commercial quantities. The number of crops has increased significantly to 200 crops identified in 1970 and 350 in 2000 (CCLRS 1971,

CDFA 2001). A growing in-state population, together with national and export demands, and changing consumer preferences for fresh fruits and vegetables, nuts and wine, underlie growth in the number of commodities produced and of changing shares toward higher-valued products. It helps that now one out of every eight U.S. consumers is a resident of California. Population projections suggest that in 2040 as many as one out of every six U.S. consumers will be a Californian. The flipside of this good news is that the agricultural industry will face increased competition for its land resources and its share of available water supplies by continual strong population growth.

Demand-driven commodity growth, the existence of a multitude of physical resource combinations (soils, climate, and water), and requirements of a growing population and industry, underlie the existence of micro-markets for the 27.7 million acres of land currently inventoried as "land in farms" in California.

California Farm Real Estate

The mosaic of diverse climatic conditions, soils, water availabilities, and 350 commercial crop opportunities cannot be portrayed in simple statewide or regional average land-value statistics. Small (micro) markets emerged as California's agriculture evolved from the cattle- and wheat-dominated agricultural economies of the 1800s, to one producing a full line of demand-driven, high-valued crops and products by the end of the Twentieth Century.

Not just land. Farm real estate consists of the portfolio of lands associated with farms enumerated in periodic Census surveys.[iii] Annual farmland values for the aggregate United States and individual states (USDA/NASS 2002) are made with reference to the previous Census benchmark, but then are often revised following tabulation of the next five-year census. By definition, land includes those objects permanently affixed to the land—buildings, perennial tree and vine plantings, and permanent irrigation systems. The value includes investments made to level land for irrigation and drainage as well as investments made for the development costs of tree and vine plantings. The 1997 Census (USDA/NASS 1999a) reveals that orchards and vineyards now account for 2.6 million acres (30 percent) of irrigated California cropland area.

Not just for agricultural use. Every conceivable motivation for owning farm real estate exists in California. Economic principles relate both to real-estate marketability and to real-estate productivity (ASFMRA 2000: 12–24). Principles strongly in play for agricultural use, as well as for the conversion of California land to other uses, include:

- Principle of Substitution (Value is determined by the cost of acquiring an equally desirable substitute property),
- Principle of Anticipation (Value is determined by expected economic gain – flow of income and capital gain), and
- Principle of Contribution (Value is the combined contribution of land and objects permanently affixed to land, e.g., buildings and improvements).

Competition for land has displaced agriculture from many temperate coastal areas (e.g., Los Angeles basin and Santa Clara/Silicon valley), shifting production "over the hill" to the Central Valley. There are very few orange trees remaining in Orange County.

Value of California Farm Real Estate, 1950-2002

How has the value of agricultural land changed over time? Figure 20.4 shows the time-series data for California and for U.S. land values. Both showed modest appreciation through the 1950s and 1960s, rapid escalation during the 1970s into the early 1980s, and significant declines through 1988, before resuming their upward trends.

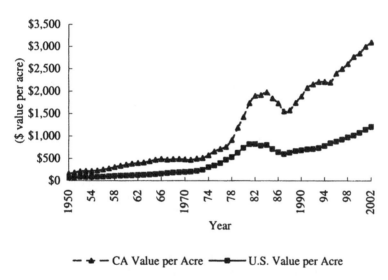

Figure 20.4 Average Value of Farm Real Estate, United States and California, 1950 to 2000

The general trends are similar. The California time series is particularly influenced by the development of new irrigated lands in the San Joaquin Valley (served by the new state and federal surface systems constructed in the 1960s and early 1970s), by on-going increases in perennial plantings (e.g., vineyards and orchards)[iv], and by population also spilling "over the hill" into once predominately agricultural areas of the Central Valley. Not only were there additional plantings of high-valued crops, but shifts in production patterns of two types occurred: 1) significant agricultural activity moved "over the hill" (e.g., fresh fruit from the Santa Clara valley moved to the Central Valley, dairy from southern California moved to the southern San Joaquin Valley, vegetables for short shoulder seasons moved into the west side of the San Joaquin Valley).

2) production actually relocated from the Sacramento Valley to newer, more productive areas of the San Joaquin Valley (e.g., almonds and processing tomatoes) attracted by the combination of new water supplies, better climate, rich alluvial soils, large undeveloped parcels, and new cultural and management innovations (e.g., higher yielding new varieties and irrigation systems). (Johnston 1990, McCalla and Johnston 2003).

Value of California Farm Real Estate in 2002

The average per-acre value of California farm real estate, including all land and buildings, was U.S. $3,100 as of 1 January 2002 (Table 20.1), which was an increase of 65 percent from 1990 (USDA/NASS 2002). The same source also gives values for general types of farmland, which gives credence to the notion that the average per-acre value of California farmland is a weighted average of values in major uses.

Table 20.1 Average Value of California Farm Real Estate, 2002

Types of agricultural land use	Average per-acre value, 2002	1997 Agricultural land use
	U.S. dollars	Million acres
All farm real estate	3,100	Land in farms: 27.7
Pasture	1,060	Pastureland, excluding woodland pasture: 15.6
Cropland	6,510	Total cropland: 10.8
- Irrigated Cropland	7,000	Irrigated Land: 8.7
- Non-Irrigated Cropland	1,420	Cropland, not irrigated: 2.1

Source: USDA/NASS (2002); USDA/NASS (1999a)

Retirement and Residential/Lifestyle Farms. A portion of California farm real estate (perhaps one-quarter or more) reflects ownership objectives that are not primarily focused on agricultural production.

A farm, for statistical purposes, is any "place" from which U.S. $1,000 or more of agricultural products were sold or normally would have been sold.[v] In 1997, there were 74,126 farms enumerated in the Census, ranging from very small retirement and residential farms to operations with sales in the millions of dollars. While the average size of farm was 374 acres, 28 percent were less than 9 acres in size and 60 percent were less than 50 acres in size. Only 53 percent of farms identified "farming" as the principal occupation of the operator.

A recently developed farm typology based primarily on annual sales and based on the occupation of owners identified 54 percent of U.S. farms in 1997 as retirement (14 percent) or residential/lifestyle (40 percent) farms (USDA/ERS 2001). Retirement and residential/lifestyle motivations are evident in many agricultural areas of California. Among farms enumerated in the 1997 Census, 32,487 (44 percent) had farm sales of less than U.S. $9,999 and an additional 10,267 farms (also with sales of less than U.S. $9,999) were subsequently identified as under-enumerated.[vi] Those that were enumerated in the 1997 Census were identified with 19 percent of the total land area in farms and accounted for 13 percent of agricultural marketings.

The upshot is that as much as one-quarter or more of California farmland may be made up of farms with ownership motives that are predominantly other than agricultural. Their existence creates differentiated micro-markets for small tracts and parcels that are substantially different from larger agricultural-use parcels with in the same geographical areas.

AGRICULTURAL PRODUCTION REGIONS OF CALIFORNIA

There are nine very different agricultural production regions in California: North Coast, Central Coast, South Coast, North Mountain, Sacramento Valley, San Joaquin Valley, Northeast Mountain, Sierra Nevada, and South Desert (Figure 20.5). Selected characteristics, mainly from the 1997 Census of Agriculture (USDA/NASS 1999a), depict diversity in the structure of agriculture and exposure to population pressures (Table 20.2). Most irrigated land is located in the

Figure 20.5 Agricultural Production Regions of California

Table 20.2 Characteristics of the Agricultural Production Regions of California.

	Total for California	North	Central Coast	Sacramento Valley	San Joaquin Valley	Sierra Nevada	South Coast	South Desert
1. Land Area								
1,000 acres	100,207	20,860	10,148	7,166	175,251	155,291	8,758	20,219
2. Land in Farms								
1,000 acres	27,699	3,526	5,269	3,967	9,764	1,423	1,827	1,923
Percent of total land area in farms	28	17	52	55	56	9	21	10
3. Cropland								
1,000 acres	10,804	718	1,182	2,901	5,339	209	468	797
Percent of land in farms that is cropland	39	20	22	53	55	15	26	41
4. Irrigated Land								
1,000 acres	8,713	486	563	1,712	4,793	136	325	699
Percent of cropland that is irrigated	31	68	48	82	90	65	69	88
5. Number of Farms								
Number of Farms	74,126	4,521	11,803	10,329	27,489	3,709	11,165	5,060
Average farm size (acres)	374	771	446	384	355	284	164	380
6. Average Value of Farm Products Sold								
$1,000 per farm	331	95	314	194	424	33	233	497
$ per acre	832	123	702	506	1,193	87	1,425	1,308
7. Average Value of Land & Buildings								
$ per acre	2,519	1,059	2,581	2,484	2,936	1,549	3,989	2,298

(Continued)

Table 20.2 Characteristics of the Agricultural Production Regions of California (Continued)

	Total for California	North	Central Coast	Sacramento Valley	San Joaquin Valley	Sierra Nevada	South Coast	South Desert
8. Top 5 commodities	Dairy	Cattle/Calves	Wine Grapes	Rice	Dairy	Cattle/Calves	Flowers & Foliage	Dairy
	Grapes	Wine Grapes	Lettuce	Nut Crops	Grapes	Hay	Nursery	Vegetables
	Nursery	Nursery	Nursery	Prunes & Peaches,	Cotton	Pasture & Range,	Vegetables	Cattle/Calves
	Lettuce	Dairy	Broccoli	Processed Tomato	Poultry	Wine Grapes	Strawberries	Alfalfa hay
	Cattle & Calves	Alfalfa Hay	Strawberries	Wine Grapes	Almonds		Avacados	Nursery
9. Larger towns and cities		Eureka	Santa Rosa	Red Bluff	Stockton	Quincy	Santa Barbara	San Bernardino
		Yreka	San Francisco	Chico	Modesto	Bishop	Los Angeles	Riverside
		Alturas	San Jose	Sacramento	Fresno		San Diego	Ontario
			Salinas		Bakersfield			

Source: Lines 1-2: County Supervisors Association of California, California County Fact Book. Lines 3-7: Census of Agriculture, 1997. Line 8: CDFA, 2001

Central Coast region and in the Sacramento and San Joaquin regions of the Central Valley. The crops and products most important to each region (Table 20.2: Line 8) range from the extensive (cattle and calves, hay, pasture and range) in the North Coast, North Mountain, Northeast Mountain, and the Sierra Nevada regions, to high-valued specialty crops in the North, Central, and South Coast regions where urban populations are also concentrated (Table 20.2: Line 9). The average value of farm products sold ranges from only about U.S. $100 per acre (i.e., the North Coast, North Mountain, Northeast Mountain, and the Sierra Nevada regions) to more than U.S. $1,000 per acre in the San Joaquin, South Coast, and South Desert regions.

Average per-acre land values (Table 20.2: Line 7) reflect the existent portfolio of agricultural lands in the region, ranging from only U.S. $1,059 in the North Coast, North Mountain, Northeast Mountain, and the Sierra Nevada regions, to nearly U.S. $4,000 on average in the South Coast region

Micro-Market Agricultural Land-Value Data, 1958 to 2003

Three organizations have been primarily responsible for publishing farm real-estate estimates for California special land uses—the U.S. Department of Agriculture (from 1958 to 1984), the California Department of Food and Agriculture through 1991, and the California Chapter of the American Society of Farm Managers and Rural Appraisers to the present. This section reveals the origins of published micro-market data for the more significant micro-markets.

U. S. Department of Agriculture (USDA)

The USDA published micro-market, land-value statistics beginning in 1958, gradually expanding coverage in a series that continued through 1984. The USDA first attempted to report special (other than statewide) estimated land values for California in the May issue of *Current Developments in The Farm Real Estate Market* (USDA/ARS 1958: 7–8). It noted that:

> The average value of all farmland in California, which was estimated at $282 per acre as of March 1, is a composite of many types and grades of land. Values range from less than $50 per acre for some grazing land to more than $5,000 per acre for certain irrigated orchards.

The average per-acre value and the range in value estimates were reported initially for 13 fruit and nut crops and 3 field-crop groups – truck and vegetables, intensive field crops and extensive field crops. This and subsequent annual reports in the series contain statistics and excellent narratives of factors influencing agricultural land values.

In 1960, the USDA reported sales of farmland for five types of intended, nonfarm uses. This component would not appear in subsequent issues. (USDA/ARS 1960: 13).

In 1962, the USDA first reported regional estimates of orchards, vineyards, and groves for the four major agricultural production regions—Central Coast, Sacramento Valley, San Joaquin Valley, Southern California. (USDA/ERS 1962: 18–21).

In 1965, the USDA extended coverage to include regional values estimates for the 3 field-crop groups—truck and vegetables, plus intensive and extensive field crops. (USDA/ERS 1965: 15–19).

In 1966, the USDA expanded crop coverage to differentiate between irrigated and non-irrigated field crop, pasture, and rangeland. (USDA 1966: 19)

In 1977, the land value estimates were reported in a manner designed to acknowledge the existence of micro-markets for orchards, vineyards, and groves, even within agricultural regions. Standard deviations of estimated average values were also published. (USDA 1977: 35)

The August 1994 issue of *Farm Real Estate Market Developments* was the last USDA publication of California micro-market land values (USDA/ERS 1994). Regional coverage for cropland, pasture, and rangeland had, by that time, been expanded to include a fifth region to cover northern California. Fruit and nut coverage had expanded to 18 tree and vine crops.

California Department of Food and Agriculture

The California Department of Agriculture had long included USDA land values in its annual reports, beginning in the mid 1960s. It successor, the California Department of Food and Agriculture, continued publishing micro-market data for a short period of time covering 1985 through 1990 (CDFA 1991). Crop/commodity coverage was expanded to include additional perennial crops.

The data were consistent with those published in the earlier USDA reports, but were available only through the annual state publication. All USDA and USDA-supported data collection of micro-market data terminated with the final 1990 estimates.

California Chapter of the American Society of Farm Managers and Rural Appraisers

The void in micro-market data published by federal and state agencies has been offset by efforts of the California Chapter of ASFMRA (Cal ASFMRA 2002). The publication, *Trends in Agricultural Land & Lease Values*, results from member efforts reporting an annual survey of sales and lease information in seven regions.

Annual surveys are not comparable to previous efforts and, thus, cannot be spliced to provide continuity in the state-federal series that extended 1958 to 1990. Regions do not correspond exactly to the agricultural production regions used by the public statisticians, nor does the list of agricultural land uses match those used in previous efforts. Regional committees only report ranges in land values. Average land values are not reported.

The new effort does provide richer information about micro-markets for land. Some are as small as a county in size, and may be further differentiated by soil quality, by water source, or by other important determinants of value. The 2001 report on land values has coverage of 167 micro-markets, and reflects the major productive uses of agricultural lands throughout the state (CalASFMRA 2002). The annual narrative report also provides excellent descriptions of market behavior in selected market areas.

Factors Differentiating Micro-Markets

Some examples from the most recent CalASFMRA survey results reveal important factors affecting land values in selected micro-markets.

Wine grape vineyards

Table 20.3 summarizes regional conclusions about wine-grape vineyards. In general, coastal vineyards are more highly valued compared to interior valley plantings; rootstock is important. It is obvious that the North Coast counties—Napa, Sonoma,

Table 20.3 Ranges of Values Per-Acre for Wine-Grape Vineyards, 2001

Agricultural land micro-market	U.S. Dollar Per-acre value	Market activity/trend
NORTH COAST – Resistant Root-stock		
Napa County	85,000 to 180,000	Limited / Stable
Sonoma County	85,000 to 105,000	Moderate / Stable
Mendocino County	35,000 to 65,000	Moderate / Stable
Lake County	23,000 to 32,000	Limited / Stable
NORTH COAST – AXR-1 Root-stock		
Napa County	55,000 to 120,000	Limited / Stable
Sonoma County	65,000 to 95,000	Moderate / Stable
Mendocino County	25,000 to 35,000	Limited / Stable
Lake County	12,000 to 20,000	Limited / Stable
CENTRAL & SOUTH COAST		
Monterey County	17,700 to 25,700	Decreasing
San Luis Obispo/Santa Barbara Counties	22,000 to33,500	Strong / Stable
SACRAMENTO VALLEY	16,000 to 20,000	Very Limited
SAN JOAQUIN VALLEY		
San Joaquin County	14,000 to 22,000	Limited / Stable to Slightly Decreasing
Central San Joaquin area	4,000 to 8,000	Limited / Decreasing
Madera County	3,500 to 5,500	Limited / Decreasing

Source: CalASFMRA (2002).

and Mendocino—have stronger market attraction/recognition even in land markets. Wide ranges in values within a market reflect vineyard characteristics (e.g., variety and age) and physical determinants affecting yield and quality.

Location in one of 74 official U.S. wine appellation areas is an important determinant of land value. Each location has a strong micro-market identification, particularly in premium North Coast counties. Napa County alone has 11 viticultural areas. Market activity in the face of an apparent wine glut, is "moderate" to "limited" and the trend has shifted from "increasing" in 2001 to "stable" in 2002 in the premium areas. Vineyards in more marginal areas have "limited" to "very limited" market activity and the trend in values is mostly "decreasing."

The desire for the limited home-site possibilities among vineyards in the premium regions is reflected in recent sales transactions—U.S. $2 million in one Sonoma County transaction and U.S. $3.5 million in one Napa County transaction.

Central Valley irrigated cropland

Table 20.4 demonstrates variation among micro-markets for open cropland. In the Sacramento Valley, location that reflects soils-type, climate, and water-supply variability within the valley, helps determine the relative value of cropland. The two northern subregions have fewer cropping alternatives (rice is the major field crop), while the southern subregions have expanding plantings of grapes and walnuts, as well as development and growth pressures from the San Francisco Bay Area.

Table 20.4 Ranges of Values per Acre for Irrigated Cropland, 2002

Agricultural land micro-market	U.S. Dollar Per acre value	Market activity/trend
SACRAMENTO VALLEY		
Northwestern counties	1,500 to 2,300	Stable
Yuba-Sutter area	2,200 to 3,000	Stable
Southern counties	2,000 to 5,000	Stable / Slightly Upward
SAN JOAQUIN VALLEY		
San Joaquin County - typical	4,000 to 5,500	Average / Stable
San Joaquin County – choice	5,500 to 9,000	Average / Stable
Stanislaus County – well & USBR water	3,000 to 6,000	Stable / Stable to Slightly Increasing
Stanislaus County – district water	5,000 to 8,500	Average / Stable
Merced County – well water	1,750 to 3,800	Limited / Stable to Slightly Increasing
Merced County – district water	2,500 to 5,000	Limited / Stable to Slightly Increasing
Merced County – west county	2,500 to 6,000	Stable to Slightly Increasing
Fresno County – USBR west side	1,000 to 3,000	Limited / Stable
Fresno County - USBR exchange	2,300 to 3,500	Moderate / Stable
Fresno County - district water	3,000 to 5,500	Limited / Decreasing
Fresno County – well water	1,000 to 3,500	Limited / Decreasing

Source: CalASFMRA (2002).

San Joaquin Valley markets show the strong influence of water supply and soil quality in several areas. Values for the most northern area, San Joaquin County, is reported in "typical" and "choice" categories. Pressures for urban development are evident with small parcels for rural residences ranging from U.S. $10,000 to U.S. $25,000 per acre. Demand continues for "choice" cropland with potential for permanent crop development, while "typical" cropland sales include those for forage production from adjacent dairy operations. In Stanislaus County, values are higher in areas served by district water supplies and lower in higher cost well areas and U.S. Bureau of Reclamation Water (USBR) areas where water supplies are less reliable from year-to-year and also most expensive. The same is also apparent in Merced County and Fresno County, where USBR water supplies have been severely curtailed below contracted quantities for the past decade.

Irrigated vegetable and row cropland

Vegetable-crop production in California has grown considerably in response to consumer demands for fresh produce. Premium production areas on the Central and South Coast regions are very limited due to urban development pressures. Large established vegetable growers appear willing to add to their land base by purchasing the limited amount of good quality farmland when offered for sale, ignoring land offerings in more outlying areas where more variable climatic factors contribute to diminished demand.

The Salinas Valley in Monterey County is the nation's salad bowl, producing vegetables of all kinds, in addition to those for which it is best known—e.g., lettuce and leafy salad greens, broccoli and strawberries. The most valuable land is in the Salinas area where the climate is coolest (Table 20.5).

Table 20.5 Ranges of Values per Acre for Irrigated Vegetable and Row-Crop Land, 2002

Agricultural land micro-market	U.S. Dollar Per-acre value	Market activity/trend
MONTEREY COUNTY		
Salinas	20,000 to 39,000	Stable / Increasing
Chualar – Gonzalez	13,000 to 17,500	Stable / Increasing
Soledad - Greenfield	7,000 to 17,000	Stable / Increasing
King City	9,000 to 19,000	Stable / Increasing
SAN LUIS OBISPO and SANTA BARBARA COUNTIES	11,000 to 21,000	Stable
VENTURA COUNTY	30,000 to 49,000	Strong / Stable

Source: CalASFMRA (2002).

Land values decrease as one moves from the northern to the most southern region where different vegetable crops are grown under progressively warmer conditions. Market activity in 2001 was "stable" and the price trend was reported to be "increasing." South coast areas also produce seasonal vegetables and strawberries.

Conclusion – California Agricultural Land Markets

California farmland is a mosaic of 27 million acres of irrigated and non-irrigated field cropland, row cropland, orchards, vineyards, groves, pastureland, and rangeland. Diversity in soils, climate, and water supply; the wide diversity in crop production possibilities (350 commercially significant crops); and the specialized nature of much of the state's agricultural production are the basis for describing some farm real-estate markets as micro-markets. For some agricultural uses, micro-markets may be very small, even restricted, in area (e.g., a small viticultural appellation region). About one-quarter, or more, of land currently in farms may be in uses best described as retirement and residential/lifestyle farms and/or in small farms with minor, incidental quantities of agricultural marketings. Markets for these may be sharply differentiated from those parcels characterized as having a predominant agricultural use with operators whose primary occupation is farming.

Ultimate highest and best use of California's farmlands is also influenced by zoning, public ordinances and regulations; and by growing municipal, industrial, and environmental demands. In particular, competing uses for land and water arise from large nonagricultural populations in major urban areas, chiefly in coastal regions, but also increasingly in major cities of the Central Valley region where population growth is expected to be highest in the Twenty-first Century.

CHAPTER SUMMARY—MICRO-MARKETS FOR FARMLAND: THE CASE OF FLORIDA AND CALIFORNIA

Market realities in these two agriculturally important states include significant interfaces and conflicts between agricultural-production opportunities and growing municipal, industrial and environmental demands for land currently identified as "in farms." Diversity in resource endowments and production alternatives exist within major agricultural production regions giving rise to the existence of micro-markets for farm real estate.

Evidence of the existence of micro-markets is in the form of recognizing types of real estate that are differentiated from other farmlands, even in the same agricultural region. In the short space available to us, we have but revealed a bit of the complexity that exists. While the coverage of our discussion may not be exhaustive, it is certainly suggestive. The bottom line, in our opinion, is that local markets in the traditional sense are more dominant generally in Florida or in California as they are in agricultural states with lesser transitional nonagricultural use alternatives for farmland.

REFERENCES

ASFMRA (American Society of Farm Managers and Rural Appraisers). (2000) *The Appraisal of Rural Property*, 2nd Edition. Chicago: Appraisal Institute.

Barnard et al. (2003) See Chapter 18, this volume.

Brown, M.J. (1996) *Forest Statistics for Florida, 1995.* Southern Research Station Resource Bulletin SRS-6. Washington: USDA/Forestry Service.

CalASFMRA (California Chapter of the American Society of Farm Managers and Rural Appraisers). (2002: 52) *Trends in Agricultural Land & Lease Values–2001.* Visalia: CalASFMRA.

CASS (California Agricultural Statistical Service). (2002) *Summary of County Agricultural Commissioners' Reports, 2001.* Sacramento: CASS. (15)

CCLRS (California Crop and Livestock Reporting Service). (1971: 16) *California Agriculture 1970.* Oakland: University of California, Division of Agriculture and Natural Resources.

CDFA (California Department of Food and Agriculture). (1991: 24) *California Agriculture Statistical Review 1990.* Sacramento: CDFA.

_____. (2001: 176) *Agricultural Resource Directory 2001.* Sacramento: CDFA.

Johnston, W.E. (1997) "Cross Sections of a Diverse Agriculture: Profiles of California's Production Regions and Principal Commodities." In *California Agriculture: Issues and Challenges*, Jerome B. Siebert, ed., (63–100). Giannini Foundation Information Series No. 97-1. Berkeley: Giannini Foundation (August).

_____. (1990) "Land: Competition for a Finite Resource—Flexibility or Irreversibility?" In *California's Central Valley—Confluence of Change*, Harold O. Carter and Carole Frank Nuckton, eds., (75–86). University of California, Agricultural Issues Center, Davis.

Johnston, W.E. and H.O. Carter. (2000) "Structural Adjustments, Resources, Global Economy to Challenge California Agriculture." *California Agriculture* 54(4): 16–22.

McCalla, A.F. and W.E. Johnston. (2003) "Whither Goes California Agriculture: Up, Down or Out? Some Thoughts About the Future." Giannini Foundation of Agricultural Economics. Berkeley: Giannini Foundation (Forthcoming).

Raup, P.M. (2003) See Chapter 2, this volume.

Reynolds, J.E. (2002) "Agricultural Land Values Increase as Citrus Land Values Decrease: 2002 Survey Results." University of Florida, Florida Food and Resource Economics No. 150. Gainesville: Food and Resource Economics (July–August).

_____. (2000) "Florida Rural Land: Competition Between Agricultural and Urban Uses." Soil and Crop Science Society of Florida Proceedings 59: 94–98.

_____. (1995) "The Florida Land Value Survey: A Decade of Results." University of Florida, Florida Food and Resource Economics No. 122. Gainesville: Food and Resource Economics (January–February).

_____. (1986) "Analytical Studies of Factors Affecting Farm Real Estate Values." In *The Farm Real Estate Market: Proceedings of a Regional Workshop.* Southern Rural Development Center SRDC No. 83. Mississippi State: Mississippi State University (29–41).

Reynolds, J.E. and A. Regalado. (2002) "The Effects of Wetlands and Other Factors on Rural Land." *The Appraisal Journal* 70(2): 182–190.

Reynolds, J.E. and D.L. Tower. (1978) "Factors Affecting Rural Land Prices in an Urbanizing Area." *Review of Regional Studies* 8(Winter): 23–34.

Shi, J.Y., T.T. Phipps, and D. Colyer. (1997) "Agricultural Land Values under Urbanizing Influences." *Land Economics* 73(1): 90–100.

Shonkwiler, J.S. and J.E. Reynolds. (1986) "A Note on the Use of Hedonic Price Models in the Analysis of Land Prices at the Urban Fringe." *Land Economics* 62(1): 58–61.

Siebert, Jerome B. (ed.). (1997) *California Agriculture: Issues and Challenges.* Giannini Foundation Information Series No. 97-1. Berkeley: Giannini Foundation.

USDA/ARS (U.S. Department of Agriculture, Agricultural Research Service). (1960) *Current Developments in The Farm Real Estate Market, November 1959—March 1960.* ARS 43-126 (CD-55). Washington, DC: USDA/ARS.

_____. (1958) *Current Developments in The Farm Real Estate Market, November 1957—March 1958.* AR S43-74 (CD-49). Washington, DC: USDA/ARS.

USDA/ERS (U.S. Department of Agriculture, Economic Research Service). (2001) Economic Research Service, Agricultural Information Bulletin No. 768. Washington, DC: USDA/ERS.

_____. (1994) *Farm Real Estate Market Developments.* Economic Research Service CD-69. Washington, DC: USDA/ERS.

_____. (1993) *Agricultural Resources: Agricultural Land Values and Markets, Situation and Outlook Report.* Economic Research Service AR-31. Washington, DC: USDA/ERS.

_____. (1977) *Agricultural Resources: Agricultural Land Values and Markets, Situation and Outlook Report.* Economic Research Service AR-31. Washington, DC: USDA/ERS.

_____. (1966) *Farm Real Estate Market Developments.* Economic Research Service CD-68. Washington, DC: USDA/ERS.

_____. (1965) *Farm Real Estate Market Developments.* Economic Research Service CD-67. Washington, DC: USDA/ERS.

_____. (1962) *Farm Real Estate Market Developments.* Economic Research Service CD-61. Washington, DC: USDA/ERS.

USDA/ NASS (U.S. Department of Agriculture, National Agricultural Statistics Service). (2002) *Agricultural Land Values.* USDA/NASS Sp Sy 3 (02). Washington, DC: U.S. Government Printing Office.

_____. (1999a) *1997 Census of Agriculture, California State and County Data, Volume 1, Geographic Area Studies, Part 5.* Washington, DC: U.S. Government Printing Office.

USDA/NASS (U.S. Department of Agriculture, National Agricultural Statistics Service). (1999b) *1997 Census of Agriculture, Florida.* Washington, DC: U.S. Government Printing Office.

_____. (1998) *Agricultural Land Values.* USDA/NASS Sp Sy 3 (98). Washington, DC: U.S. Government Printing Office.

ENDNOTES

[i] Other states may also have important agricultural areas subject to similar pressures, but they are likely not to be as extensive as those found in Florida or in California.

ii In contrast, California's production was dominated by extensive field crops and cattle and calves as recently as 1955. The shares were: livestock and livestock products, 35 percent; field crops, 28 percent; fruit and nut crops, 21 percent; and vegetable crops, 16 percent; with unreliable marketings from nursery and greenhouse production. Cash receipts from farm marketings amounted to U.S. $2.6 billion. (Johnston 1997).

iii Section 25 of the 1997 Census of Agriculture asks respondents to "Report your best ESTIMATE of the CURRENT MARKET VALUE OF LAND and BUILDINGS."

iv For example, the selection of Stags Leap Cabernet Sauvignon as the winner of a prize in a blind tasting competition in Paris in 1976 spurred expansion of premium wines and growing export markets for tree fruits and nuts attracted investments.

v This Census of Agriculture definition has been used in a consistent manner since 1974.

vi See Appendix C, 1997 Census of Agriculture.

Chapter 21

Effects of Government Restrictions on Land Ownership: The Saskatchewan Case

Jared Carlberg and Hartley Furtan
University of Manitoba
University of Saskatchewan

INTRODUCTION

Governments place the responsibility of farmland-ownership restrictions on certain economic agents in society. These agents are often policymakers who oversee the implementation of the government's farmland-ownership restrictions. The reasons for these farmland-ownership restrictions are seldom stated publicly, but often they are the direct result of political rent seeking by the local land buyers who want the opportunity to capture anticipated land rents in the absence of other possible market participants (Laband 1984). This interpretation of farmland-ownership restrictions builds on Tullock's (1989: 19) view that politicians engage in information obfuscation because direct transfers to them "would be just too raw. The voters would not buy it."

One example of a political farmland-ownership restriction is the striking of the Saskatchewan Farm Security Act (FSA) implemented in the Canadian province of Saskatchewan in 1974, in which restrictions were placed on the quantity of land that could be owned by non-Saskatchewan residents and by nonagricultural corporations. There were two stated political reasons for the restrictions. The first reason was to protect the family farm by lowering Saskatchewan land prices, thereby giving residents (particularly young farmers) the opportunity to buy land more economically. The second reason was to reduce the exodus of farmers from Saskatchewan, because policymakers believed that nonresidents would own the land and would not live on the farms, which would increase the consolidation of farms and depopulate the rural communities. The purpose of this chapter is to determine whether or not these two FSA objectives have been achieved.

We hypothesize that the reduction in the number of Saskatchewan farms occurred at a rate more slowly in Saskatchewan under the FSA than it did in Alberta and Manitoba, provinces that did not restrict ownership of farmland by non-provincial residents.

We test this hypothesis using two separate methods. In the first method, we construct an econometric model that resembles previous land-pricing models but which incorporates the FSA as a permanent fundamental component of the land price time series. Any effect of the FSA on Saskatchewan land prices should be reflected in a dummy variable that is significant and reflects the regulatory change. In this econometric model, we use a present-value (PV) framework that calculates land values as the discounted sum of all future payments to the land.

Under the second method, we test our hypothesis by comparing the ratio of the value of land both before and after the enactment of the FSA in Saskatchewan, to the value of land in the neighboring provinces of Alberta and Manitoba, which did not implement land restrictions. (Using the provinces of Alberta and Manitoba as controls, however, is imperfect because the intent of the FSA was to slow the exodus of Saskatchewan farm numbers. Its intent was not to keep the exodus below the levels occurring in either Alberta or Manitoba.) Nevertheless, comparing Saskatchewan to its neighboring provinces helps to illustrate the effects, if any, of the FSA legislation.

If the FSA were effective in its objectives, the ratio of the value of Saskatchewan farmland to the value of Albertan and Manitoban farmland should have decreased after Saskatchewan's implementation of the FSA.

We find that the FSA has no statistically significant effect on land values in Saskatchewan, indicating that the effect of FSA is too small to be measured accurately. We also find that the number of farms in Saskatchewan declined faster than those in either Alberta or Manitoba between 1976 and 1996. Clearly, the objectives of the FSA policymakers have not been achieved.

The immediate contribution of this chapter is to assist the 2002 policy debate by determining whether or not the FSA should be removed as a protection device for Saskatchewan residents from foreign competition in the land market. Results suggest that Saskatchewan land prices are determined by local market conditions and are not affected by the influx of nonresident capital. As long as Saskatchewan residents do not face a credit constraint, they will bid as much for land as nonresidents do. The second contribution of this chapter is to demonstrate that ownership restrictions do not stop the exodus of labor from the agricultural sector. The presence of larger labor saving farm equipment has more to do with the reduction of farm numbers than does land ownership per se.

HISTORY OF LIMITED FARM OWNERSHIP IN SASKATCHEWAN[i]

The concept of limiting the nonresident and corporate ownership of Saskatchewan's agricultural lands was an outcome of at least two major events on the

Canadian Prairies. First, farm incomes were low in the late 1960s and early 1970s because of depressed grain sales. As a result of these depressed grain sales, the Canadian Wheat Board (CWB), which markets most of the farmers' grain (including wheat, durum, barley, and oats), encountered slow sales due to excess supplies on the world market. Second, the price of wheat rose to record levels in 1974 because of large Canadian grain purchases by the Union of Soviet Socialist Republics (USSR). Land prices in Canada followed the increase in the price of wheat and reached record levels by 1976 (Schmitz 1995). Suddenly, both farmers and nonfarmers saw land as a valuable asset, and each attempted to acquire additional acreage. Thus, rent-seeking activities were exercised by involved farmers and by the community at large, which made the punitive legislation politically palatable.

The formal proposal to limit farmland ownership first appeared in March 1973 when the report of the Special Committee on the Ownership of Agricultural Lands (Committee) was finalized (GOS, 1973: 6). The mandate of the Committee was to "investigate the effects of the purchase and ownership of agricultural lands by nonresident, foreign, and corporate persons." Also of concern to the 1973 Committee was the plight of Saskatchewan's young farmers, who were leaving Saskatchewan for other parts of Canada (particularly to the provinces of Alberta and Ontario), in search of nonfarm employment opportunities. According to the Committee (GOS 1973: 13):

> The need for action is clear. We cannot hope for a single 'cure-all,' but we must take every measure that will increase opportunities for young people to get into farming with good chances for success. The cycle of declining population leading to cutbacks of social amenities and local businesses leading to further loss of population must be stopped.

The conclusion the Committee reached was that young farmers must be encouraged and supported to ensure the stability of Saskatchewan's agricultural sector. To help preserve the family-farm unit, the Committee had twelve recommendations, some of which formed the basis of The Farm Ownership Act of 1974 (GOS 1973).

The Farm Ownership Act of 1974

The Farm Ownership Act of 1974 affected two distinct groups—nonresidents and nonagricultural corporations. Subject to certain exemptions, nonresidents of Saskatchewan were limited in the amount of agricultural land they could own such that it was an aggregate landholding capacity of CDN $15,000 of municipal assessment. (Aggregate landholding refers to the holdings of an individual, his or her spouse, and dependent children.) The only instances in which nonresidents were permitted to hold or to acquire more than CDN $15,000 of community assessment in Saskatchewan landholdings were when:

- The nonresident had acquired the landholding or the right to own the land prior to 31 March 1974.
- The nonresident, who was a Saskatchewan farmer and had farmed the landholding for five years prior to 31 March 1974, held the landholding.
- The nonresident, who had received a transfer of land, was a designated relative of a one-time farmer.
- The nonresident, who intended to become a resident of Saskatchewan within three years, was granted an exemption from the CDN $15,000 limit by the Farm Land Security Board.

The Farm Ownership Act of 1974 defined an agricultural corporation as one that was primarily engaged in the business of farming, and one that was at least 60-percent owned and controlled by resident Saskatchewan farmers. Thus agricultural corporations were not restricted in the amount of land they could hold. On the other hand, nonagricultural Saskatchewan corporations were limited to an aggregate landholding of 160 acres, and those in excess of 160 acres were required to dispose of the excess by 1 January 1994. Certain exemptions also applied to corporate holdings. For example, corporations acquiring an interest by way of realization of a quit-claim settlement or foreclosure of a mortgage were given two years to dispose of any land in excess of the permitted 160 acres. Also, if a trust company held land in trust for a Saskatchewan resident, it was exempt from the application of the legislation. The Farm Land Security Board was also empowered to grant exemption for a nonagricultural corporation to acquire an interest in land in excess of 160 acres.

Subsequent Amendments

In 1977, several amendments were introduced to the Farm Ownership Act of 1974. The most significant of these was the lowering of the limit on nonresident acquisitions from the CDN $15,000 assessed-value limit to an acreage limit of 160 acres of agricultural property. Canadian residents (excluding, of course, Saskatchewan residents) continued to be included in the broad definition of a nonresident. In the late 1970s, however, concerns arose over the exploitation of the 160-acre limit that was put in place for both nonresidents and nonagricultural corporations. For example, nonresidents who were eligible could acquire tracts of land and could place each 160 acres of land in a different family member's name. Alternatively, an individual who was eligible could set up numerous corporations, and could acquire 160 acres in each corporate name. As a consequence to this exploitation, purchases made through such loopholes were blamed for escalated land prices in certain Saskatchewan rural municipalities.

In response to this trend, the Farm Ownership Act was further amended in 1980. The amount of land nonresidents and nonagricultural corporations were allowed to acquire was not to exceed 10 acres. Canadians (excluding

Saskatchewan residents) continued to be classified in the same manner as foreign nonresidents. Thus the Farm Ownership Act of 1980 also prohibited a Saskatchewan resident from acquiring a landholding on behalf of a nonresident or a nonagricultural corporation when that acquisition would place the individual or corporation in contravention of the legislation.

The 1980 legislation required the agricultural corporation to be engaged in the business of farming only if 60 percent of all voting and nonvoting shares were legally and beneficially owned by resident farmers. The Farm Ownership Act was amended again in 1988 to define an agricultural corporation as one that was primarily engaged in the business of farming if the majority of all issued shares were legally and beneficially owned by resident farmers. Thus the definition of an agricultural corporation was also changed.

In 1988, the Farm Ownership Act was replaced with the Saskatchewan Farm Security Act (FSA). A number of significant changes occurred at that time. Nonresidents continued to be limited to an interest in 10 acres of land, but a special provision was added to the legislation allowing Canadian residents to acquire an interest in up to 160 acres of farmland. Similarly, while nonagricultural corporations continued to be restricted to a 10-acre interest, nonagricultural corporations in which the majority of issued voting shares was held by Saskatchewan residents was allowed to acquire an interest in up to 320 acres of agricultural property.

The definition of an agricultural corporation was again changed in 1988. It was redefined as a corporation that was engaged in the business of farming that had the majority of issued voting shares legally and beneficially owned by resident producers. [A producer was defined as an individual who was engaged in the business of farming, including both the actual physical labor (active farming) and/or the management of farm property (passive farming)]. We disclose the changes made to the Farm Ownership Act/Farm Security Act from 1974 to 1988 (Table 21.1).

Table 21.1 Changes to Saskatchewan Farmland Ownership Regulations (1974 to 1988)

Owner type	1974	1977	1980	1988
Saskatchewan residents	unrestricted	unrestricted	unrestricted	unrestricted
Canadian residents	CDN $15,000 assessment	160 acres	10 acres	320 acres
Non-Canadian residents	CDN $15,000 assessment	160 acres	160 acres	10 acres
Agricultural corporations	unrestricted	unrestricted	unrestricted	unrestricted
Saskatchewan nonagricultural corporations	160 acres	160 acres	10 acres	320 acres
Other nonagricultural corporations	160 acres	160 acres	10 acres	10 acres

LAND-PRICE THEORY

Two types of auctions, the oral (English) and the first-price sealed-bid auction, are used to sell Saskatchewan farmland. In an oral (English) auction the auctioneer begins with a price for a parcel of land and then lowers the price until a bid is received. If another bidder wishes to offer a higher bid, he or she does so, and the bidders compete for a parcel of land until only one buyer remains. Bidders in this type of auction may make multiple bids. Alternatively, in a first-price sealed-bid auction, bids are submitted in a sealed envelope to the vendor until a deadline date and time is reached. At the deadline date and time, with no bidders aware of anyone else's bid, the envelopes are opened. Participants in this type of auction make only one bid, and the buyer is revealed when the bids are opened and examined. This type of auction is the one most often used for Saskatchewan farmland. According to the revenue-equivalence theorem, the oral and first-price sealed-bid auctions yield the same price on average (Vickrey 1961), but the oral auction requires the bidders to be present at the auction. If it is expensive or inconvenient for bidders to attend the auction, their numbers may be recorded and the winning bid may be lowered. This could be the reason behind the preference of first-price sealed-bid auctions for farmland sales in Saskatchewan (Milgrom 1989).

Brannman et al. (1987) and McAfee and McMillan (1987) find in their research that the lower the number of bidders at an auction, the lower the winning bid becomes, regardless of the type of auction. Any regulation that restricts the number of bidders can therefore be expected to lower the amount of the winning bid, all else being equal. If bidders placing a higher value on the parcel of farmland are excluded by regulation, bids will be lowered still further because bids are functions of the participants' evaluation of the item for sale (McAfee and McMillan 1987). Lapping and Lecko (1983) note that a tax advantage, for example, could be a reason that nonresidents place a higher value on farmland than residents do. Another such reason could be the use of land for recreational or other nonfarming purposes, which would cause a nonresident to assign a higher value to a parcel of land than would a resident. (Land required for a retirement home, for instance, could be valued more highly than a similar parcel would be that is used only for agricultural purposes.)

There are competing theories as to how farmland purchasers arrive at a bidding price for land. Melichar (1979) initially suggests that the PV model is an appropriate method of valuing farmland. Since then, numerous authors either confirm (Alston 1986; Burt 1986; Pongtanakorn and Tweeten 1986) or deny (Featherstone and Baker 1987; Falk 1991; Clark et al. 1993; Just and Miranowski 1993; Tegene and Kuchler 1993; Schmitz 1995; Chavas and Thomas 1999) the model's usefulness in the evaluation of farmland. Falk and Lee (1998) break farmland price-time series down into three uncorrelated components and find that deviations of farmland price from the predictions of the PV model are not important in the long run. Given this result, and given that the PV model is the most often used for similar Canadian studies (e.g., Baker et al. 1991), it is adopted in this chapter as the benchmark model for the competitive case of Saskatchewan land-price valuation.

In the PV model, the price of land is the sum of its discounted future returns

$$V_t = \sum_{S=1}^{S} R_{t+S} / (1+d)^S,$$ (1)

where V_t is the value of land at time t, R_t is the rent to the land in time t, d is the discount rate, and S is the number of discounting periods.

In the PV model, the role of expectations must be considered explicitly since bidders use expected rents in their land-valuation process. Weisensel et al. (1988) and Veeman et al. (1993) recognize that, due to the uncertain nature of commodity prices and government subsidies, rational expectations cannot be assumed for bidders, and instead they employ adaptive expectations as a basis for their land valuation. In our models, we use the Weisensel et al. (1988) and Veeman et al. (1993) PV framework for our analysis.

In an adaptive-expectations model, the dependent variable is determined by the expected rather than by the current values of the independent variables (Kennedy 1998; Greene 2000). This can be written as

$$V_t = \beta_0 + \beta_1 R_t^* + \varepsilon_t,$$ (2)

where R_t^* is the expected value of rent in time t, with expectations formed in time period $(t-1)$, and ε_t as the error term.

Since future rents are not known, a simple rule can be used to base expectations on past forecast errors. Specifically, expectations on the independent variable are formed by adding to the previous period's expected value a constant proportion λ of the previous period's forecast error. This yields

$$R_t^* = R_{t-1}^* + \lambda(R_t - R_{t-1}^*).$$ (3)

Equation (2) can be rearranged to show that $R_t^* = (V_t - \beta_0 - \varepsilon_t)/\beta_1$, so $R_{t-1}^* = (V_{t-1} - \beta_0 - \varepsilon_{t-1})/\beta_1$. Therefore, past values for rents depend on past values of land, which in turn depend on the past values of rent since forecasts are made in time period $t-1$, and so on. The lag lengths must then be truncated for practical purposes. The final form of the model consists of only known values

$$V_t = f(V_{t-1,t-2,...t-i}; R_{t-1,t-2...t-j}),$$ (4)

where i and j are the lag lengths on land values and rents.

DATA

Farmland value is approximated as the value of land and buildings as reported by Statistics Canada's CANSIM service (Statistics Canada 2001). This is reasonable, since the land value includes the value of structures erected on it. There is no data available for cash rents in Saskatchewan, so cash receipts from farm products are used as approximations for the 1950 to 1970 period and total cash receipts are used for the 1971 to 1999 period.[ii] Veeman et al. (1993) employ a similar measure, total cash receipts, to approximate rents in their application of the PV model to Canadian farmland. In this study, the value of farmland and the rent series both were converted to real values using the real CPI (1992 = 100).

The data for the number of farms was taken from the 1976 and 1996 Canada Census on Agriculture (Statistics Canada 2001). Adjacent Alberta counties and Saskatchewan rural municipalities were matched, as were adjacent Manitoba and Saskatchewan municipalities. (Rural municipalities are the most micro units of measurement for farm numbers.) From 1976 to 1996, the percentage change in farm numbers was calculated for each of the paired 56 rural municipalities.

ESTIMATION PROCEDURE FOR LAND PRICES

The impact of the regulatory changes of Saskatchewan's FSA on land prices in that province is our focus in this chapter. We use land prices for the provinces of Alberta and Manitoba as controls. Since neither of Saskatchewan's western nor eastern neighbor restricts farmland ownership on the basis of Canadian residency,[iii] land values in those provinces should not have been measurably affected by Saskatchewan's regulatory change.

Stationarity in first differences for land values and rents is a key element of the applicability of PV models for land valuation (Falk 1991, and Clark et al. 1993). We applied the Dickey-Fuller test to the time-series data for each of the three provinces. Our test results indicate that the first-difference stationarity condition holds.

We use two separate methods to determine whether the regulations of the FSA affected farmland values in Saskatchewan. The first method involves the calculation of a coefficient on the dummy variable representing the FSA in the adaptive-expectations PV model. If the FSA affected land values in Saskatchewan but not in Alberta or Manitoba, then the dummy variable representing the regulatory change should have a negative sign and should be statistically significant for Saskatchewan but not for either of the other two provinces. Equation (4) implies that there could be long lag lengths on both land values and rents in the adaptive-expectations PV land model. Lag lengths are truncated for estimation purposes when further lagged values are not statistically significant. The resulting equations for Saskatchewan, Alberta, and Manitoba land values are

$$V_t^S = \beta_0 + \beta_1 V_{t-1}^S + \beta_2 V_{t-2}^S + \beta_3 R_t^S + \beta_4 R_{t-1}^S + \beta_5 FSA + \varepsilon_t, \tag{5}$$

$$V_t^A = \alpha_0 + \alpha_1 V_{t-1}^A + \alpha_2 V_{t-2}^A + \alpha_3 R_t^A + \alpha_4 R_{t-1}^A + \alpha_5 FSA + u_t, \text{ and} \tag{6}$$

$$V_t^M = \delta_0 + \delta_1 V_{t-1}^M + \delta_2 V_{t-2}^M + \delta_3 R_t^M + \delta_4 R_{t-1}^M + \delta_5 FSA + w_t \tag{7}$$

where land value and land-rent variables are as previously defined, *FSA* is the indicator variable representing the regulatory change; S, A, and M are the provinces of Saskatchewan, Alberta, and Manitoba, respectively; and ε_t, u_t, and w_t are the error terms, respectively.

If separate equations were affected by common factors influencing their disturbances, it would be appropriate to treat the equations as a system (Johnson and DiNardo 1997). Pongtanakorn and Tweeten (1986) include numerous factors that exert minor influences on the price of land that could affect the disturbance terms in the equations for land values in all three provinces.[iv] One method for the estimation of the set of equations above provided there are no codependent regressors is seemingly unrelated regressions (SUR). By estimating the equations as a system, we enhance the efficiency of the estimates by taking the cross-equation correlations into account (SAS Institute Inc. 1999).

Autocorrelated residuals in models with lagged dependent variables can cause all desirable estimator properties to be lost. A Durbin *h*-test reveals that autocorrelation is present in the residuals for each of Equations (5) to (7). Stepwise autoregression is then used to determine the order of the autoregressive model for the equations and we conclude that the equations for both provinces follow a first-order autoregressive AR(1) error model. The AR(1) procedure in SAS allows for the estimation of a SUR system with autocorrelation, and is used to estimate Equations (5) through (7) in double-log form, incorporating an AR(1) error structure. The Shapiro-Wilk *w*-test statistics for Saskatchewan, Alberta, and Manitoba are 0.97, 0.98, and 0.96, respectively, with *p*-values of 0.49, 0.75, and 0.10, respectively. As such, the null hypothesis of normality in the residuals is not rejected. A Henze-Zirkler *t*-test statistic of 1.83 with a *p*-value of 0.067 indicates that normality is not rejected for the SUR system as a whole. Godfrey's *LM*-test statistic for serial correlation is 1.60 (with a *p*-value of 0.21) for Saskatchewan and 3.26 (with a *p*-value of 0.07) for Alberta, indicating that serial correlation is not present in the residuals of either equation. The analogous values for the Manitoba equation are 5.85 and 0.02, indicating that serial correlation is present in the residuals. However, this affects neither the results of the estimation of the overall system nor the variables of interest in any important way.

We conduct a Chow test for structural change, yielding an *f*-value of 0.11 with a *p*-value of 0.99, which indicates that the null hypothesis of no structural change as a result of the FSA is not rejected. A modified Breusch-Pagan test is used to check for homoscedasticity of the error terms. Using the full set of regressors, the *t*-statistics for Saskatchewan, Alberta, and Manitoba are 20.74, 18.32, and 18.65, with *p*-values of 0.07, 0.15, and 0.13, respectively. The null hypothesis of homoscedastic residuals is not rejected for any of the equations.

A second method for estimating the effects of the regulatory changes of the FSA on farmland values in Saskatchewan considers the ratio of Saskatchewan farmland values to those in Alberta and Manitoba. The land-value ratios between the provinces should not have changed if the legislation did not have an effect on land values. If the ratio of land values between Saskatchewan and one of its neighboring provinces were considered to be a function of the ratio of cash receipts and new ownership restrictions in Saskatchewan, the model

$$V_t^S / V_t^A = \phi_0 + \phi_1(R_t^S / R_t^A) + \phi_2 FSA + u_t, \tag{8}$$

would hold, where V_t^S and V_t^A are the value of land and buildings in Saskatchewan and Alberta, respectively, and R_t^S and R_t^A are the analogous variables for rent, respectively. The same equation can be written for the Saskatchewan or the Manitoba case

$$V_t^S / V_t^M = \theta_0 + \theta_1(R_t^S / R_t^M) + \theta_2 FSA + z_t. \tag{9}$$

We used ordinary least squares (OLS) regressions to estimate Equations (8) and (9). Autocorrelation is found in the residuals for each equation, and stepwise autoregression indicates an AR(1) model is appropriate. A q-statistic test and an LM test both reject the null hypothesis of homoscedasticity. As such, corrections for both autocorrelation and heteroscedasticity are required. The generalized autoregressive conditional heteroscedasticity (GARCH) model is one method for addressing heteroscedasticity in time-series models. This method allows a long memory process that is appropriate in this case since the *LM*-tests for heteroscedasticity are significant at long-lag lengths (SAS Institute, Inc., 1999). We combine an AR(m) process with a GARCH(p,q) process to model a time series with an autoregressive-error structure when heteroscedasticity is present. In most cases, a GARCH(1,1) will suffice. For present purposes an AR(1) and GARCH(1,1) are employed, using the method of maximum likelihood.

RESULTS ON LAND PRICES

We give the results of SUR estimation of Equations (5) through (7) (Table 21.2). The FSA indicator variable is not significant for any of the three provinces, and does not have the hypothesized negative sign for Saskatchewan. Equality of coefficients on the dummy variable for the respective provinces is tested with a Wald test and not rejected, indicating that the effect of the legislation was not different in any of the three provinces. The coefficient on the dummy variable representing the legislation for the Saskatchewan model is 0.043. Since the mean on the dependent variable (the logarithm of the value of land and buildings) is 14.727, by dividing that number by 0.043 we see that the FSA can be interpreted as generating a 0.29 percent increase in Saskatchewan land values.
Based on the average value of provincial land and buildings, this translates into an increase of CDN \$23.5 million. The coefficient for the FSA in Alberta is 0.056, which is a 0.41 percent increase when given a mean dependent variable of 14.765. The effect of the FSA in Alberta therefore is CDN \$77.3 million dollars, influence of the FSA. For Manitoba, the percent increase is 0.31 when given a coefficient on the dummy variable of 0.043 and a mean value on the dependent variable of 13.725. This translates into an effect of approximately CDN \$90.2 million but, as in the Alberta case, this is not due to the FSA.

Table 21.2 Seemingly Unrelated Regressions for Value of Land and Buildings, Saskatchewan and Alberta (1952–99)

Variable	Province		
	Saskatchewan	Alberta	Manitoba
Intercept	0.233	0.266*	0.290*
	(0.194)	(0.149)	(0.152
Land value, lagged one period	1.595**	1.444**	1.460**
	(0.094)	(0.082)	(0.102)
Land value, lagged two periods	−0.667**	−0.571**	−0.614**
	(0.087)	(0.074)	(0.088)
Rent	0.258**	0.504**	0.362**
	(0.061)	(0.072)	(0.067)
Rent, lagged one period	−0.197**	−0.385**	−0.217**
	(0.064)	(0.074)	(0.067)
FSA	0.043	0.056	0.030
	(0.051)	(0.043)	(0.043)
Adjusted R^2	0.9987	0.9987	0.9988
N	48	48	48

Note: Standard errors are given in parentheses. Double and single asterisks indicate significance at the 5-percent and 10-percent level, respectively.

If we were to assume that the value of land and buildings in Saskatchewan would have changed by an average of the same proportion as in the other provinces due to the regulatory change (i.e., if Saskatchewan land values would have increased by 0.36 percent compared to Manitoba's 0.31 percent and Alberta's 0.41 percent in absence of the legislation), the effect of the FSA could be calculated. We find that Saskatchewan's land values increased 0.07 percent less than did the average of the others because of FSA legislation. This translates into an FSA legislation effect for the province of around CDN $565,000. Based on a real mean-value of land and buildings in the province of nearly 9 billion dollars, this effect is not very large.

We show the GARCH model estimates for Equation (8) for Saskatchewan/Alberta and (9) for Saskatchewan/Manitoba (Table 21.3). The dummy variable for the FSA is not significant in either case, suggesting that FSA legislation did not change the ratios of land values in Saskatchewan more significantly than did the legislation that was applied to the Alberta or Manitoba land values. In addition, the dummy variable for the FSA does not have the expected sign, as in the SUR model presented above.

Results of both methods that we used for determining the effects of the FSA legislation on land values in Saskatchewan therefore imply that the impact of the change in ownership restrictions is negligible. No evidence is found that the FSA lowered the value of farmland in Saskatchewan relative to either of the control provinces of Alberta or Manitoba. The evidence in this chapter cannot support claims that farmland prices have been lowered as a result of the FSA legislation.

RESULTS ON FARM NUMBERS

The second policy objective of the Farm Security Act of 1974 was to slow down the exodus of farm families from the sector. Policymakers believed that by placing a restriction on who could own Saskatchewan farmland, the number of family farms would remain more stable than if the land market remained open to all.

We test the hypothesis that the Farm Security Act of 1974 did in fact reduce the exodus of farmers in Saskatchewan by comparing the percentage rate of change in farm numbers in Saskatchewan with that in Alberta and Manitoba. We calculate the percentage change in farm numbers in the 56 rural municipalities along the Alberta-Saskatchewan and Saskatchewan-Manitoba border for the 1976 to 1997 period. The rural municipalities in Saskatchewan are paired to the adjacent rural municipalities in Alberta and Manitoba to capture the effect of similar weather, soil type, and settlement patterns.

The first test we use determines if the paired means of the two data series are significantly different. This is done using a simple t-test for a paired data series. The $H_0 : \mu_0 = \mu_1$ where μ_0 is the mean of the change in Alberta and Manitoba farm numbers and μ_1 is the mean of the change in Saskatchewan farm numbers. The calculated $t = 17.23$ with a critical value of 2.00, thus the hypothesis that the means are equal is rejected. This indicates that the change in farm numbers is occurring at a different rate in the two series (i.e., Alberta-Manitoba and Saskatchewan).

The change in farm numbers in the Alberta and Manitoba regions is specified to be a function of the change in farm numbers in the paired Saskatchewan regions. There is a potential variation in farm numbers due to the difference in farming conditions along the provincial border. As an example, soil types, settlement pattern, and weather conditions may be different among the Northern, Central, and Southern parts of the provinces. We inserted a dummy variable for the Northern and Central regions of the provinces with the intercept capturing the Southern region. The authors chose the regions. The estimated regression equation with t-statistics is shown in parentheses in Equation (10).

We reveal the plot of the regression line for the change in farm numbers in the Alberta and Manitoba regions (Figure 21.1). If the estimated coefficient on the *SASK* variable were not significantly different from one and with a zero intercept, the percentage rate of change in farm numbers would be equal among Saskatchewan and the other two provinces. However, the estimated coefficient

$$Y = -15.00 + 0.373SASK + 8.931Dn + 15.07Dm$$
$$(-4.38) \quad (4.15) \qquad (1.87) \qquad (3.28) \qquad\qquad (10)$$
$$\text{Adjusted } R^2 = 0.26 \quad F = 7.59 \text{ Prob. } F = 0.000266$$

on the *SASK* variable indicates that the change in farm numbers occurred faster in Saskatchewan than it did in Alberta and Manitoba.

Table 21.3 Parameter Estimates from Generalized Autoregressive Conditional Heteroscedasticity Model, for Ratio of Saskatchewan-to-Alberta and Saskatchewan-to-Manitoba Value of Land and Buildings, 1950 to 1999

Variable	Province	
	Alberta	Manitoba
Intercept	0.898*	2.219*
	(0.117)	(0.990)
Cash-receipts ratio	0.213*	0.020
	(0.052)	(0.148)
FSA	0.042	0.281
	(0.074)	(1.128)
N	50	50

Note: Standard errors are given in parentheses. An asterisk indicates significance at the 5-percent level.

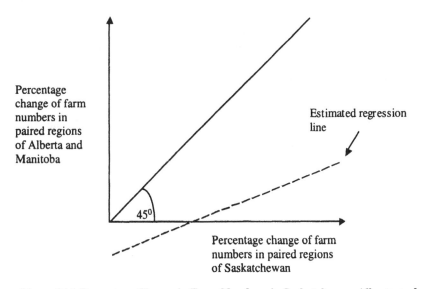

Figure 21.1 Percentage Change in Farm Numbers in Saskatchewan-Alberta and Saskatchewan-Manitoba 1976 to 1996

The intercept term is significantly lower in Saskatchewan for its Southern and Central regions. This indicates that Saskatchewan had more farmers in these two regions than Alberta and Manitoba had in their respective adjacent regions. In the Northern region, all provinces began in 1976 with the same number of farmers, but the percentage rate of decline in the number of farmers was most rapid in Saskatchewan.

This analysis of the percentage change in farm numbers suggests that the second major policy objective of the Farm Security Act of 1974 was not achieved. Farm numbers declined at a faster rate in Saskatchewan than they did in either Alberta or Manitoba. The reason for the more rapid decline remains an open-ended question, the answer to which may have had nothing to do with farmland-ownership restrictions.

SUMMARY AND CONCLUSIONS

The purpose of this chapter was to determine the effect of the Farm Security Act of 1974 on land values and on the change in farm numbers in Saskatchewan. The legislation introduced restrictions on the amount of land that could be owned by individual Saskatchewan nonresidents. No evidence was found that the FSA caused land values in Saskatchewan to decrease. One possible economic explanation for this result is that Saskatchewan residents faced no credit constraints and were willing to bid away all possible future land rents. As such, the FSA regulations were not effective at lowering Saskatchewan farmland prices for resident farmers.

We found evidence that the exodus of farmers from Saskatchewan exceeded that from both Alberta and Manitoba. The hypothesis that the rate of exodus was slower in Saskatchewan after the Farm Security Act of 1974 was introduced was not tested. However, all tests will suffer the problem of identifying an appropriate control environment for comparison.

The rent-seeking hypothesis was not formally modeled or tested in this chapter. Land rents are well understood in an agrarian economy and more groups than farmers seek to capture them. Government policy may be endogenous in this case with the trade off occurring between rent seeking for land rents and support for the government. This is an area for future research.

The results of this study are subject to some considerable limitations. Better data may have led to more precise results from the research. The data for cash receipts and the value of land and buildings are highly aggregated, making the effects of the FSA on individual land transactions difficult to discover. Additionally, the use of farm cash receipts as a proxy for cash rent is not perfect, although it has been used in Canadian land-value studies by Veeman (1993) and is the only reliable approximation available. Also, having only fifty observations in the time series limits the ability to draw strong inferences from the results of our study.

REFERENCES

Alston, J.M. (1986) "An Analysis of Growth of U.S. Farmland Prices, 1963–82." *American Journal of Agricultural Economics* 68(1): 1–9.

Baker, T.G., E.H. Ketchabaw, and C.G. Turvey. (1991) "An Income Capitalization Model for Land Value with Provisions for Ordinary Income and Long-Term Capital Gains Taxation." *Canadian Journal of Agricultural Economics* 39(1): 69–82.

Brannman, L., J.D. Klein, and L.W. Weiss. (1987) "The Price Effects of Increased Competition in Auction Markets." *Review of Economics and Statistics* 69(1): 24–32.

Burt, O.R. (1986) "Econometric Modeling of the Capitalization Formula for Farmland Prices." *American Journal of Agricultural Economics* 68(1): 10–26.

Chavas, J.P. and A. Thomas. (1999) "A Dynamic Analysis of Land Prices." *American Journal of Agricultural Economics* 81(4): 772–84.

Clark, J.S., M. Fulton, and J.T. Scott, Jr. (1993) "The Inconsistency of Land Values, Land Rents, and Capitalization Formulas." *American Journal of Agricultural Economics* 75(1): 147–55.

Falk, B. (1991) "Formally Testing the Present-Value Model of Farmland Prices." *American Journal of Agricultural Economics* 73(1): 1–10.

Falk, B. and B. Lee. (1998) "Fads versus Fundamentals in Farmland Prices." *American Journal of Agricultural Economics* 80(3): 696–707.

Farm Land Security Board. (2002) "Saskatchewan's Farm Ownership Legislation." Unpublished report, Government of Saskatchewan, Regina.

Featherstone, A.M. and T.G. Baker. (1987) "An Examination of Farm Sector Real Asset Dynamics." *American Journal of Agricultural Economics* 69(3): 532–46.

GOS (Government of Saskatchewan). (1973) "Final Report of the Special Committee on the Ownership of Agricultural Lands." Unpublished report. Regina: Government of Saskatchewan.

Greene, W.H. (2000) *Econometric Analysis, Fourth Edition.* Upper Saddle River: Prentice Hall.

Johnson, J. and J. DiNardo. (1997) *Econometric Methods 4th Edition.* New York: McGraw-Hill.

Just, R.E. and J.A. Miranowski. (1993) "Understanding Farmland-Price Changes." *American Journal of Agricultural Economics* 75(1): 156–68.

Kennedy, P. (1998) *A Guide to Econometrics 4th Edition.* Cambridge: The MIT Press.

Laband, D.N. (1984) "Restriction of Farm Ownership as Rent-Seeking Behavior: Family Farmers Have It Their Way." *American Journal of Economics and Sociology* 43(2): 179-89.

Lapping, M.B. and M. Lecko. (1983) "Foreign Investment in U.S. Agricultural Land: An Overview of the Issue and a Case Study of Vermont." *American Journal of Economics and Sociology* 42(3): 291–304.

McAfee, R.P. and J. McMillan. (1987) "Auctions and Bidding." *Journal of Economic Literature* 25(2): 699–738.

Melichar, E. (1979) "Capital Gains versus Current Income in the Farming Sector." *American Journal of Agricultural Economics* 61(5): 1085–92.

Milgrom, P. (1989) "Auctions and Bidding: A Primer." *Journal of Economic Perspectives* 3(1): 3–22.

Pongtanakorn, C. and L. Tweeten. (1986) *Determinants of Farmland Price and Ratio of Net Rent to Price.* Agricultural Experiment Station, Oklahoma State University, Research Report P-878, Stillwater (May).

SAS Institute, Inc. (1999) *SAS / ETS User's Guide, Version 8.* Cary: SAS Institute.

Schmitz, A. (1995) "Boom/Bust Cycles and Ricardian Rent." *American Journal of Agricultural Economics* 77(5): 1110–25.

Statistics Canada. (2001) Internet Website:
http://www.statcan.ca/english/CANSIM/cansim1.htm

Tegene, A. and F. Kuchler. (1993) "A Regression Test of the Present Value Model of U.S. Farmland Prices." *Journal of Agricultural Economics* 44(2): 223–36.

Tullock, G. (1989) *The Economics of Special Privilege and Rent Seeking.* Boston: Kluwer Academic Publishers.

Veeman, M.M., X.Y. Dong, and T.S. Veeman. (1993) "Price Behaviour of Canadian Farmland." *Canadian Journal of Agricultural Economics* 41(1): 135–43.

Vickrey, W. (1961) "Counterspeculation, Auctions, and Competitive Sealed Tenders." *Journal of Finance* 16(1): 8–37.

Weisensel, W.P., R.A. Schoney, and G.C. Van Kooten. (1988) "Where Are Saskatchewan Farmland Prices Headed?" *Canadian Journal of Agricultural Economics* 36(1): 37–50.

ENDNOTES

[i] This section draws from the report entitled *Saskatchewan's Farm Ownership Legislation* (Farm Land Security Board, 2002).

[ii] Statistics Canada changed its method of reporting cash receipts to farm operators in 2001. Since potential bidders on a parcel of land are aware of past returns to the land, a change in how those returns are measured by a government agency will not affect their expectations unless the changes are accompanied by an amended government policy. The ratio of Saskatchewan rents to those of Alberta or Manitoba will not be affected either since the 2001 Statistics Canada change applied to all provinces.

[iii] Alberta and Manitoba, like Saskatchewan, restrict ownership by non-Canadian residents. However, we assumed that the number of prospective non-Canadian land buyers is much smaller than is the number of prospective Canadian land buyers.

[iv] Pongtanakorn and Tweeten (1986) list factors including interest rates on farm loans, population density, and stock market returns.

Index

Printed and bound by CPI Group (UK) Ltd, Croydon, CR0 4YY

16/04/2025

14658603-0002